Molecular Structure, Electronic and Vibrational Spectra Theoretical Calculations in Materials Sciences

Molecular Structure, Electronic and Vibrational Spectra Theoretical Calculations in Materials Sciences

Guest Editor

Georgiy V. Girichev

MDPI

Basel • Beijing • Wuhan • Barcelona • Belgrade • Novi Sad • Cluj • Manchester

Guest Editor
Georgiy V. Girichev
Department of Physics
Ivanovo State University of
Chemistry and Technology
Ivanovo
Russia

Editorial Office
MDPI AG
Grosspeteranlage 5
4052 Basel, Switzerland

This is a reprint of the Special Issue, published open access by the journal *International Journal of Molecular Sciences* (ISSN 1422-0067), freely accessible at: https://www.mdpi.com/journal/ijms/special_issues/molecular_calculations.

For citation purposes, cite each article independently as indicated on the article page online and as indicated below:

Lastname, A.A.; Lastname, B.B. Article Title. *Journal Name* **Year**, *Volume Number*, Page Range.

ISBN 978-3-7258-5479-0 (Hbk)
ISBN 978-3-7258-5480-6 (PDF)
https://doi.org/10.3390/books978-3-7258-5480-6

Contents

About the Editor

Georgiy V. Girichev

Georgiy V. Girichev is a Full Professor of Physics at the Faculty of Inorganic Chemistry and Technology, the Ivanovo State University of Chemistry and Technology (ISUCT), Ivanovo, Russia. He was a Head of the Physics Department at the ISUCT for more than three decades (1989 to 2024). His main research interests lie in the field of molecular structure and thermodynamics of chemical compounds. In 1971 at the Ivanovo Institute of Chemistry and Technology, he created a gas-phase electron diffraction laboratory, where more than fifty doctoral dissertations have been defended to date. He has several titles, including Honored Worker of Higher Education of the Russian Federation, Honored Worker of Higher Professional Education of the Russian Federation, Full Member of the Russian Academy of Natural Sciences, and Foreign Member of the Norwegian Academy of Science and Letters. In 2024, he was selected as Laureate (winner) of the national award "Professor of the Year".

Preface

This Reprint includes studies that successfully combine experimental and theoretical research methods to determine the structure and spectra of a number of substances in gaseous and condensed states. The presented articles exemplify the self-sufficiency of quantum chemical methods and their ability to replace costly experiments in studying the structure, nature of chemical bonds, electronic and vibrational spectra, and magnetic properties of chemical compounds. This Reprint is intended for readers interested in both the physical chemistry of the specific compounds discussed in the articles and those pursuing the advances of modern quantum chemistry.

Georgiy V. Girichev
Guest Editor

International Journal of
Molecular Sciences

MDPI

Editorial

Special Issue "Molecular Structure, Electronic, and Vibrational Spectra Theoretical Calculations in Materials Sciences"

Georgiy V. Girichev

Department of Physics, Ivanovo State University of Chemistry and Technology, Sheremetevsky Avenue 7, 153000 Ivanovo, Russia; g.v.girichev@mail.ru; Tel.: +7-4932-35-9874

1. Introduction

The dialectic of the development of science is expressed in the fact that today's experiments must give way to tomorrow's theory, which makes these experiments unnecessary, but which, in turn, must give way to a new, more accurate experiment the day after tomorrow, and so on along the path of progress. In the case of structural studies, the problem along this path is the different physical meaning of the molecular parameters obtained in the experiment and in quantum chemical calculations, where the researcher is faced with the task of comparing them with each other or using the calculated data to explain the properties of a real object.

Experimental data refer to a real physical object. For example, structural parameters, obtained by diffraction methods, are averaged over electronic, vibrational, and in some cases rotational levels populated under the experimental conditions, or they bear the imprint of collective interaction. The latter often manifests itself in the results of XRD and many spectroscopic studies.

At the same time, the structural and spectroscopic parameters obtained in routine quantum chemical calculations refer to the unobservable hypothetical equilibrium structure of the molecule. In this regard, the problem of comparing and mutually complementing experimental and calculated data, which are in principle incompatible due to different physical meanings, arises, the solution of which requires further development.

Currently, the most important technique for theoretical calculations in the field of chemistry and materials science is the use of quantum chemical methods of varying levels of complexity.

Quantum chemistry in its development appears in the form of stages associated with both the improvement of the theoretical level and the expansion of the range of research objects. The beginning of quantum chemistry as a special branch of physical chemistry was laid thanks to the concept of the French physicist Louis de Broglie, according to whom the key elementary particle responsible for the chemical properties of a substance, the electron, has both corpuscular and wave properties; this was presented in his article [1] (this phenomenon subsequently received a special name—wave–particle duality). The article mentioned was published in 1925, and this year marks exactly 100 years since its publication. Based on the concept of Louis de Broglie, the Austrian physicist Erwin Schrödinger in his classic works [2,3] formulated the so-called the wave equation that bears his name and describes the state of an electron in an atom; namely, it became the basis of quantum chemistry and retains its significance to this day.

In its development, quantum chemistry has gone through a number of stages, expanding its application from the analytical (theoretical) description of the object of study to its numerical modeling, consistently increasing the theoretical level of description of the system under study.

By now, two main approaches to calculating molecular characteristics have been formed—*ab initio* methods and DFT methods. The latter methods are less demanding on computer resources for calculations, and they have today received a significantly larger scale of practical application than *ab initio* methods—Hartree–Fock and post-Hartree–Fock methods. Nevertheless, the "gold standard" of quantum chemistry for small molecules is the CCSD(T) coupled cluster method, which belongs to the *ab initio* method class.

Returning to the beginning of this article, which generally examines the interaction of experiment and theory as the basis of the dialectic of scientific development, it is important to note the main pragmatic purpose of theory—replacing an expensive experiment while simultaneously ensuring the accuracy of predicting the properties being studied, which is not inferior in accuracy to experimental data.

At present (if not in principle at all), the key to the successful development of theoretical methods is the presence of metrologically reliable and clear in physical meaning experimental data, which can be relied upon when testing new theoretical constructs.

Among spectroscopic methods, microwave spectroscopy is most widely used to determine the exact geometric structure of free molecules. However, the application of this method is limited to molecules consisting of a relatively small number of atoms and possessing a dipole moment.

A valuable method for establishing molecular structure is NMR. However, the accuracy of determining internuclear distances through this method is 1–2 orders of magnitude lower than the accuracy of microwave and diffraction methods. It should be noted that the obtained structural data, as a rule, relate to the condensed state of the substance or solution and largely concern clarifying issues of the chemical identification of the sample, conformational analysis, internal influence effects, etc.

Valuable information about the geometric and electronic structure of molecules can be obtained by studying their vibrational (IR and Raman) and electronic (emission, absorption, etc.) spectra.

The most frequently used method for determining the structure of substances in the solid state is single-crystal XRD. If the crystal is molecular, the geometric structure of separate molecules can be found. Thanks to favorable statistics, namely, a very wide sample (the number of reflections recorded during X-ray structural analysis of single crystals can reach tens of thousands), this method allows us to determine interatomic distances with very high metrological accuracy. However, compared to the structure of an isolated (individual) molecule, the molecular structure obtained in X-ray diffractometry is distorted by the collective (intermolecular) interaction in the crystal. This distortion can significantly exceed the error of the diffraction experiment and in some cases can lead to erroneous conclusions when using X-ray structural data to describe the steric and electronic effects that determine the structure of an isolated molecule. Another factor that must be considered when using most XRD structural data is that the scattering of X-rays occurs on the electron shells of atoms in a molecule. Therefore, it is not the internuclear distances that are determined, but the distances between the centers of gravity of the electron density on the atoms, which can differ significantly from the internuclear ones due to the specifics of the chemical bond that is formed. Examples of such situations are the distances between light atoms that have lone electron pairs or the distances between atoms, one of which is a hydrogen atom. In the latter case, special geometric techniques are used that formally define the positions of hydrogen atoms.

The listed circumstances should be taken into account when formulating conclusions based on a comparison of the XRD and quantum chemical structure.

The main experimental method for determining the structure of polyatomic (up to 100 atoms or more) molecules free from collective interaction is gas electron diffraction. However, the molecules under study must be transferred to the gas phase. In some cases, the vaporization process is accompanied by the incongruent evaporation of the sample or oligomerization, its interaction with the inlet system material, and in some cases this requires an effusion experiment at elevated temperatures (up to 2000 K and higher), which significantly complicates both the experiment and the interpretation of its results. The issue of identifying the object under study was resolved in the works [4,5], in which the authors combined electron diffraction and mass spectrometric experiments, thereby obtaining the ability to continuously monitor the composition of the vapor under study throughout the diffraction experiment. However, the different physical meaning of the structure obtained by electron diffraction (it is a nuclear configuration averaged over all electronic, vibrational and rotational states populated under the experimental conditions) and the structure from quantum chemical calculations (it corresponds to a minimum on the potential energy surface and is unobservable in the experiment) greatly complicates the task and requires that the specifics of intramolecular dynamics, based on the calculated quantum chemical force field of the molecule, are taken into account when deciphering the diffraction pattern of molecules.

Considering the vibrational effects and the effects of centrifugal distortion when interpreting the data of a gas-phase electron diffraction experiment allows us to determine the equilibrium molecular structure, which in its physical meaning coincides with the calculated quantum chemical structure. This further allows us to use it equally with the calculated structure to establish structure–property correlations of real objects or when testing quantum chemical methods for calculating molecular structure.

Thus, the question of the physical meaning of the obtained molecular characteristics is in many cases the most crucial key if we want to explain the subtle effects of the relationship between the structure and physicochemical properties of substances.

This introduction does not claim to be an exhaustive examination of all methods capable of providing information on the structure of molecules, but it merely draws the attention of researchers to the validity and quantitative side of conclusions that are based on a comparison of the structural and energy characteristics of molecules obtained by different methods.

This Special Issue "Molecular structure, electronic and vibrational spectra. Theoretical calculations in materials science" represents a characteristic cross-section of the current situation in the literature, with the help of which one can indirectly conceptualize the nature of the interaction of experiment and theoretical calculations and the direction of development of the problem under discussion.

2. An Overview of the Published Articles

Contribution 1. The work of Adabekyan and co-authors is a classic example of the successful complementarity of experimental and computational methods for the purpose of deep penetration into the physicochemistry of the compounds studied. In this work, the photoprocesses in 1,4-diazadistyrylbenzene and 1,3-diazadistyrylbenzene derivative diperchlorates in MeCN were studied by absorption, luminescence, and kinetic laser spectroscopy. The comprehensive experimental measurements were added by DFT/PBE0/6-31+G(d,p)/PCM(MeCN) calculations of *cis* and *trans* isomer structure diazadistyrylbenzenes and their electrocyclization products (dihydrophenanthrenes DHP-1 and DHP-2) in the ground S0 and singlet excited S1 states. The absorption and emission spectra were calculated by TDDFT with the same functional, basis set, and solvent model. The information obtained in the work on photoprocesses in diazadistyrylbenzenes can be implemented in the construction of photoactive supramolecular systems based on their use of macrocyclic compounds.

Contribution 2. The work of Giricheva and co-authors has demonstrated an even closer connection between experimental and theoretical methods. The main goal of the work was to establish the possibility of the existence of cyclic dimers of 4-n-propyloxybenzoic acid formed by hydrogen bonds in the gas phase. The authors used a combined method, including gas electron diffraction, mass spectrometry, and DFT calculations. The calculated structural parameters were refined during the LS structural analysis of the experimental diffraction pattern. In addition, the calculated force field was used both to calculate the starting values of the root mean square vibration amplitudes and vibrational corrections to internuclear distances in order to give them the physical meaning of values close to equilibrium ones. Thus, the experimental and calculated structural parameters were reduced to a single scale, thereby ensuring their correct comparison. The energy of hydrogen bonds to the cyclic fragment of the found dimer was calculated quantum chemically and the bonding nature was described within the framework of NBO analysis.

Contribution 3. The work of Sagan and co-authors is a completely theoretical study of chemical bonding in the associates "cyclopropenylidene—MX_2" and "imidazol-2-ylidene—MX_2", where M = Be, Mg, Zn; X = H, B. The physical nature of these "carbene...MX_2" interactions has been quantitatively characterized by the joint use of a topological QTAIM-based Interacting Quantum Atoms (IQA) decomposition scheme, the molecular orbital-based Extended Transition State—Natural Orbitals for Chemical Valence (ETS-NOCV) charge, and a Local Energy Decomposition scheme (LED) based on the Domain-based Local Pair Natural Orbital Coupled Cluster with Single, Double, and perturbative Triple excitations (DLPNO-CCSD(T)) method, which was used to minimize the error introduced by the theoretical level of calculations.

The characteristic feature of the considered associates is the bent structure of the MX2 molecule. The formation of the C...M bond leads to the significant elongation of the M–X bond and a greater opening of the L–C–L angle in the carbene molecule. It is also worth noting that for the bond length M–X, the relation Be < Zn < Mg applies so that for a given X, the magnesium atom forms the longest M–X bond.

Contribution 4. The work of Tverdova and co-authors reports the results of studying the geometric structure, vibrational frequencies, electronic characteristics, and thermodynamic characteristics of conformers, as well as the structure and energy of transition states of 4-(4-tritylphenoxy)phthalonitrile (TPPN), which can be used in material science as a precursor in the synthesis of phthalocyanines with bulky substituents. A set of complementary methods include gas electron diffraction, mass spectrometry, IR spectroscopy, and quantum chemical calculations (DFT/B3LYP, PBE, CAM-B3LYP, and B97D). It was found that the molecule TPPN has four conformers. All vibrational bands in the experimental IR spectrum of TPPN have been assigned by symmetry and the distribution of normal vibration energy over internal coordinates. A slight difference was noted in the theoretical IR spectra of the conformers, which, however, made it possible to determine that TPPN consists of a mixture of conformers in the condensed state. The structure of transition states between conformers is determined. The close calculated values of the total energy of the conformers and high transition barriers suggest that the conformational diversity of TPPN is formed during the synthesis of the compound.

Contribution 5. The work of Tiunova and co-authors describes the synthesis of cationic Mn(III) complexes with ligands 5-Hal-sal2323 (Hal = Cl, Br) and a paramagnetic divalent counterion $[ReCl_6]^{2-}$ as follows: $[Mn(5\text{-}Cl\text{-}sal_2323)]_2[ReCl_6]$ (1) and $[Mn(5\text{-}Br\text{-}sal_2323)]_2[ReCl_6]$ (2). The crystal structures and magnetic properties of both isostructural two-component ionic compounds were studied at six temperatures (from 100 to 423 K).

The phenomenon of different spin behavior of Mn atoms in stoichiometrically identical fragments of [Mn(5-Cl-sal2323)] within 1 and 2 was discovered. A thermally induced spin transition at high temperature, associated with the cationic moiety, and a field-induced slow magnetic relaxation of magnetization at cryogenic temperature, associated with the anionic moiety of the complexes, were detected.

Contribution 6. In the work of Merkel and co-authors, FTIR spectroscopy in combination with DFT/B3LYP/6-311 G (p, d) calculations were applied to study the geometrical structure and intermolecular interaction of two mesogenic groups based on cyanobiphenyl (CB) with thioether-linking groups (C-S-C) termed $CBSC_nSCB$ (n = 5, 7) and the asymmetric system with the ether- and thioether-linking groups, termed $CBSC_nOCB$ (n = 5, 7). Changes in the behavior of intermolecular interactions were observed by significant differences in the values of transition dipole moments for selected vibrational bands. One significant finding of this study was evidence that the intermolecular forces evolved on the transition from the nematic phase to the twist–bend phase. A model for the packing of the dimer molecules in the twist–bend phase was developed due to the overlapping of the rigid cyanobiphenyl cores stabilizing this phase and thus affecting the helical pitch.

Contribution 7. In the work of Ferjani and co-authors, new salt of the antifungal drug fluconazole, $(H_2Fluconazole){\cdot}SnCl_6{\cdot}2H_2O$, was synthesized and characterized by single-crystal XRD, Hirshfeld surface analysis, Raman, FTIR, UV-visible spectroscopies, and magnetic property measurements. To understand and predict the chemical behavior of $(H_2Fluconazole){\cdot}SnX_6{\cdot}2H_2O$; (X = F, Cl, Br, and I), global chemical reactivity descriptors were used, including E_{HOMO}, E_{LUMO}, ionization potential, electron affinity, hardness, softness, chemical potential, electrophilicity, and electronegativity, and these were calculated at the theory level HSE1PBE/6-311++G**,LANL2DZ(Sn). TDDFT calculations were performed at the optimized geometries using the same levels of theory.

Contribution 8. In the work of Loska and co-authors, FTIR spectroscopy and DFT/B3LYP/6-311 G (p, d) calculations have been used to study the structure and molecular interactions in the nematic and twist–bend phases of symmetrical and asymmetrical dimers with the cyanobiphenyl mesogenic groups linked in two mesogenic cores by methylene, thioether, and ether fragments. The presence of conformers and the knowledge of their structural and energetic characteristics are important for calculating the orientation order in liquid crystals. In this connection, the dihedral angle distributions and conformations in terms of potential energy functions for torsional motion were the objects of special attention. Another distinctive feature of this work is the presence of detailed data on the theoretical and experimental frequencies, dichroism values, relative intensity, direction of the transition dipole moment, and approximate band assignments for all dimers investigated.

Contribution 9. In the work of Qi Gong and Guiling Zhang, a state-of-the-art quantum mechanics-based program, CASTEP module of Materials Studio 2020 package (Accelrys Inc., Materials Studio version 2020.08, San Diego, CA, USA), specifically designed for solid-state materials science was applied to study of new 2D topological insulators (TIs) with a large bulk bandgap. It was identified through first-principles calculations that HSb_2XH_n and HBi_2XH_n form a new family of giant bandgap 2D TIs. On the examples of HSb_2XH_n and HBi_2XH_n, spin–orbit coupling electronic structures of Janus (asymmetrically functionalized by CH_3 or OH) Sb and Bi monolayers with nontrivial topology, Rashba splitting, and valley-contrast circular dichroism are proposed and demonstrated in the present study.

Contribution 10. In the work of Ryzhov and co-authors, the electronic and geometric structures of metal-free tetrapyrazinoporphyrazine (TPyzPA), octachlorotetrapyrazinoporphyrazine ($TPyzPACl_8$), and their complexes with Al, Ga, and In were studied at the

DFT/PBE0-D3/def2TZVP theory level. The main objective was to study the effect of chlorination on the structure and properties of these macrocycles.

It has been established that the structural parameters of the macrocyclic ligand undergo minimal changes when replacing a central metal atom or introducing chlorine atoms at the periphery.

The substitution of H-atoms for Cl leads to a decrease in the energy of the frontier MOs of M(Cl)TPyzPACl_8 as compared to M(Cl)TPyzPA. The electron absorption spectrum of TPyzPACl_8 contains three intense bands, while only two bands were found for TPyzPA. The calculated IR spectra of TPyzPACl_8 have a significant difference in comparison with TPyzPA due to heavy Cl atoms involved in the vibrations that form medium and strong bands in the region of 1000–1500 cm^{-1}.

Contribution 11. The work of Klyukin and co-authors is devoted to the combined theoretical and experimental study of the protonation of $[2,6\text{-}B_{10}H_8O_2CCH_3]^-$. This is a typical example of the successful use of quantum chemical calculations to understand the mechanism of a chemical reaction for the synthesis of a compound, in particular, the $[2,6\text{-}B_{10}H_8O_2CCH_3{*}H_{fac}]^0$ complex. The structure of this complex was obtained using NMR experiments and DFT calculations. The 11B-1H NMR spectra of this complex were found to be complex and difficult to interpret. Notably, DFT/B97/IGLO-III calculations reproduced the experimental NMR spectrum well.

Contribution 12. The work of Ziyi Wang and co-authors demonstrates the effectiveness of combining experimental spectroscopy and quantum chemical calculations for interpreting vibrational spectra. The initiative of this work relates to the problem of the corrosion of technological equipment under the influence of organic sulfur compounds. However, the problem of corrosion itself is not specifically considered here, and the study is limited to a detailed analysis of the structure, as well as vibrational spectra and vibrational behaviors, of two sulfur-containing compounds, dibenzyl disulfide and dibenzyl sulfide (DBDS, DBS), as well as bibenzyl (BiBz). The obtained IR and Raman spectra of these compounds were interpreted using DFT calculations. The conformational composition was calculated according to the Boltzmann distribution in relative Gibbs free energy. The analysis of the distribution of potential energy of normal vibrations along internal coordinates was used in the assignment of experimental spectra. The Noncovalent Interaction Analysis (NCI) method was applied to reveal the nature of dominant conformations with low free energy.

No comments were made regarding the validity of the comparison of vibrational spectra, since these were measured for the solid phase and calculated for the gas.

Contribution 13. In the work of Dan Deng and co-authors, the phosphorescence mechanism of the (E)-3-(((4-nitrophenyl)imino)methyl)-2H-thiochroman-4-olate-BF_2 compound has been theoretically investigated. The calculated radiative and nonradiative rate constants confirm the predominant relaxation scheme of the excited state S2 in the form of a chain of transitions S0→S2→S1→T2→S0, indicating a violation of the Kasha rule.

3. Conclusions

The articles presented in this Special Issue attest to the increasingly deep complementarity of experiments and quantum chemical calculations. In some cases, calculations act as an independent research method, claiming to eliminate the need for experiment and demonstrating the availability of obtaining the wider spectrum of properties of the compounds studied compared to experiments. Improvements in calculation methods eventually eliminate the need for an experiment in cases where this replacement cost is effective. In science circles, the following quote is known: "There is nothing more practical than theory". However, it is necessary to remember that the effectiveness of theoretical methods is tested by comparing their predictions with the results of experiments. In this regard, the following quote of Nobel Prize Laureate, Professor Pyotr L. Kapitsa, is substantiated: "Theory is a good thing, but well done experiment lasts forever" [6]. In this regard, works that contain experimental and theoretical results that are both complementary and mutually verifiable are of particular value.

Funding: This research received no external funding.

Data Availability Statement: Not applicable.

Acknowledgments: The author expresses gratitude to O.M. Mikhailov for the useful discussions.

Conflicts of Interest: The author declares no conflicts of interest.

List of Contributions:

1. Atabekyan, L.S.; Freidzon, A.Y.; Ionov, D.S.; Nuriev, V.N.; Medved'ko, A.V.; Vatsadze, S.Z.; Gromov, S.P.; Chibisov, A.K. Photoprocesses in Derivatives of 1,4- and 1,3-Diazadistyryldibenzenes. *Int. J. Mol. Sci.* **2022**, *23*, 15346. https://doi.org/10.3390/ijms232315346.
2. Giricheva, N.I.; Bubnova, K.E.; Krasnov, A.V.; Girichev, G.V. Cyclic Dimers of 4-n-Propyloxybenzoic Acid with Hydrogen Bonds in the Gaseous State. *Int. J. Mol. Sci.* **2022**, *23*, 15079. https://doi.org/10.3390/ijms232315079.
3. Sagan, F.; Mitoraj, M.; Jabłoński, M. Nature of Beryllium, Magnesium, and Zinc Bonds in Carbene$\cdots$$MX_2$ (M = Be, Mg, Zn; X = H, Br) Dimers Revealed by the IQA, ETS-NOCV and LED Methods. *Int. J. Mol. Sci.* **2022**, *23*, 14668. https://doi.org/10.3390/ijms232314668.
4. Tverdova, N.V.; Giricheva, N.I.; Maizlish, V.E.; Galanin, N.E.; Girichev, G.V. Molecular Structure, Vibrational Spectrum and Conformational Properties of 4-(4-Tritylphenoxy)phthalonitrile-Precursor for Synthesis of Phthalocyanines with Bulky Substituent. *Int. J. Mol. Sci.* **2022**, *23*, 13922. https://doi.org/10.3390/ijms232213922.
5. Tiunova, A.V.; Kazakova, A.V.; Korchagin, D.V.; Shilov, G.V.; Aldoshin, S.M.; Dmitriev, A.I.; Zhidkov, M.V.; Zakharov, K.V.; Yagubskii, E.B. Spin-Crossover and Slow Magnetic Relaxation Behavior in Hexachlororhenate(IV) Salts of Mn(III) Complexes $[Mn(5\text{-Hal-sal2323})]_2[ReCl_6]$ (Hal = Cl, Br). *Int. J. Mol. Sci.* **2022**, *23*, 11449. https://doi.org/10.3390/ijms231911449.
6. Merkel, K.; Loska, B.; Arakawa, Y.; Mehl, G.H.; Karcz, J.; Kocot, A. How Do Intermolecular Interactions Evolve at the Nematic to Twist–Bent Phase Transition? *Int. J. Mol. Sci.* **2022**, *23*, 11018. https://doi.org/10.3390/ijms231911018.
7. Ferjani, H.; Bechaieb, R.; Alshammari, M.; Lemine, O.M.; Dege, N. New Organic–Inorganic Salt Based on Fluconazole Drug: TD-DFT Benchmark and Computational Insights into Halogen Substitution. *Int. J. Mol. Sci.* **2022**, *23*, 8765. https://doi.org/10.3390/ijms23158765.
8. Kocot, A.; Loska, B.; Arakawa, Y.; Mehl, G.H.; Merkel, K. Study of the Experimental and Simulated Vibrational Spectra Together with Conformational Analysis of Thioether Cyanobiphenyl-Based Liquid Crystal Dimers. *Int. J. Mol. Sci.* **2022**, *23*, 8005. https://doi.org/10.3390/ijms23148005.
9. Gong, Q.; Zhang, G. Rashba Splitting and Electronic Valley Characteristics of Janus Sb and Bi Topological Monolayers. *Int. J. Mol. Sci.* **2022**, *23*, 7629. https://doi.org/10.3390/ijms23147629.
10. Ryzhov, I.V.; Eroshin, A.V.; Zhabanov, Y.A.; Finogenov, D.N.; Stuzhin, P.A. DFT Study of Molecular Structure, Electronic and Vibrational Spectra of Tetrapyrazinoporphyrazine, Its Perchlorinated Derivative and Their Al, Ga and In Complexes. *Int. J. Mol. Sci.* **2022**, *23*, 5379. https://doi.org/10.3390/ijms23105379.
11. Klyukin, I.N.; Kolbunova, A.V.; Novikov, A.S.; Nelyubin, A.V.; Selivanov, N.A.; Bykov, A.Y.; Klyukina, A.A.; Zhdanov, A.P.; Zhizhin, K.Y.; Kuznetsov, N.T. Protonation of Borylated Carboxonium Derivative $[2,6\text{-}B_{10}H_8O_2CCH_3]^-$: Theoretical and Experimental Investigation. *Int. J. Mol. Sci.* **2022**, *23*, 4190. https://doi.org/10.3390/ijms23084190.
12. Wang, Z.; Song, R.; Chen, W.; Wang, J.; Wang, P.; Zhang, Z.; Zhang, X.; Wan, F. Vibrational Spectra and Molecular Vibrational Behaviors of Dibenzyl Disulfide, Dibenzyl Sulphide and Bibenzyl. *Int. J. Mol. Sci.* **2022**, *23*, 1958. https://doi.org/10.3390/ijms23041958.

13. Deng, D.; Suo, B.; Zou, W. New Light on an Old Story: Breaking Kasha's Rule in Phosphorescence Mechanism of Organic Boron Compounds and Molecule Design. *Int. J. Mol. Sci.* **2022**, *23*, 876. https://doi.org/10.3390/ijms23020876.

References

1. De Broglie, L. Recherches sur la théorie des Quanta. *Ann. Phys.* **1925**, *10*, 22–128. Available online: https://theses.hal.science/tel-00006807 (accessed on 29 July 2025). (In French) [CrossRef]
2. Schrödinger, E. An Undulatory Theory of the Mechanics of Atoms and Molecules. *Phys. Rev.* **1926**, *28*, 1049–1070. [CrossRef]
3. Schrödinger, E. Quantisierung als Eigenwertproblem; von Erwin Schrödinger. *Ann. Phys.* **1926**, *384*, 361–377. (In German) [CrossRef]
4. Bohatka, S.; Berecz, I.; Tremmel, J.; Hargittai, I. Combined Quadrupole Mass Spectrometric and Electron Diffraction Experiment for Molecular Structure Deteremination. In Proceedings of the 7th International Vacuum Congress & 3rd International Conference on Solid Surfaces, Viena, Austria, 12–16 September 1977; pp. 193–196.
5. Girichev, G.V.; Shlykov, S.A.; Petrova, V.N.; Subbotina NYu Lapshina, S.B.; Danilova, T.G. Equipment and methods for a joint electronographic and mass spectrometric experiment and their application to the study of titanium trihalide molecules. *Izv. VUZ. Chem. Chem. Tech.* **1988**, *31*, 46–51. (In Russian)
6. Kapitsa, P.L. *Experiment, Theory, Practice*; Nauka: Moscow, Russia, 1981. (In Russian)

International Journal of
Molecular Sciences

MDPI

Article

Photoprocesses in Derivatives of 1,4- and 1,3-Diazadistyryldibenzenes

Levon S. Atabekyan [1,*], Alexandra Y. Freidzon [1,2,*], Dmitry S. Ionov [1], Vyacheslav N. Nuriev [1,3], Alexey V. Medved'ko [1,4], Sergey Z. Vatsadze [1,3], Sergey P. Gromov [1,3] and Alexander K. Chibisov [1]

1 Photochemistry Center of RAS, FSRC "Crystallography and Photonics", Russian Academy of Sciences, Novatorov Str. 7A-1, 119421 Moscow, Russia
2 Institute of Nanoengineering in Electronics, Spintronics and Photonics, National Research Nuclear University MEPhI, Kashirskoye Shosse, 31, 115409 Moscow, Russia
3 Department of Chemistry, M. V. Lomonosov Moscow State University, Leninskie Gory 1-3, 119991 Moscow, Russia
4 N.D. Zelinsky Institute of Organic Chemistry RAS, Leninsky Prosp. 47, 119991 Moscow, Russia
* Correspondence: levat51@mail.ru (L.S.A.); freidzon.sanya@gmail.com (A.Y.F.)

Abstract: Photoprocesses in 1,4-diazadistyrylbenzene (**1**) and 1,3-diazadistyrylbenzene derivative (**2**) diperchlorates in MeCN were studied by absorption, luminescence, and kinetic laser spectroscopies. For compound **1**, *trans-cis*-photoisomerization and intersystem crossing to a triplet state are observed. For compound **2**, photoelectrocyclization is suggested. Quantum chemical calculations of diazadistyrylbenzene structures in the ground and excited states were carried out. The schemes for photoreactions were proposed.

Keywords: diazadistyrylbenzene; electrocyclization; fluorescence; laser photolysis; triplet state; *trans-cis*-isomerization; density functional theory

1. Introduction

Distyrylbenzenes are promising photoactive compounds for the design of organic light-emitting diodes, materials for solar cells, nonlinear optical materials, and chemical sensors [1–5]. The presence of two C=C double bonds in a molecule of distyrylbenzene offers additional opportunities for their phototransformations. Distyrylbenzenes have been the object of a sufficient number of studies [6–14], but there were no data on photoprocesses of diazadistyrylbenzenes and their derivatives. Earlier, we studied the photonics of tetramethoxy-1,4-distyrylbenzenes [15], and also bis-15-crown-5- and 18-crown-6-containing 1,3- and 1,4-distyrylbenzenes [16–21] and their ability to form supramolecular complexes with metal cations.

The present study is focused on the derivatives of 1,4- and 1,3-diazadistyrylbenzenes (**1** and **2**, respectively, Figure 1). The aim of this work was a comparative study of the photonics of these compounds in MeCN by absorption, fluorescence, and kinetic laser spectroscopies, detecting the triplet state and intermediate products of photoreactions. In parallel with experimental studies, quantum chemical calculations were carried out to elucidate the structure of the chromophore in the course of photoexcitation, as well as to interpret the experimental absorption and fluorescence spectra.

The established ability of isomeric 1,4- and 1,3-derivatives of distyrylbenzenes to undergo various types of photochemical transformations, including those involving triplet states, has not been studied previously and is of considerable interest. This ability makes them promising for use in photoactive materials. The photoprocesses of diazadistyrylbenzenes under study can also be implemented in the construction of photoactive supramolecular systems on their basis with the use of macrocyclic compounds.

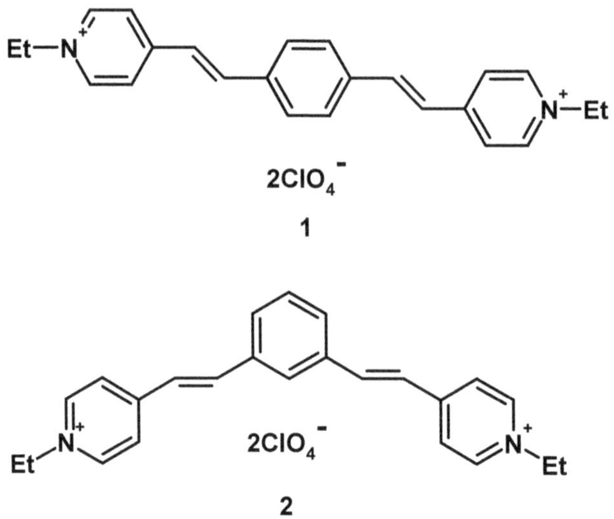

Figure 1. Structures of 4,4′-{1,4-phenylenebis[(E)-ethene-2,1-diyl]}bis(1-ethylpyridinium diperchlorate (**1**) and 4,4′-{1,3-phenylenebis[(E)-ethene-2,1-diyl]}bis(1-ethylpyridinium) diperchlorate (**2**).

2. Results and Discussion

2.1. Absorption and Fluorescence Spectra

Figure 2 shows the absorption spectra, and Table 1 represents the corresponding absorption band maxima and extinction coefficients for compounds **1** and **2**. The absorption spectra of **1** and **2** differ both in shape and in the position of the absorption band maxima. A significant blue (40 nm) shift is observed for compound **2** as compared with **1**. In this case, the extinction coefficients of both compounds are virtually identical.

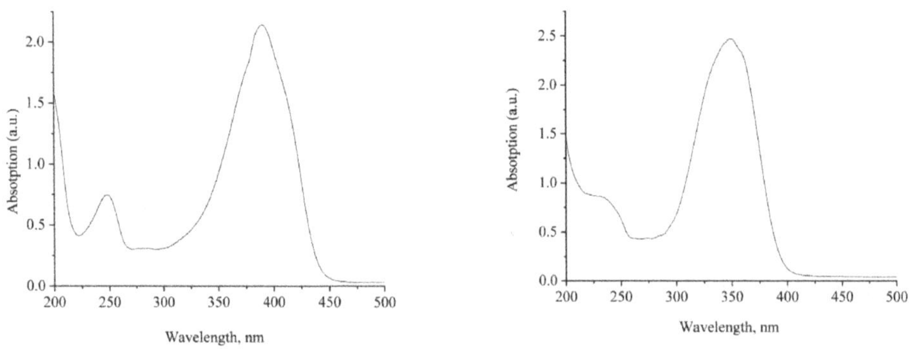

(**a**) (**b**)

Figure 2. Absorption spectra of compounds (**a**) **1** and (**b**) **2**. The concentration of compounds is 4×10^{-5} (**a**) and 5×10^{-5} (**b**) mol/L.

The calculation shows that the absorption of **1** originates from a single delocalized electronic transition with high oscillator strength, while the absorption of **2** originates from four localized electronic transitions with lower oscillator strength, which sum up to almost the same intensity as the absorption of **1**.

Table 1. Absorption band maxima and extinction coefficients.

Compound	Experiment		Calculation	
	λ_{max}, nm	ε, $L \cdot mol^{-1} \cdot cm^{-1}$	λ_{max}, nm	*f*
1	390	5.4×10^4	420	2.07
2	350	5.0×10^4	372 [1] 350 [1] 348 [1] 314 [1]	0.59 [1] 0.58 [1] 0.75 [1] 0.23 [1]

f is the oscillator strength; [1] four absorption bands of **2**.

Figure 3 shows the fluorescence spectra of compounds **1** and **2**, and Table 2 represents their absorption band maxima, fluorescence quantum yields, and fluorescence lifetimes.

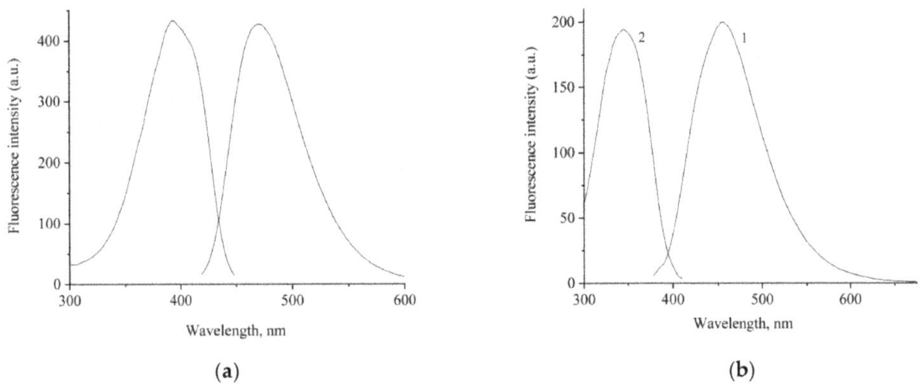

Figure 3. Fluorescence emission (1) and excitation (2) spectra of compounds (**a**) **1** and (**b**) **2**. The concentration of compounds **1** and **2** is 5×10^{-6} mol/L.

Table 2. Emission band maxima, fluorescence quantum yields φ and fluorescence lifetimes.

Compound	Experiment						Calculation		
	φ	λ_{max}, nm	$\lambda_{0\text{-}0}$, nm	λ_{exc}, nm	Δλ, nm	τ_{fl}, ps	λ_{max}, nm	*f*	τ_r, ps
1	0.05	470	434	393	77	120	454 [1] 472 [2]	2.18 1.34	1420 2500
2	0.02	455	395	346	109	160	409 [1] 422 [2]	0.30 0.19	8350 13700

Δλ is a Stokes shift between the fluorescence emission and excitation maxima; [1] *trans* isomer; [2] *cis* isomer.

A weak fluorescence of compounds **1** and **2** was detected. Fluorescence excitation spectra coincide with the corresponding absorption spectra. At the same time, compound **2** demonstrates a large Stokes shift of the fluorescence band (109 nm) and a significant (by 2.5 times) decrease in the fluorescence quantum yield compared with compound **1**, which can be explained by a lower degree of conjugation and a significant deformation of the structure upon photoexcitation of **2**. The measured fluorescence lifetime is 120 ps for **1** and 160 ps for **2**. The fluorescence kinetics for compound **2** was found to have a biexponential character, which can confirm the presence of another form of **2** in the excited state with a structure significantly different from the initial one.

The calculated radiative lifetimes are 20–50 times longer than the measured ones, which is consistent with the low fluorescence quantum yield ($\varphi = \tau/\tau_r$). In the absence of a quencher and the occurrence of the intersystem crossing to the triplet state, the fluorescence lifetime is determined as $\tau = 1/(k_r + k_{nr} + k_{ic})$. The nonradiative lifetime τ_{nr} for *trans* isomer

1 is 126 ps, for **2,** it is 163 ps, whereas the radiative lifetime is 1420 and 8950 ps, respectively, thus being in agreement with the experimental values.

2.2. Kinetic Spectroscopy

Upon laser flash excitation at 355 nm, a change in the absorption spectrum of compound **1** is observed, as illustrated in Figure 4. The corresponding kinetic curves measured in the maxima of the photoinduced absorption bands are shown in Figure 5.

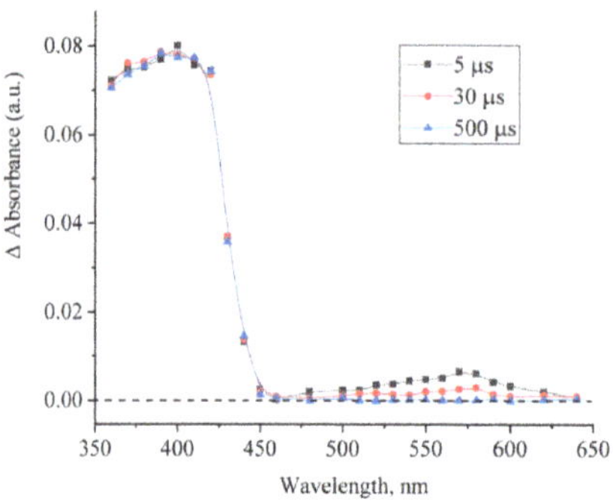

Figure 4. Difference photoinduced absorption spectra of deoxygenated solution of compound **1** measured at 5, 30, and 300 μs after the laser pulse. The concentration of compound **1** is 5×10^{-5} mol/L.

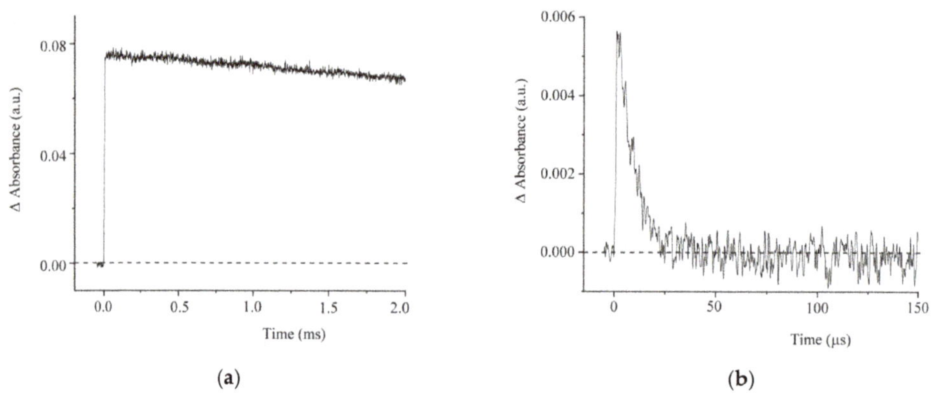

Figure 5. Kinetic curves of the absorbance change in the deoxygenated solution of compound **1** measured at λ = 400 nm (**a**) and 570 nm (**b**). The concentration of compound **1** was 5×10^{-5} mol/L.

The photoinduced absorbance change at 350–450 nm can be attributed to *trans-cis*-isomerization of **1** to give a relatively long-lived *cis*-isomer (Figure 5a). In this case, the rate of decrease in absorption at 400 nm does not depend on the ambient oxygen content in the solution and characterizes the process of *cis-trans* isomerization. The absorption band in the region of 500–650 nm is observed only for the deoxygenated solution and can be associated with the T-T absorption. The triplet state lifetime of **1** is 30 μs (Figure 5b).

The absence of bleaching in the principal absorption band of compound **1** can be attributed to the similarity of the absorption spectra of the *trans* and *cis* isomers of the compound.

For compound **2**, the difference in the photoinduced absorption in the visible spectral range is negligible and more than an order of magnitude smaller than for compound **1**. Figure 6 illustrates the different spectrums of photoinduced absorption of compound **2** under the accumulation of absorption curves.

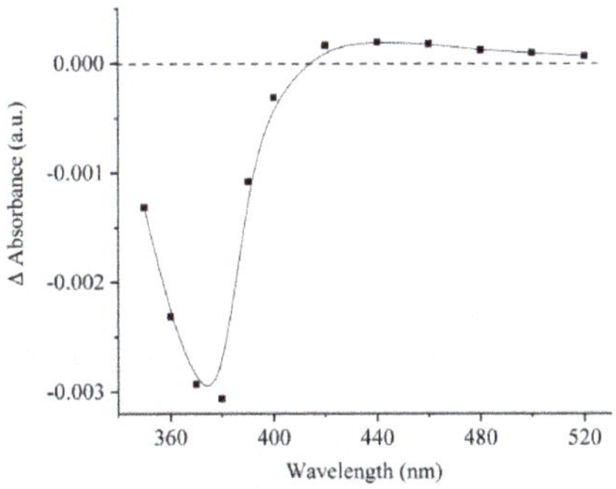

Figure 6. Photoinduced absorption spectrum of compound **2** measured at 5 μs after the laser pulse. The concentration of compound **2** was 5×10^{-5} mol/L.

The observed photoinduced changes in the absorbance of compound **2** can be explained by assuming that these changes occur mainly in the short-wavelength region of the spectrum, which made it impossible to detect them by kinetic laser spectroscopy. In addition to spectral measurements performed upon laser excitation, the absorption spectra of compounds **1** and **2** were measured upon stationary UV irradiation. Figure 7 shows the obtained data.

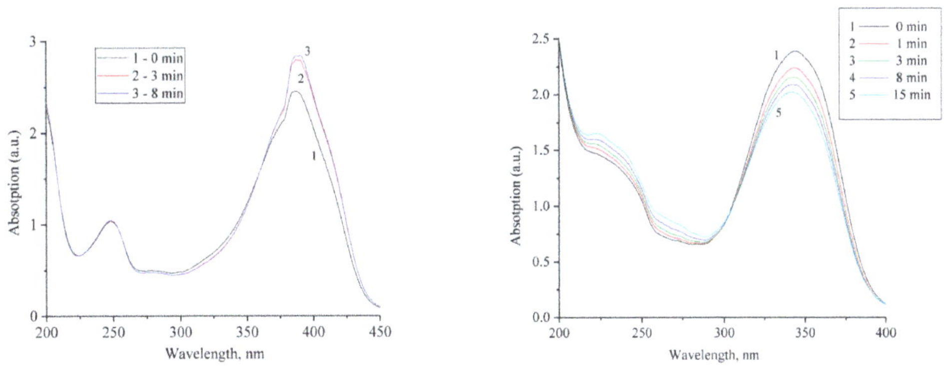

(a) (b)

Figure 7. Absorption spectra of compounds **1** (**a**) and **2** (**b**) measured upon UV irradiation at various times as indicated in the insets. The concentration of compounds **1** and **2** was 5×10^{-5} mol/L.

Upon UV irradiation, an increase in optical density of the solution of compound **1** was observed, which corresponds to a photoinduced change in absorption measured upon laser excitation (Figure 4) and attributed to the formation of a long-lived *cis* isomer of **1**, which has an absorption spectrum similar to that of the *trans* isomer and a significant extinction coefficient.

On the contrary, UV irradiation of compound **2** results in a decrease in absorption of the principal absorption band and an increase in absorption in the short-wavelength region of the spectrum, which is consistent with the experimental data obtained upon laser excitation. It can be assumed that for compound **2**, photoelectrocyclization takes place to give the final product, which absorbs in the short-wavelength region of the spectrum. This conclusion is supported by the literature data confirming the participation of 1,3-distyrylbenzenes in the electrocyclization reaction [1–3,8]. Photocyclization of 1,3-diazadistyrylbenzene seems to occur in the singlet excited *cis* isomer, which has a short lifetime. The possible scheme of electrocyclization of compound **2** is depicted in Figure 8.

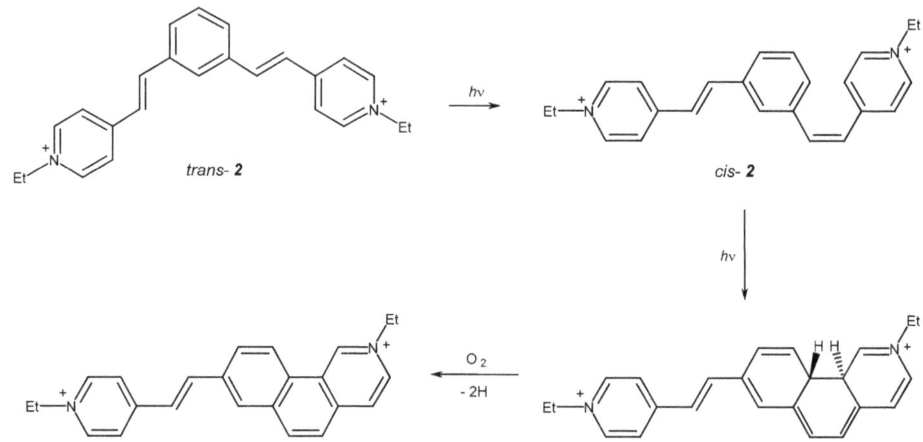

Figure 8. Scheme of electrocyclization of compound **2**.

2.3. *Quantum Chemical Calculations*

2.3.1. Model

In general, the processes of laser flash photolysis experiments can be viewed as follows.

The reactions of *trans-cis* and *cis-trans* isomerization proceed via a conical intersection (CI), and the minimum energy paths from the local *trans* or *cis* minima on the S_1 potential energy surface to the CI include saddle points, which limit the photoisomerization rate. The energy of the transition state relative to the *trans* or *cis* local minimum can be considered as the activation energy of *trans-cis* photoisomerization.

A laser pulse (355 nm) gives the molecule (*trans*-**1** or *trans*-**2**) excessive energy, which may be sufficient to overcome the barriers of *trans-cis* photoisomerization. After that, the molecules can easily reach the region of the conical intersection and relax nonradiatively either to the *trans* or to the *cis* form in the ground state. This explains why the *trans-cis* isomerization takes place in the nanosecond timescale under the laser flash photolysis conditions.

When photoinduced absorption spectra are measured (after at least 5 μs after the laser pulse), the excess energy is already dissipated either radiatively or nonradiatively, and the mixture consists of the initial molecule in the ground state and its primary photoproducts in their lowest states (S_0 for singlet products and T_1 for triplet products). In order to measure photoinduced absorption, the molecules in the mixture are excited directly to their absorption bands without pumping excessive energy. In this case, the barriers for the excited-state reactions can be overcome only through thermal activation, and the reaction rates can be estimated using the Arrhenius formula, given that the corresponding excited state is sufficiently long-lived.

Therefore, after the laser pulse, the mixture consists of both *trans* and *cis* of **1** or **2** in their ground state and, where possible, their triplet forms. Upon measuring the photoinduced absorption spectra, the triplet forms absorb at ~500–600 nm, and the singlet forms absorb at ~350–400 nm, undergo thermally activated or barrierless phototransformations, or emit light (fluoresce).

2.3.2. Results

The structures of 1,4-diazadistyrylbenzene (**1**) and its electrocyclization product calculated by PBE0/6-31+G(d,p)/PCM(MeCN) are shown in Figure 9. The *trans* isomer is planar both in the ground and in the excited states. The *cis* isomer is nonplanar; the C-C=C-C torsion angle is ~9° in the ground state and ~23° in the excited state. These torsions are close to the same angles in the DHP (~8 and 18° in the ground and excited states, respectively). The ground-state DHP is a local minimum, while the excited-state DHP is a conical intersection (CI). *Trans-cis* photoisomerization proceeds via another conical intersection at

~90° C-C=C-C torsion angle. The minimum energy path from both *trans* and *cis* minima to the CI includes two saddle points at ~121 and 66° C-C=C-C torsion angle.

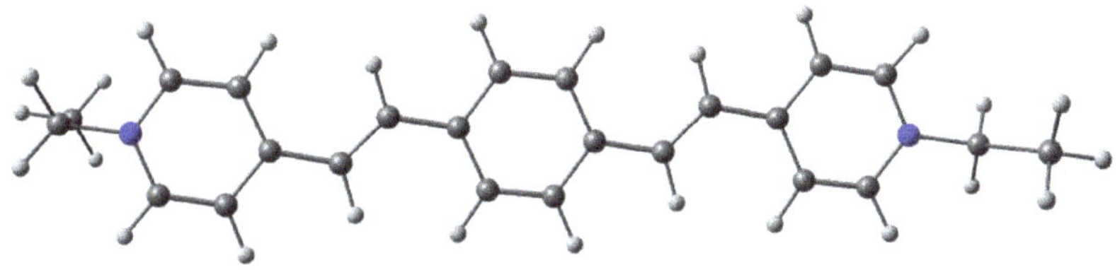

(**a**)

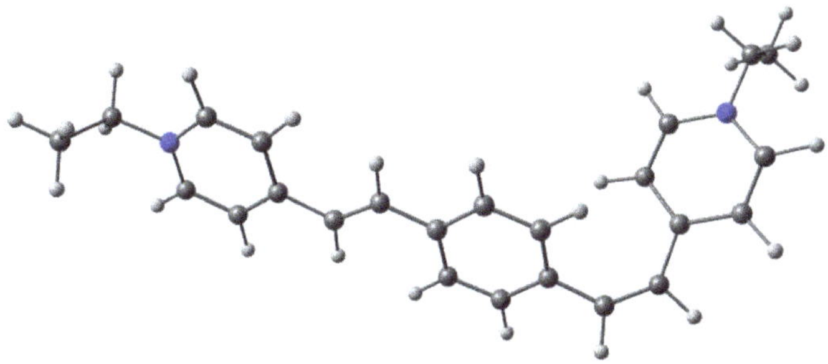

(**b**)

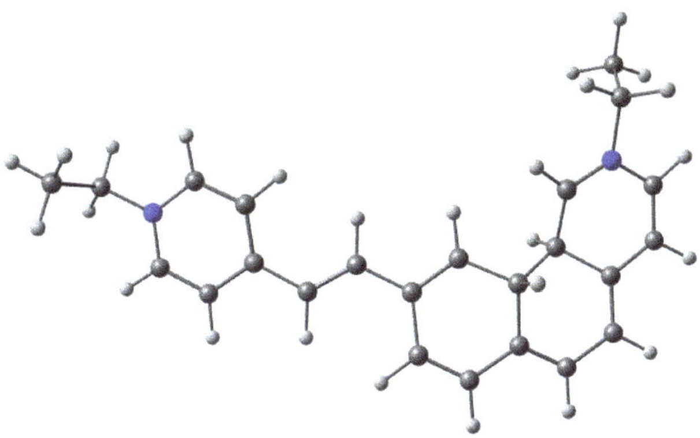

(**c**)

Figure 9. Structures of (**a**) *trans*-**1** and (**b**) *cis*-**1** and (**c**) its electrocyclization product dihydrophenanthrene DHP-**1**.

Figure 10 shows the energy level diagram of diazadistyrylbenzene **1** (a), its possible phototransformations (b), and the absorption maxima of the phototransformation products (c). On the basis of the calculated data, we may infer the following picture of the photoprocesses in **1**.

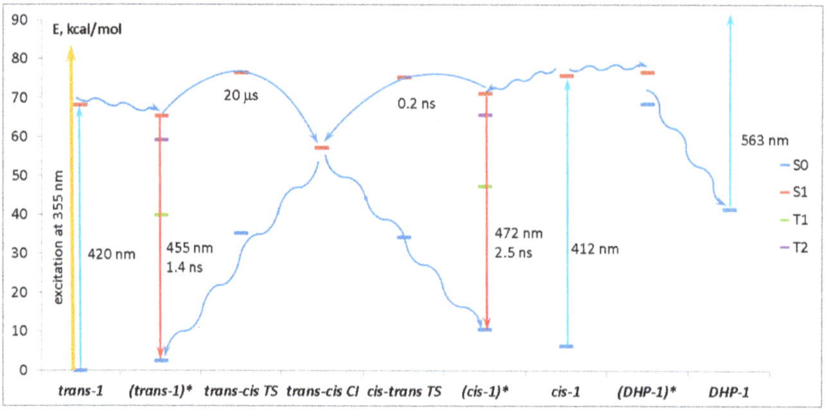

(**a**)

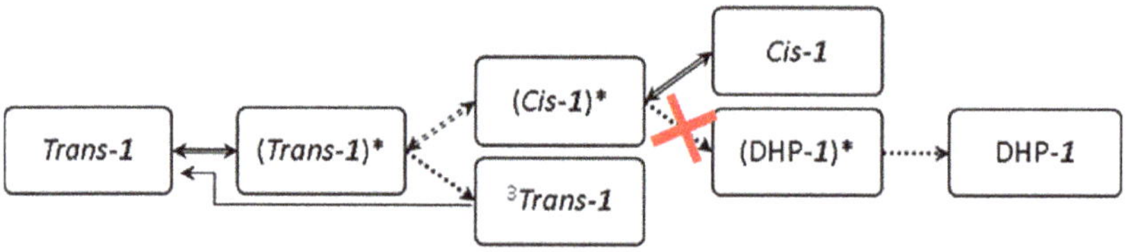

(**b**)

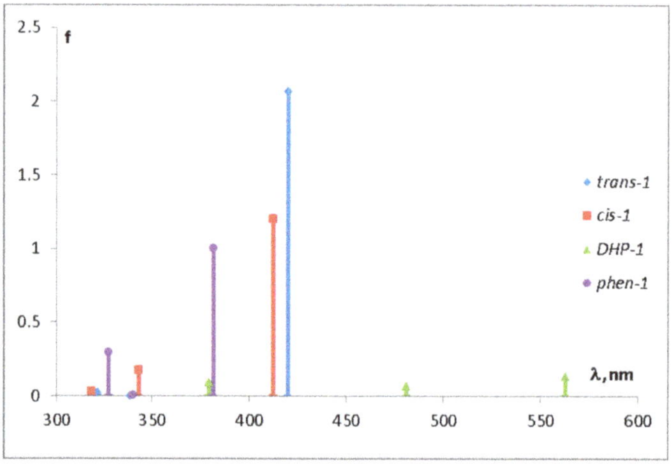

(**c**)

Figure 10. (**a**) Energy level diagram of diazadistyryl benzene **1**, (**b**) scheme of its possible phototransformations, and (**c**) absorption maxima of the phototransformation products. Straight arrows on the energy diagram denote radiative processes, and wavy arrows denote nonradiative processes.

As mentioned above, after the laser pulse, the mixture consists of *trans*-**1**, *cis*-**1**, and, possibly, ^{3}*trans*-**1**. When photoinduced absorption is measured, *trans*-**1** absorbs at 420 nm (390 nm in the experiment) with further relaxation to the minimum of the S_1 state ((*trans*-**1**)*). From the minimum, the molecule can emit light at 455 nm (470 nm in the experiment), convert to the triplet state (^{3}*trans*-**1**) through S_1-T_2 intersystem crossing with further phosphorescence and nonradiative relaxation, or undergo a *trans-cis* isomerization. Without excess excitation energy provided by the laser pulse, this process proceeds in the microsecond timescale and, therefore, is sufficiently slow. The reverse *cis-trans* isomerization in the excited state proceeds much faster, namely, in the nanosecond timescale and can take place even under the conditions of steady-state photoexcitation. Transformation to DHP-**1** from any of the S_1 minima is unfavorable (denoted by red cross in Figure 10b) because the excited (DHP-**1**)* is ~5.5 kcal/mol higher than the excited (*cis*-**1**)* minimum, and the excess energy received by the molecule upon primary flash excitation is dissipated.

The absorption bands of *cis*-**1** and *trans*-**1** are very close and overlap when measured (calculated 420 and 412 for *trans*-**1** and *cis*-**1**, respectively). The emission proceeds in the nanosecond timescale, and the fluorescence spectra of both *cis* and *trans* isomers also overlap (calculated at 455 and 472 nm for *trans*-**1** and *cis*-**1**, respectively).

The rate of phototransformation to the triplet state was not evaluated. However, it is worth noticing that T_2 lies only ~6 kcal/mol lower than S_1 both in the *trans*- and *cis*- S_1 minima, and the nature of the T_2 state is different from the nature of the S_1 state (see Figure S3 of Supplementary Materials). According to the El-Sayed rule, ISC is efficient in this case, leading to further evolution of the triplet molecule.

Since DHP-**1** does not accumulate, its oxidation product, phenanthrene-**1** (Phen-**1**), which could be detected by absorption at ~380 nm, does not form.

For the 1,3-diazadistyrylbenzene (**2**) and its electrocyclization product, the corresponding structures calculated by the same method are shown in Figure 11. The *trans* isomer is planar both in the ground and in the excited states. The *cis* isomer is nonplanar; the C-C=C-C torsion angle is ~8° in the ground state and ~14° in the excited state. These torsions are even closer to the same angles in the DHP (~8 and 16° in the ground and excited states, respectively) than in **1**. The ground-state DHP is a local minimum, while the excited-state DHP is a conical intersection. *Trans-cis* photoisomerization proceeds via another conical intersection at ~90° C-C=C-C torsion angle. The minimum energy path from both *trans* and *cis* minima to the CI includes two saddle points at ~128 and 57° C-C=C-C torsion angle.

Figure 12 shows the energy level diagram of diazadistyrylbenzene **2** (a), its possible phototransformations (b), and the absorption maxima of the phototransformation products (c). On the basis of the calculated data, we may infer the following picture of the photoprocesses in **2**.

After the laser pulse, the mixture consists of *trans*-**2**, *cis*-**2**, and, possibly, ^{3}*trans*-**2**. When photoinduced absorption is measured, *trans*-**2** absorbs at 314–372 nm (350 nm in the experiment) with further relaxation to the minimum of the S_1 state ((*trans*-**2**)*). From this minimum, the molecule can either emit light at 409 nm (455 nm in the experiment), transform to the triplet state through the S_1-T_3 intersystem crossing with further phosphorescence or nonradiative relaxation, or undergo a *trans-cis* isomerization. Without excess excitation energy provided by the laser pulse, this process proceeds in the microsecond timescale and, therefore, is sufficiently slow. The reverse *cis-trans* isomerization proceeds slower by order of magnitude, which may result in the accumulation of the *cis*-**2** isomer.

From the (*cis*-**2**)* minimum, the molecule can transform to the excited (DHP-**2**)*, which has almost the same energy as the excited (*cis*-**2**)* minimum. The (DHP-**2**)* is a conical intersection and serves as a funnel for the nonradiative relaxation of (*cis*-**2**)*. The ground-state DHP-**2**, which absorbs in the visible region, is not accumulated due to its fast oxidation to the corresponding phenanthrene-**2** (Phen-**2**), whose absorption band (calculated 358–361 nm) overlaps with those of the *trans*-**2** and *cis*-**2**.

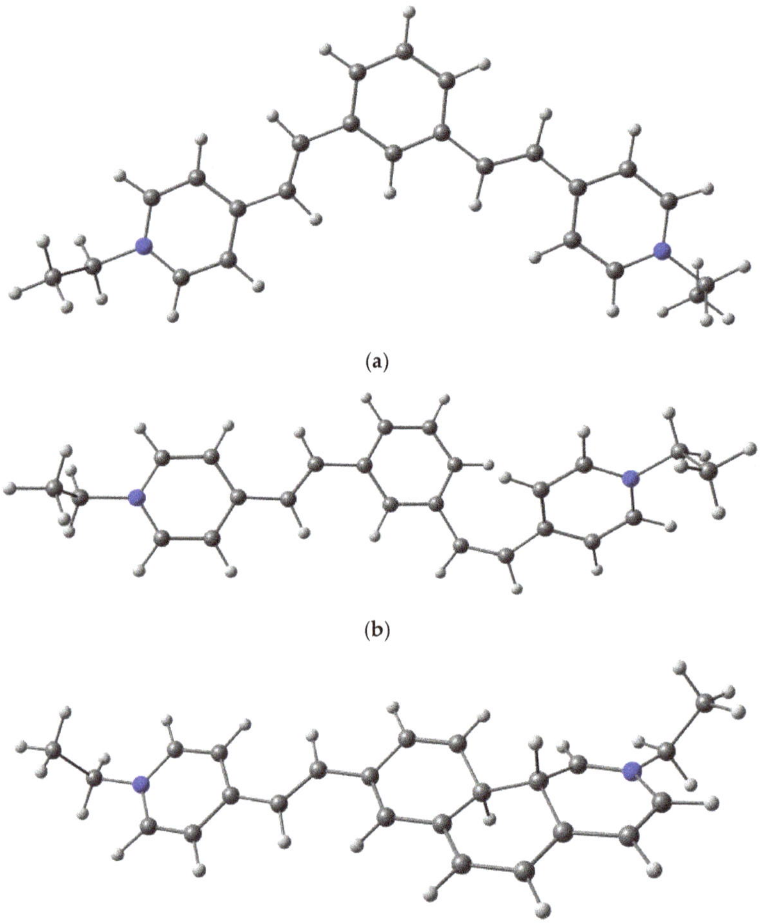

(a) (b) (c)

Figure 11. Structures of (**a**) *trans*-**2** and (**b**) *cis*-**2** and (**c**) its electrocyclization product dihydrophenanthrene DHP-**2**.

ISC in the (*trans*-**2**)* can proceed from the S_1 to the closest triplet level, which is T_3. However, both S_1 and T_3 states are formed by the same orbitals (see Figure S3 of Supplementary Materials), and this makes ISC inefficient according to the El-Sayed rule (denoted by red cross in Figure 12b). This explains the absence of photoinduced absorption at 500–600 nm in **2**.

The absorption bands of *cis*-**2** and *trans*-**2** are even closer than in **1** and overlap (calculated 372 and 368 nm for *trans*-**2** and *cis*-**2**, respectively). The emission proceeds in the nanosecond timescale and the fluorescence spectra of both *cis* and *trans* isomers also overlap (calculated at 409 and 422 nm for *trans*-**2** and *cis*-**2**, respectively).

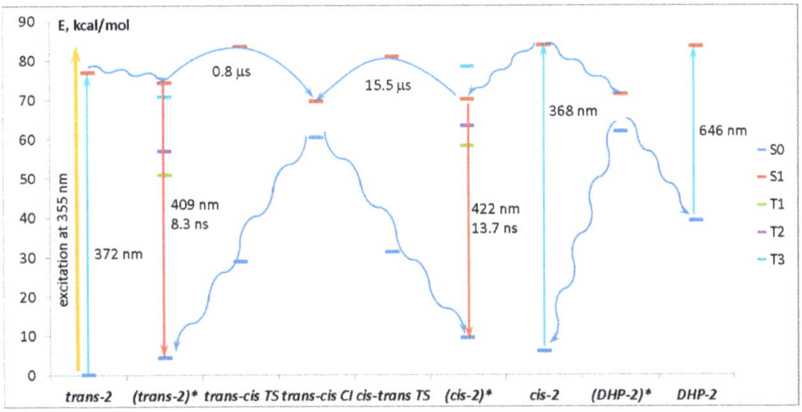

(**a**)

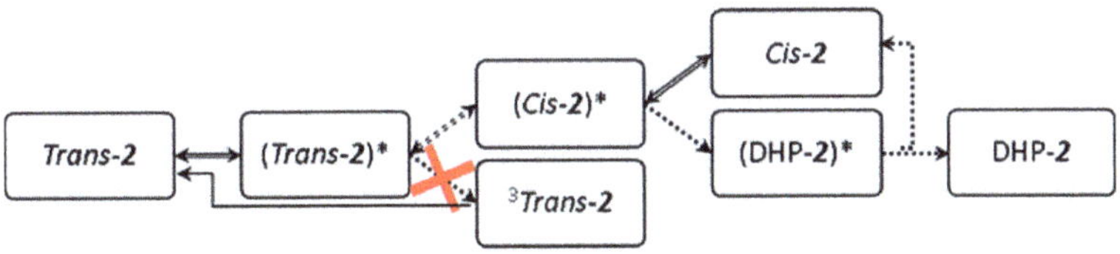

(**b**)

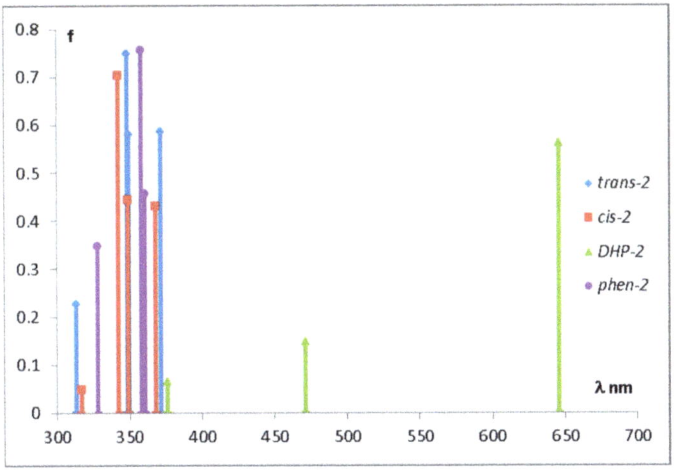

(**c**)

Figure 12. (**a**) Energy level diagram of diazadistyrylbenzene **2**, (**b**) scheme of its possible phototransformations, and (**c**) absorption maxima of the phototransformation products. Straight arrows on the energy diagram denote radiative processes, whereas wavy arrows denote nonradiative processes.

3. Materials and Methods

The synthesis of perchlorates of 1,4- and 1,3-diazadistyrylbenzenes (**1** and **2**) is beyond the scope of this study and will be published elsewhere. ^{1}H-NMR spectra of **1** and **2** can be found in Supplementary Materials, Figures S1 and S2.

The absorption spectra were recorded on the Agilent 8453 spectrophotometer. The luminescence spectra were recorded on the Varian Eclipse spectrofluorimeter. Fluorescence lifetimes were measured on the Fluotime 300 spectrofluorimeter. The difference absorption spectra of photoproducts and kinetics decay were measured by means of the nanosecond laser flash photolysis apparatus [22]. Irradiation was performed using the third harmonic of the Nd-YAG laser Solar (λ = 355 nm). The dissolved oxygen was removed by the argon bubbling of the solution. Stationary irradiation of diazadistyrylbenzenes solutions was performed using a DKSSh-150 lamp with a UFS-6 glass filter. For the measurement of fluorescence quantum yields, 9,10-diphenylanthracene (quantum yield 0.92 in ethanol [23]) was used as a standard. The accuracy of quantum yield measurement was 10%. All measurements were carried out in MeCN (high purity grade, grade 0, Cryochrome) at room temperature.

Kinetic spectroscopy allows us to draw conclusions about the presence of certain photoreaction products from indirect data (photoinduced absorption spectra and kinetic curves). Since the concentration of the diazadistyrylbenzenes is very low, it is not possible to accumulate and isolate (in general, unstable) photoreaction products in their pure form.

Calculation

The structures of the *cis* and *trans* isomers of diazadistyrylbenzenes **1** and **2** and their electrocyclization products (dihydrophenanthrenes DHP-**1** and DHP-**2**) in the ground S_0 and singlet excited S_1 states, as well as the transition states of the *trans-cis* photoisomerization on the S_1 potential energy surfaces and DHP oxidation products, phenanthrenes Phen-**1** and Phen-**2**, were optimized by the DFT with PBE0 functional and 6-31+G(d,p) basis set using FireFly program package [24] partially based on GAMESS(US) code [25]. The solvent effect (MeCN) was included through the dielectric polarizable continuum model (D-PCM) [26].

The absorption and emission spectra were calculated by TDDFT with the same functional, basis set, and solvent model. The absorption spectra were calculated after geometry optimization of the ground state, while the emission spectra were calculated after geometry optimization of the π–π* excited state. The vibrational frequencies and normal modes were calculated for all the studied stationary points of the ground and S_1 excited states. The triplet energies were calculated by TDA [27].

The radiative lifetime τ_r was calculated as follows:

$$k_r = \frac{2}{3} f_{0i} \nu_{i0}^2; \qquad \tau_r = \frac{1}{k_r} \tag{1}$$

where k_r is the fluorescence rate constant, f_{0i} is the oscillator strength, and ν_{i0} is the frequency (cm^{-1}) of the transition.

The *trans-cis* and *cis-trans* photoisomerization times τ_{tc} were calculated using the Arrhenius equation:

$$k_{tc} = c\omega_0 \exp\left(\frac{-E_A}{kT}\right); \qquad \tau_{tc} = 1/k_{tc} \tag{2}$$

where E_A is the activation energy calculated for the corresponding transition state on the S_1 potential energy surface, c is the speed of light, and ω_0 is the frequency of the reorganizational mode [28–30].

Since the energy profile of the isomerization and subsequent electrocyclization goes through conical intersection regions, the applicability of TDDFT needs special discussion. It is known that TDDFT is inapplicable to cases with quasidegeneracy. In the quasidegeneracy region, the excited state geometry optimization procedure fails because the gap between the ground and excited state (or the excitation frequency) becomes zero, and Casida equations do not have positive eigenvalues [31]. However, for our qualitative conclusions, we neither need the exact energy of the minimum energy CI nor its exact structure. For our purposes, it is sufficient to know that both (DHP-**1**)* and (DHP-**2**)* lie in the region of the conical intersection. If this intersection is peaked [32], it acts as a reaction funnel. By TDDFT, we may roughly demonstrate that the near-CI region for (DHP-**1**)* lies higher than the local *cis* and *trans* minima of the S_1 state, while for (DHP-**2**)*, this region has lower energy and, therefore, a barrierless path from the *cis* form to the CI is possible in **2**.

The calculations were also repeated using BHandHLYP functional to demonstrate that the results are qualitatively the same and only slightly depend on the weight of exact exchange in the functional (Figure S4 of Supplementary Materials)

This methodology has been benchmarked against multireference (XMCQDPT2/CASSCF) calculation in [33].

4. Conclusions

It was found that the photoprocesses of 1,4- and 1,3-isomers of diazadistyrylbenzenes differ substantially. The phototransformation of the 1,4-diazadistyrylbenzene derivative involves *trans-cis* isomerization and intersystem crossing to the triplet state. In this case, the formation of DHP-**1** is energetically unfavorable. For 1,3-diazadistyrylbenzene, *trans-cis* isomerization is followed by the electrocyclization from the *cis*-form to generate DHP-**2** and further fast oxidation with atmospheric oxygen to give phenanthrene, while intersystem crossing is not efficient. The considered photoprocesses of diazadistyrylbenzenes can be implemented in the construction of photoactive supramolecular systems on their basis with the use of macrocyclic compounds.

Supplementary Materials: The following Supplementary Material can be downloaded at https://www.mdpi.com/article/10.3390/ijms232315346/s1.

Author Contributions: L.S.A.: Investigation, Formal analysis, Visualization, Writing—Original Draft, Writing–Review and Editing. A.Y.F.: Formal analysis, Visualization, Writing—Original Draft, Writing—Review and Editing, Methodology. D.S.I.: Investigation. V.N.N.: Resources. A.V.M.: Resources. S.Z.V.: Resources. S.P.G.: Conceptualization, Project administration, Funding acquisition. A.K.C.: Writing—Review and Editing. All authors have read and agreed to the published version of the manuscript.

Funding: Synthesis and steady-state spectroscopy of diazadistyrylbenzenes were financially supported by the Russian Science Foundation (Project No. 22-13-00064). Laser pulse photolysis and quantum chemical calculations were financially supported by the Ministry of Science and Higher Education of the Russian Federation within the State Assignment to the Federal Research Center "Crystallography and Photonics", Russian Academy of Sciences (no grant number).

Institutional Review Board Statement: Not applicable.

Informed Consent Statement: Not applicable.

Data Availability Statement: Not applicable.

Acknowledgments: The authors thank the Russian Science Foundation and the Ministry of Science and Higher Education of the Russian Federation for financial support. The calculations were performed using the computational facilities of the Joint Supercomputer Center of the Russian Academy of Sciences.

Conflicts of Interest: The authors declare no conflict of interest.

References

1. Kim, M.; Whang, D.R.; Gierschnerb, J.; Park, S.Y. A distyrylbenzene based highly efficient deep red/near-infrared emitting organic solid. *J. Mater. Chem. C* **2015**, *3*, 231–234. [CrossRef]
2. Cavazzini, M.; Quici, S.; Orlandi, S.; Sissa, C.; Terenziani, F.; Painelli, A. Intimately bound coumarin and bis(alkylaminostyryl)benzene fragments: Synthesis and energy transfer. *Tetrahedron* **2013**, *69*, 2827–2833. [CrossRef]
3. Chaieb, A.; Khoukh, A.; Brown, R.; Francois, J.; Dagron-Lartigau, C. Characterization of model luminescent PPV analogues with, donating or withdrawing groups. *Opt. Mater.* **2007**, *30*, 318–327. [CrossRef]
4. Motoyoshiya, J.; Fengqiang, Z.; Nishii, Y.; Aoyama, H. Fluorescence quenching of versatile fluorescent probes based on strongly electron-donating distyrylbenzenes responsive to aromatic chlorinated and nitro compounds, boronic acid and Ca^{2+}, Spectrochim. *Spectrochim. Acta Part A Mol. Biomol. Spectrosc.* **2008**, *69*, 167–173. [CrossRef]

5. Pond, S.J.K.; Tsutsumi, O.; Rumi, M.; Kwon, O.; Zojer, E.; Bredas, J.L.; Marder, S.R.; Perry, J.W. Metal-Ion Sensing Fluorophores with Large Two-Photon Absorption Cross Sections: Aza-Crown Ether Substituted Donor-Acceptor-Donor Distyrylbenzenes. *J. Am. Chem. Soc.* **2004**, *126*, 9291–9306. [CrossRef]
6. Morgan, D.D.; Horgan, S.W.; Orchln, M. Photocyclization of stilbene analogs I. The oxidative photocyclization of 1,3-distyrylbenzene. *Tetrahedron Lett.* **1970**, *11*, 4347–4350. [CrossRef]
7. Laarhoven, W.H.; Cuppen, T.J.H.M.; Nivard, R.J.F. Photodehydrocyclizations in stilbene-like compounds—II: Photochemistry of distyrylbenzenes. *Tetrahedron* **1970**, *11*, 1069–1083. [CrossRef]
8. Zertani, R.; Meier, H. Photochemie des 1,3-Distyrylbenzols—Ein neuer Weg zu syn-[2.2](1,3)Cyclophanen. *Chem. Ber.* **1986**, *119*, 1704–1715. [CrossRef]
9. Marri, E.; Galiazzo, G.; Mazzucato, U.; Spalletti, A. Excited state properties of cross-conjugated 1,2- and 1,3-distyrylbenzene and some aza-analogues. *Chem. Phys.* **2005**, *312*, 205–211. [CrossRef]
10. Marri, E.; Elisei, F.; Mazzucato, U.; Pannacci, D.; Spalletti, A. Triplet-sensitized photobehaviour of the three stereoisomers of 1,4-distyrylbenzene and some aza-analogues. *J. Photochem. Photobiol. A Chem.* **2006**, *177*, 307–313. [CrossRef]
11. Ginocchietti, G.; Galiazzo, G.; Pannacci, D.; Mazzucato, U.; Spalletti, A. Effect of stereoisomerism on the radiative and reactive relaxation channels of two thio-analogues of distyrylbenzene. *Chem. Phys.* **2006**, *331*, 164–172. [CrossRef]
12. Ginocchietti, G.; Mazzucato, U.; Spalletti, A. Excited state behaviour of some thio-analogues of 1,3-distyrylbenzene. *J. Photochem. Photobiol. A* **2008**, *196*, 233–238. [CrossRef]
13. Kalanoor, B.S.; Bisht, P.B.; Annamalai, S.; Aidhen, I.S. A new distyrylbenzene derivative with Weinreb amide functionality: An efficient laser dye and nonlinear optical material. *J. Lumin.* **2009**, *129*, 1094–1098. [CrossRef]
14. Toyoshima, T.; Yoshida, S.; Watanabe, S. Synthesis of an alkylthio-substituted dibenz[a,j]anthracene with improved solubility via the oxidative photocyclization of 1,3-distyrylbenzene derivatives. *Tetrahedron* **2013**, *69*, 1904–1911. [CrossRef]
15. Atabekyan, L.S.; Avakyan, V.G.; Nuriev, V.N.; Vatsadze, S.Z.; Gromov, S.P. Photonics of tetramethoxy-1,4-distyrylbenzene. *Russ. Chem. Bull.* **2018**, *67*, 2016–2024. [CrossRef]
16. Vedernikov, A.I.; Nuriev, V.N.; Fedorov, O.V.; Moiseeva, A.A.; Kurchavov, N.A.; Kuz'mina, L.G.; Freidzon, A.Y.; Pod'yacheva, E.S.; Medved'ko, A.V.; Vatsadze, S.Z.; et al. Synthesis, structure and complexation of biscrowncontaining 1,4-distyrylbenzenes. *Russ. Chem. Bull.* **2016**, *11*, 2686–2703. [CrossRef]
17. Nuriev, V.N.; Fedorov, O.V.; Moiseeva, A.A.; Freidzon, A.Y.; Kurchavov, N.A.; Vedernikov, A.I.; Medved'ko, A.V.; Pod'yacheva, E.S.; Vatsadze, S.Z.; Gromov, S.P. Synthesis, Structure, Spectral Properties, and Electrochemistry of Bis(crown ether) Containing 1,3-Distyrylbenzenes. *Russ. J. Org. Chem.* **2017**, *53*, 1726–1737. [CrossRef]
18. Atabekyan, L.S.; Avakyan, V.G.; Chibisov, A.K.; Gromov, S.P.; Vatsadze, S.Z.; Nuriev, V.N.; Medved´ko, A.V. Bis(15-crown-5)-1,4-distyrylbenzene and its complexes with metal perchlorates: Photonics and structure. *Russ. Chem. Bull.* **2019**, *68*, 2053–2064. [CrossRef]
19. Atabekyan, L.S.; Avakyan, V.G.; Chibisov, A.K.; Gromov, S.P.; Vatsadze, S.Z.; Nuriev, V.N.; Medved'ko, A.V. Photonics of bis(18-Crown-6)-1,4-Distyrylbenzene and Its Complexes with Metal Perchlorates. *High Energy Chem.* **2019**, *53*, 115–124. [CrossRef]
20. Atabekyan, L.S.; Freidzon, A.Y.; Chibisov, A.K.; Gromov, S.P.; Vatsadze, S.Z.; Nuriev, V.N.; Medvedko, A.V. Photoprocesses in bis(18-crown-6)-1,3-distyrylbenzene and its complexes with metal perchlorates. *High Energy Chem.* **2021**, *184*, 108773–108782. [CrossRef]
21. Atabekyan, L.S.; Avakyan, V.G.; Chibisov, A.K.; Nuriev, V.N.; Medvedko, A.V.; Koshkin, A.V.; Gromov, S.P. Photoprocesses in bis(15-crown-5)-1,3-distyrylbenzene and its complexes with metal perchlorates. *J. Photochem. Photobiol. A* **2021**, *414*, 113293–113302. [CrossRef]
22. Atabekyan, L.S.; Chibisov, A.K. Photoprocesses in Aqueous Solutions of 9-Ethylthiacarbocyanine Dyes in the Presence of Surfactants. *High Energy Chem.* **2007**, *41*, 91–96. [CrossRef]
23. Montalti, M.; Credi, A.; Prodi, L.; Gandolfi, M.T. *Handbook of Phochemistry*, 3rd ed.; Taylor and Fransis: London, UK, 2006; p. 574.
24. Granovsky, A.A. Firefly, version 8.2.4; 2017. Available online: http://classic.chem.msu.su/gran/firefly/index.html (accessed on 1 September 2022).
25. Schmidt, M.W.; Baldridge, K.K.; Boatz, J.A.; Elbert, S.T.; Gordon, M.S.; Jensen, J.H.; Koseki, S.; Matsunaga, N.; Nguyen, K.A.; Su, S.; et al. General Atomic and Molecular Electronic Structure System. *J. Comput. Chem.* **1993**, *14*, 1347–1363. [CrossRef]
26. Tomasi, J.; Mennucci, B.; Cammi, R. Quantum Mechanical Continuum Solvation Models. *Chem. Rev.* **2005**, *105*, 2999–3093. [CrossRef] [PubMed]
27. Peach, M.J.G.; Tozer, D. Overcoming Low Orbital Overlap and Triplet Instability Problems in TDDFT. *J. Phys. Chem. A* **2012**, *116*, 9783–9789. [CrossRef]
28. Eyring, H.; Walter, J.; Kimball, G.E. *Quantum Chemistry*; Wiley: New York, NY, USA, 1944.
29. Bazilevskii, M.V.; Faustov, V.I. Modern theories of chemical reactions in condensed phases. *Russ. Chem. Rev.* **1992**, *61*, 651–672. [CrossRef]
30. Henriksen, N.E.; Hansen, F.Y. *Theories of Molecular Reaction Dynamics: The Microscopic Foundation of Chemical Kinetics*; Oxford University Press: Oxford, UK, 2008.
31. Casida, M. *Time-Dependent Density Functional Response Theory for Molecules in Recent Advances in Density Functional Methods*; Chong, D.P., Ed.; World Scientific: Singapore, 1995; pp. 155–192. [CrossRef]
32. Atchity, G.J.; Xantheas, S.S.; Ruedenberg, K. Potential energy surfaces near intersections. *J. Chem. Phys.* **1991**, *95*, 1862–1876. [CrossRef]
33. Quentin, C.; Gerasimaitė, R.; Freidzon, A.; Atabekyan, L.S.; Lukinavičius, G.; Belov, V.N.; Mitronova, G.Y. Direct Visualization of Amlodipine Intervention into Living Cells by Means of Fluorescence Microscopy. *Molecules* **2021**, *26*, 2997. [CrossRef]

International Journal of
Molecular Sciences

MDPI

Article

Cyclic Dimers of 4-n-Propyloxybenzoic Acid with Hydrogen Bonds in the Gaseous State

Nina I. Giricheva [1,*], Ksenia E. Bubnova [2], Alexander V. Krasnov [2] and Georgiy V. Girichev [2]

1 Nanomaterial Research Institute, Ivanovo State University, 153025 Ivanovo, Russia
2 Department of Physics, Ivanovo State University of Chemistry and Technology, 153000 Ivanovo, Russia
* Correspondence: n.i.giricheva@mail.ru; Tel.: +7-4932-32-92-29

Abstract: A comprehensive study of saturated vapors of 4-n-propyloxybenzoic acid (POBA) by gas electron diffraction (GED) and mass spectrometric (MS) methods supplemented by quantum chemical (QC) calculations was carried out for the first time. An attempt was made to detect dimeric forms of the acid in the gaseous state. It has been established that at the temperature of GED experiment, vapor over a solid sample contains up to 20 mol.% of cyclic dimers with two O-H...O hydrogen bonds. The main geometrical parameters of gaseous monomers and dimers of POBA are obtained. The distance r(O … O) = 2.574(12) Å in the cyclic fragment of the gaseous dimer is close to that in the crystal structure (2.611 Å). In the mass spectrum of the POBA recorded the ions of low intensity with a mass exceeding the molecular mass of the monomer were detected. The presence of ions, whose elemental composition corresponds to the dissociative ionization of the dimer, confirms the results of the GED experiment on the presence of POBA dimers in the gas state. The results of GED studies of acetic acid, benzoic acid, and POBA were compared. It is shown that the COOH fragment saves its geometric structure in monomers, as well as the COOH...HOOC fragment with two hydrogen bonds in dimers of different acids. The intermolecular interaction energy in considered acid dimers was estimated using QC calculations (B97D/6-311++G **). The significant value of last (>84 kJ/mol) is the reason for the noticeable presence of dimers in the gas phase.

Keywords: 4-n-propyloxybenzoic acid; hydrogen bond; dimer; molecular structure; gas electron diffraction; DFT; mass spectrometry; intermolecular interaction

1. Introduction

It is well known that aromatic carboxylic acids are dimerized in crystalline and liquid crystalline phases [1–6]. Due to this feature, even benzoic acids with short substituents can exhibit liquid crystal properties [5].

During heating, substituted benzoic acid passes into a liquid-crystalline state, and then into an isotropic state with a gradual destruction of intermolecular hydrogen bonds, and according to IR spectroscopy, the number of dimers in the isotropic state becomes small [1].

For aromatic carboxylic acids the method of IR spectroscopy is a reliable tool for detecting dimeric forms of acids, since IR spectra contain characteristic absorption bands in the range of 2500–2700 cm^{-1} [3,4,6–8] and a band related to the frequency of the C=O stretching vibration (1680–1700 cm^{-1}), which is shifted to the low-frequency region of the IR spectrum compared with the frequency $\nu(C{=}O)_{st}$ for the monomeric form, which is usually localized at 1740 cm^{-1} [1,3,4]. Most of the IR spectra of aromatic carboxylic acids refer to samples in the crystalline state, or to a solution of acids in non-polar solvents, in which intermolecular hydrogen bonds are preserved.

An interesting question is the possibility of the existence of dimers of benzoic acids in vapor. The experimental methods that can be used divulge the answer are IR spectroscopy of the gas phase, mass spectrometry and gas electron diffraction. In this case, the temperature of the experiment and the total vapor pressure are very important, since the dimerization constant K_{dim} strongly depends on these parameters, in addition, the method of registration of dimeric forms is also important.

It is reported in [9–13] that substituted benzoic acids can be in the dimeric form in the gas phase. This conclusion is mainly based on the interpretation of the vibrational spectra of benzoic acid (BA) and its substituted ones obtained under special conditions.

For example, in [10], the IR absorption spectra of the jet-cooled BA monomer and dimer, recorded throughout the 500–1900 cm^{-1} range via ion dip spectroscopy, was obtained. In [13], the IR spectra of the BA dimer have been recorded under jet-cooled conditions using the double resonance method of fluorescence-dip IR spectroscopy. The authors [13] note that "the spectra are assuredly due exclusively to dimers in the ground-state zero-point level at a rotational temperature of 3–5 K". Even under these conditions, the dimers have remarkably broad spectra, extending from 2600 to almost 3150 cm^{-1}.

In [11,12], the conclusion about the existence of dimers was made based on the infrared spectra of benzoic acid and deuterobenzoic acid embedded in Ar matrices; the compounds were evaporated from a Knudsen effusion cell at 46 °C. The authors note that both a raise of benzoic acid concentration and a matrix annealing lead to the formation of H-bonded benzoic acid dimers.

At the same time, in NIST Chemistry WebBook, the IR spectra of gaseous BA and its substituted ones did not reveal bands characteristic of dimers. However, it should be noted that the IR spectra of the gas refer to the temperature of the experiment at which complete thermal dissociation of dimeric forms is possible, for example, for unsubstituted BA, the gas temperature is 160 °C.

Therefore, for the study of monomer-dimer equilibrium, the method of GED may be more effective, which requires lower experimental temperatures than IR spectroscopy of gaseous compounds, and when registering a diffraction pattern, there is no destructive effect on the molecular forms present in the vapor under study, such as this is the case in electron ionization mass spectrometry.

Such electron diffraction studies of BA were performed in two laboratories at different vapor temperatures (T = 288 K [14] and T = 405 K [15]). In both works, the interpretation of the diffraction pattern was made on the assumption of a monomer composition of the vapor. The possibility of the presence of dimeric forms in the vapor was not tested.

Thus, in the literature, there is no consensus about the possibility of maintaining hydrogen bonds (HBs) in gaseous aromatic carboxylic acids.

At the same time, for acetic acid (AA), the presence of cyclic dimers with two hydrogen bonds in vapor was proved by the electron diffraction method. A unique example of such a study is the work [16], in which a two-temperature electron diffraction/mass-spectrometric (GED/MS) experiment was performed. It is shown that at T = 296 K the gas phase contains dimeric forms in the amount of 31(2) mol%, while at T = 457 K, the number of dimers is negligibly small. The validity of this conclusion is confirmed by the distinct difference between the two radial distribution functions f(r) associated with the different composition of the gas phase, i.e., with different monomer-dimer ratio and its dependence on the temperature. At the same time, the mass spectra of acetic acid obtained at two temperatures in [16] did not contain ions with a mass greater than the molecular mass of the monomer.

The purpose of this work is a comprehensive study of vapors of 4-n-propyloxybenzoic acid (POBA) by electron diffraction, mass spectrometric and quantum chemical methods and an attempt to detect dimeric forms of the acid in the gaseous state. The object of study is the first representative in the series of 4-n-alkyloxybenzoic acids ($n \geq 3$), which exhibits liquid crystal properties, despite the presence of a short substituent $-OC_nH_{2n+1}$.

An encouraging example of a gaseous acetic acid dimer [16], on the one hand, inspires hope for a successful solution of the problem of establishing the monomer–dimer composition of POBA vapors, since, according to our estimates, the intermolecular interaction energy (E_{IMI}) in the POBA dimer is comparable to that for the acetic acid dimer. On the other hand, there are circumstances that complicate this task, since the POBA dimer is not an easy object to study by the GED method, since dimer contains internuclear distances exceeding 20 Å.

2. Results and Discussions

2.1. Peculiarity of GED Data Structural Analysis of POBA Vapor

The temperature 87(15) °C of GED experiment was lower than the temperature of "crystal-liquid crystal" transition ($T_{Cr \to LC}$ = 140.5 °C), i.e., POBA sublimation took place.

The theoretical function of molecular scattering sM(s) contained two terms:

$$sM(s) = \alpha \cdot sM(s)_{mon} + (1-\alpha) \cdot sM(s)_{dim} \quad (1)$$

where $sM(s)_{mon}$ is the theoretical function for monomer, $sM(s)_{dim}$ is the theoretical function for dimer, α is the coefficient that determines the contribution of the monomeric form to the function sM(s), which can take the value $0 < \alpha < 1$.

The coefficient α is related to the mole fraction of the monomer by the following relationship $\chi_{mon} = \alpha/[\alpha + (1-\alpha)/3.54]$, which takes into account the different scattering power of monomeric and dimeric forms. The scattering power of molecular species present in the vapor were estimated by the formula: $\Sigma Z_i Z_j \times k_{ij}/r_{ij}$, where Z_i are charges of atom nuclei; r_{ij} is the distance between i and j atoms; k_{ij} is the multiplicity of the (i–j) distance. The ratio of scattering power of dimer and monomer calculated in this way is 3.54/1.

The geometric models of the monomer and dimer with the numbering of atoms are shown in Figure 1. To describe their geometry, 21 independent parameters were used: five internuclear distances (C-O1, O1-H, C1-C, C2-H, C ... X), eleven bond angles (C-O1-H, O2=C-O1, C1-C-O1, C-C1-C2, C-C1-C6, C4-OR-C1R, C1-C2-H, OR-C1R-H, H-C1R-H, O2=C-X, O1-C-X), and five torsion angles (C1R-OR-C4-C5, C2R-C1R-OR-C4, C3R-C2R-C1R-OR, H-C1R-OR-C4, H-C3R-C2R-C1R). The non-equivalence of C-C and C-H bonds in benzene and propyl fragments was taken into account.

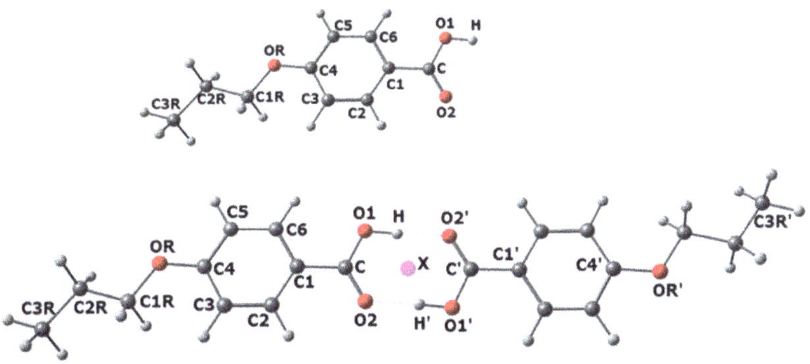

Figure 1. Geometric models of the POBA monomer and dimer with atomic numbering.

In the least squares analysis of electron diffraction data (modified KCED program [17]), the parameters obtained from the results of quantum chemical calculations (B97D/6-311++G **) were taken as the starting values of internuclear distances, bond angles and torsion angles. VibModule program [18] was applied to calculate the vibrational corrections $\Delta r = r_{h1} - r_a$ and the starting values of root-mean-square vibration amplitudes of monomer and dimer POBA at the temperature of the GED experiment using the harmonic approximation and taking into account the non-linear interrelation between internal and Cartesian vibrational coordinates. The quantum chemically calculated difference between parameters of the same type in the monomer and dimer was retained. In the least squares analysis, both 10 geometrical parameters (Table 1) along with 14 groups of vibration amplitudes of the monomer and dimer were varied simultaneously.

Table 1. Theoretical (B97D/6-311++G **) and experimental (GED) geometric parameters of the gaseous monomer and cyclic dimer POBA (R_f = 3.405%).

Parameters [a]	Monomer		Dimer	
	QC r_e Structure	GED [b] r_{h1} Structure	QC r_e Structure	GED r_{h1} Structure
$(C\text{-}C)_{av}$ (Ph)	1.404	1.405(4) p_1[c]	1.405	1.406(4)
C-O1	1.375	1.368(7) p_2	1.332	1.325(7)
C=O2	1.219	1.212(7) (p_2)	1.242	1.234(7)
O1-H	0.971	0.971	1.010	1.010
C1-C	1.483	1.484(4) (p_1)	1.484	1.484(4)
C1-C6	1.413	1.414(4) (p_1)	1.413	1.414(4)
$C\text{-}H_{av}$(Ph)	1.087	1.082(5) p_3	1.087	1.082(5)
$C4\text{-}O_R$	1.362	1.354(7) (p_2)	1.362	1.355(7)
$O_R\text{-}C1_R$	1.443	1.436(7) (p_2)	1.442	1.435(7)
$C1_R\text{-}C2_R$	1.524	1.525(4) (p_1)	1.524	1.525(4)
C-Hav (R)	1.101	1.094(5) (p_3)	1.099	1.095(5)
O2=C-O1	121.6	121.8(30) p_4	123.2	123.2(30)
C1-C-O1	112.9	114.4(13) p_5	114.6	113.1(13)
H-O1-C	105.9	105.9	110.1	110.1
C6-C1-C	123.0	122.9(4) p_6	122.1	122.0(4)
H-C6-C1	119.6	124.6(18) p_7	119.4	124.3(18)
$C4\text{-}O_R\text{-}C1_R$	118.5	117.4(23) p_8	118.5	117.4(23)
$H\text{-}C1_R\text{-}O_R$	109.3	111.6(80) p_9	109.4	111.6(80)
C2-C1-C	118.3	116.5(25) p_{10}	119.1	119.0(25)
O1-C-C1-C2	180.0	180.0	179.2	179.2
$C5\text{-}C4\text{-}O_R\text{-}C1_R$	180.0	180.0	178.6	178.7
$C3_R\text{-}C2_R\text{-}C1_R\text{-}O_R$	180.0	180.0	179.0	179.0
$H\text{-}C1_R\text{-}O_R\text{-}C4$	−59.2	−59.2	−60.0	−62.9
$H\text{-}C3_R\text{-}C2_R\text{-}C1_R$	180.0	180.0	180.0	180.0
χ, mol.fr.		0.80(10)		0.20(10)

[a] distance in Å, angle in degree. [b] Uncertainties given in parentheses were taken as: $[(2.5\sigma_{LS})^2 + \sigma^2_{scale}]^{1/2}$, σ_{scale} = 0.002r for bond distances and $3\sigma_{LS}$ for bond angles, σ_{LS} is a standard deviation in the least-squares refinement. [c] p_i—refinable independent parameter; (p_i)—parameter refined in the i-th group.

2.2. Conformers of the Monomeric and Dimeric Forms of POBA. Possibilities of Gas Electron Diffraction Method

In [2,19], we considered conformers of the POBA monomer, which differ both in the spatial structure of the OC_3H_7 substituents and in the structure of the COOH acid group. The latter can have a *cis*-(H-O1-C=O2 torsion angle is 0°) or *trans*-(H-O1-C=O2 torsion angle is 180°) conformation (Figure S1). The most energetically favorable (by ≈25 kJ/mol, B97D/6-311++G **) is the *cis*-configuration which can form cyclic dimers.

In addition, the POBA monomer has some conformers that differ in the structure of the –OC3H7 substitute. The most stable conformers, the energy difference of which does not exceed 6.7 kJ/mol, are shown on Figure S2.

The calculation results showed that, regardless of the spatial structure of the substituents, the intermolecular interaction energy, and the geometry of the –OPh-COOH···COOH-PhO-backbone in the POBA dimer remain the same.

For GED analyze the model of POBA monomer with a flat *trans*-structure of the substituent carbon backbone was chosen (Figure 1). This conformer is energetically more favorable.

The set of independent geometric parameters includes torsion angles C1R-OR-C4-C5, C2R-C1R-OR-C4, C3R-C2R-C1R-OR, H-C3R-C2R-C1R (see Section 2.1). Their variation gives effective values, which in their physical meaning correspond to thermally averaged values of torsion angles for all monomeric conformers (Figure S2) present in saturated vapor at the temperature of the GED experiment. The variation of the four indicated torsion angles does not lead to a decrease in the disagreement factor R_f but indicates a large error in their determination and shows the impossibility to determine substituent conformations in the monomeric and dimeric forms of the acid by the GED method.

It should be noted that the GED method is quite sensitive to determination of structural parameters of molecules with a set of internuclear distances in the range from 1 to 15 Å. If molecular forms have terms with internuclear distances r > 10 Å, then to them determination, it is necessary to decrease the scanning step Δs of the electron scattering intensity sM(s).

In [20], we considered the question of the relationship between the scanning step Δs and the maximum value of the internuclear distance r, as well as the reliability of determining terms corresponding to large r values. It was shown that even when the scanning step Δs is reduced to 0.1 $Å^{-1}$, instead of the traditionally used 0.2 $Å^{-1}$ or 0.25 $Å^{-1}$, the sensitivity of the GED method is very low for terms with internuclear distances r > 14 Å.

In the case of POBA, the monomeric form has an effective length of about 12 Å, the dimer is 24 Å, and the -Ph-COOH ... HOOC-Ph- dimer synthon is 12 Å. The range of distances from 1 to 12 Å also includes terms which belong to different monomers in the dimer (for example C ... C3R′, C1 ... OR′, OR ... C4′ (Figure 1)).

Thus, the set of terms determined by the GED method includes the terms of the monomer, the terms of monomer fragments in the dimer, and the terms describing the central part of the dimer. However, it is not possible to reliably determine the relative positions of the two $–OC_3H_7$ substituents in the dimer.

These limitations in studying the composition and geometry of the POBA vapor components affected the found set of geometric parameters of monomers and dimers (Table 1).

Note that two options for determining the vapor composition were used: the grid method, in which the coefficient α in equation (1) changed with a certain step when refining all other independent parameters (Figure 2), and the simultaneous variation of the structural parameters and the coefficient α associated with the concentration monomers and dimers in the framework of the least-squares analysis.

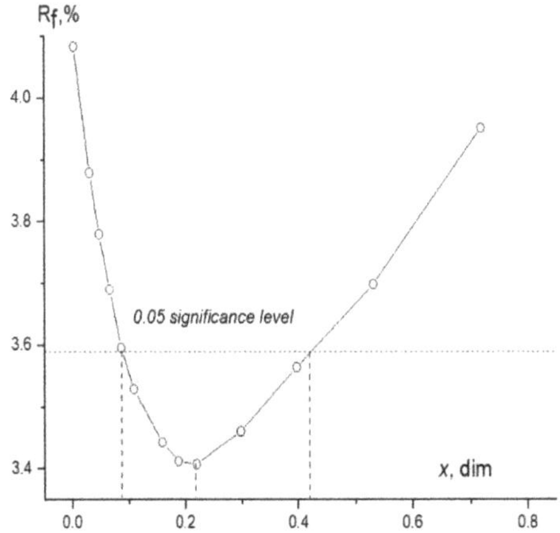

Figure 2. Dependence of the value of R_f on the mole fraction of the dimer χ_{dim} in the vapor of POBA The horizontal dashed line corresponds to the Hamilton's test [21] for uncertainties at significance level 0.05.

Both options lead to results consistent within the error (simultaneous variation of χ_{dim} = 0.20(10), R_f= 3.405%; grid method χ_{dim} = 0.22, R_f = 3.407%, Figure 2).

The experimental functions of the molecular scattering intensity sM(s) and the radial distribution f(r) together with theoretical analogs corresponding to the optimized values of the structural parameters and the optimal ratio between the monomers and dimers of POBA vapor are shown in Figures 3 and 4, respectively.

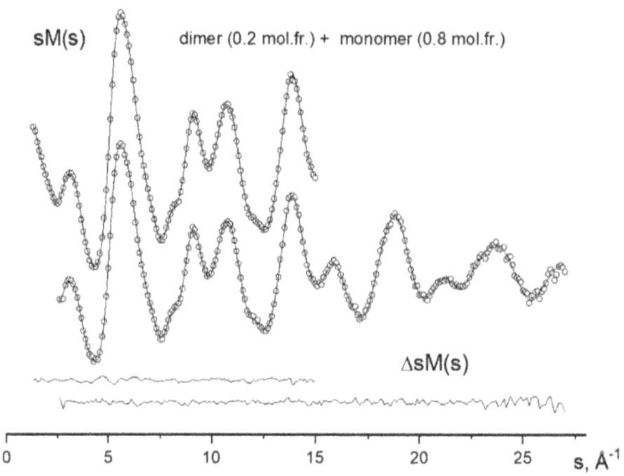

Figure 3. Experimental (points) and theoretical (lines) curves of the molecular scattering intensity sM(s) and their difference.

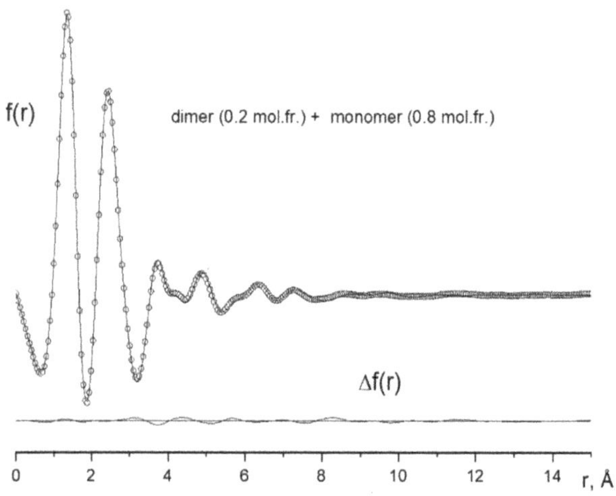

Figure 4. Experimental (points) and theoretical (lines) curves of the radial distribution f(r) and their difference Δf (r).

2.3. *Mass Spectrometric Confirmation of the Existence of Dimeric Forms in a Vapor over POBA*

To date, we are not aware of mass spectrometric data that would indicate the presence of dimeric forms of any aromatic carboxylic acids in the gas phase. The NIST database lists the mass spectrum of POBA in the range m/z = 0–180, which does not cover the mass range of ions related to the dimeric forms of the acid (Figure S3).

From the example with acetic acid, it is known that its dimers decompose upon electron impact and are not detected in the mass spectrum [16].

We have carried out a mass-spectrometric study of POBA vapors with a thorough analysis of the high mass range.

In the synchronous GED/MS experiment, the mass spectrum recorded simultaneously with the registration of the diffraction pattern in the extended mass range also did not contain intense ions belonging to dimeric form.

However, the APDM-1 mass spectrometer, which is part of complex GED/MS, has a low sensitivity. Therefore, the mass spectrometric experiment for POBA was performed on an MI-1201 device [22], which has both a higher sensitivity and resolution compared to APDM-1.

As well as in the GED/MS experiment, the ions corresponding to the monomeric form of the acid were registered in the range of 0–180 Da. The POBA monomer has a molecular weight of 180 a.m.u. In addition to them, when scanning the range of 180–400 a.m.u., heavy ions of low intensity were detected (Figure 5).

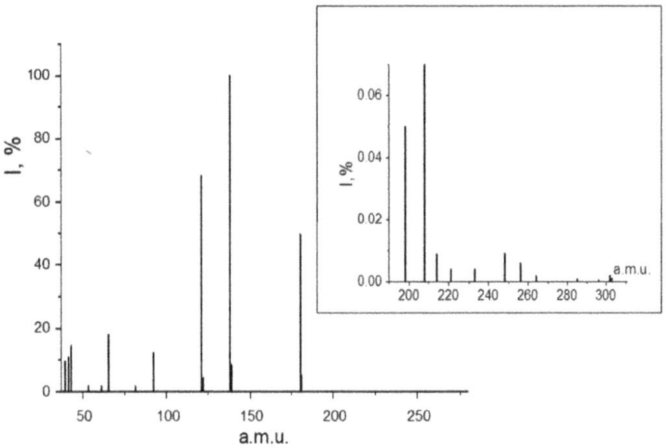

Figure 5. Mass spectrum of POBA recorded on an MI-1201 device.

The elimination of certain fragments during the ionization of molecules by electrons, known in the mass spectrometry of organic compounds [23], was used to identify elemental composition of ions.

It was supposed the elimination of the phenyl fragment -C_6H_4- and elimination of groups –CH_2–, –CH_2-CH_2–, –CH_2-CH_2-O–, –CH_2-CH_2-CH_2-O– from substituents –O-C_3H_7.

The possible elemental composition of monomer ions as well as ions with masses of more than 180 a.m.u. found in the mass spectrum of the acid are shown in Table 2. The presence of such ions corresponding to the dissociative ionization of the dimer confirms the results of the GED experiment on the presence of POBA dimers in the gas phase.

Table 2. Ions recorded in the mass spectrum of POBA in this work (MI-1201) and NIST.

Ion Mass a.m.u.	Elemental Composition of Ion [a]	I, % this Work	I, % NIST
39	C_3H_3	10	22
41–43 [b]	C_3H_n	15	21
50–53	C_4H_n	2	8
63–65	C_5H_n	18	23
81	C5H4OH	2	7
93	C_6H_4-OH	12	12
121	HOOC-C_6H_4	68	92

Table 2. *Cont.*

Ion Mass a.m.u.	Elemental Composition of Ion [a]	I, % this Work	I, % NIST
138	$HOOC\text{-}C_6H_4\text{-}OH$	100	100
180	$HOOC\text{-}C_6H_4\text{-}OC_3H_7$	50	25
198 [c]	–COOH . . . .$HOOC–C_6H_4\text{-}OCH_3$	0.05	
207	H_7C_3O–COOH . . . .$HOOC–OC_3H_7$	0.14	
214–212	H_3CO–COOH . . . .$HOOC–C_6H_4\text{-}CH_3$	0.01	
221–222	H_6C_3–COOH . . . .$HOOC–C_6H_4\text{-}CH_2$ $\text{-}C_6H_4\text{-}$**B**$\text{-}C_6H_4\text{-}CH_2\text{-}$	0.004 <0.01	
233–232	H_5C_6-**A**-C_6H_4-OH	<0.01	
248	HO-H_4C_6-**A**-C_6H_4-OH	0.01	
256–258	H_7C_3O–COOH . . . .$HOOC–C_6H_4\text{-}OCH_3$	<0.01	
264	$H_4C_2\text{-}C_6H_4$-**B**-$C_6H_4\text{-}C_2H_4$	<0.01	
285	H_7C_3O–COOH . . . .$HOOC–C_6H_4\text{-}OC_3H_7$	<0.01	
296	$H_4C_2O\text{-}C_6H_4$-**B**-$C_6H_4\text{-}OC_2H_4$	<0.01	
302–300	$H_2CO\text{-}C_6H_4$–COOH . . . .$HOOC–C_6H_4\text{-}OCH_2$	<0.01	

[a] **A** -O=C-O-C=O–; **B** -O=C-C=O–. [b] The maximum intensity of ions in the C_mH_n groups is given. [c] It was noted that the molecular ion of aliphatic carboxylic acids dimer is also not registered in the mass-spectra [24], but the M·COOH ion, which contains the COOH . . . HOOC fragment, is registered.

2.4. Comparison of the Results of GED Studies of Three Carboxylic Acids: AA—Acetic Acid, BA—Benzoic Acid, and POBA—4-n-Propyloxybenzoic Acid

Two-temperature GED studies of AA vapors were carried out in 1944, 1971, and 2020 at three GED laboratories [16,25,26].

The main conclusion of these studies is the statement that AA vapor at temperatures close to room temperature contains a significant amount of dimeric form with two HBs, and at temperatures 120–150° higher, the amount of dimers in the vapor is negligibly small.

For unsubstituted benzoic acid BA, studies were also performed at two temperatures (288 K [14] and 405 K [15]). However, they were performed in different GED laboratories on devices with different inlet systems. Moreover, the construction of new evaporator for low pressure gas electron diffraction was used in [14]. As was noted in the introduction, the interpretation of the GED data [14,15] obtained at different temperatures was performed on the assumption that only monomeric forms of BA are present in the vapor.

Similar to the results for AA, one would expect the presence of dimeric forms of BA at low temperatures [14]. However, the very small value of the structural wR factor of 2.52% indicates a good agreement between the theoretical monomer model of BA vapor and the experimental data.

Using the DFT/B97D/6-311++G ** method, we calculated the geometrical and vibrational parameters (T = 298 K) and theoretical functions sM(s) and f®of the BA monomer and dimer (Figures S4 and S5).

It can be seen the theoretical functions $sM(s)_{dim}$ and $sM(s)_{mon}$ have common visual features, which are also observed on the experimental function sM(s) [14]. At the same time, the difference curve $\Delta sM(s) = sM(s)_{d\text{-}m} - sM(s)_{mon}$ is significantly different from zero (Figure S4).

Significant differences are seen between the theoretical functions $f(r)_{mon}$ and $f(r)_{dim}$ (Figure S5). Thus, the GED method would make it possible to detect the presence of dimers in the BA vapor.

For further analysis of the results [14,15], we estimated the mole fraction of the dimer in BA vapor under the conditions of the GED experiments (T and P were taken from [14]). Using the DFT/B97D/6-311++G ** method, we calculated the Gibbs free energy of the dimerization process at different temperatures ($\Delta G_{290} = -19.5$ kJ/mol and $\Delta G_{405} = -0.75$ kJ/mol). It turned out that the equilibrium constant $k_x = x_{dim}/x^2_{mon}$ of "2·monomer↔dimer" reaction is $6 \cdot 10^{-4}$ and 0.018 at T = 290 K and T = 405 K, respectively. Small values of k_x can be an explanation for the absence of dimeric forms in BA vapor for experiments [14,15]. The low vapor pressure in the evaporator in the first experiment [14] and the high vapor temperature in the second experiment [15] contributed to the dissociation of dimers into monomers.

Table 3 compares the geometric parameters obtained by GED experiments for the monomeric forms of POBA [this work], BA [14,15], and AA [16], as well as the parameters of the cyclic COOH···HOOC fragment in POBA and AA dimers.

Table 3. The selected geometric parameters of AA, BA and POBA obtained by GED.

Parameters	POBA this Work		AA [16] [a]		BA [14,15] [a]	
	Monomer r_{h1} Structure T = 360 K	Dimer r_{h1} Structure T = 360 K	Monomer r_e Structure T = 296 K	Dimer r_e Structure T = 296 K	Monomer r_a Structure T = 288 K	Monomer r_a Structure T = 405 K
$(C_{Ph}\text{-}C_{Ph})_{av}$	1.405(4)	1.406(4)	-	-	1.393(12)	1.401(2)
C-O	1.368(7)	1.325(7)	1.350(8)	1.315(10)	1.350(10)	1.367(8)
C=O	1.212(7)	1.234(7)	1.207(10)	1.226(12)	1.205(10)	1.225(6)
$C1_{Ph}$-C	1.484(4)	1.484(4)	1.490(12)	1.493(12)	1.493(10)	1.484(7)
$C\text{-}H_{av}$(Ph)	1.082(5)	1.095(5)	1.090(17)	1.090(17)	1.103(10)	1.102(8)
O... H		1.565(8)		1.657(23)		
O... O		2.574(12)				
O–C=O	121.8(30)	123.2(30)	120.5(9)	123.0(12)	-	120.3(31)
χ_{mon}	0.80(10)		0.69(5)		1.0	1.0

[a] For a correct comparison in brackets instead of σ_{LS} [14,16] the uncertainties in internuclear distances calculated by the formula $\sigma = [(0.002r)^2 + (2.5\sigma_{LS})^2]^{1/2}$ is used.

Despite the fact that the parameters for the monomers AA, BA, and POBA are different in their physical meaning, they all agree within the experimental error. Agreement is also observed between the parameters of the cyclic fragment in the AA and POBA dimers.

Therefore, it can be concluded that the COOH fragment retains its structure in the monomers of various carboxylic acids, as does the COOH ... OOC fragment with two hydrogen bonds in the dimers.

These experimental data are confirmed by the results of quantum chemical calculations on the structure of monomeric and dimeric forms, characteristics of HB, and dimerization energy of the considered carboxylic acids (Table S1).

During the formation of cyclic dimers, as noted in [16,25,26], the C=O and O–H bonds are lengthened, while the C–O bonds are shortened, which can be clearly explained using NBO analysis.

2.5. NBO-Analysis and HB Energy

In [27], we analyze the nature of the HB in the POBA dimer and estimate the energy of the HB using the NBO analysis of the electron density distribution.

The geometric parameters of the O-H···O=C fragments and the dimerization energy correspond to a strong HB between the carboxylic acid monomers. Note that the distance r(O···O) in gaseous dimers (Table 3) is close to that in the crystal structure POBA (2.611 Å [28]).

The main donor-acceptor interactions occur between the lone electron pairs of the LP(O2) atom and the antibonding orbital σ*(O1′-H′), which lead to the formation of a bonding region between O2···H′ atoms, to a decrease in the electron density in the region of C=O2 and O1′-H′ bonds, and an increase their length.

The balance between the energy of the donor-acceptor interaction of the orbitals $E^{(2)}$ = 131.0 kJ/mol and the energy of steric repulsion E_{steric} = 79.2 kJ/mol between the electron density of the orbitals LP(O2) and the bonding orbital σ(O1′-H′) is 51.8 kJ/mol. The latter value can be considered as the energy of one HB in the POBA cyclic dimer (Table S2).

On the other hand, the HB energy can be estimated from the intermolecular interaction energy ΔE_{IMI}, calculated as the difference between the optimized dimer energy E_{dim} and the energies of the monomers with their geometry in the complex $E_{mon}(SP)$. To calculate the latter, the SP (Single Point) variant was used.

$$\Delta E_{IMI} = E_{dim} - 2E_{mon}(SP)$$

The interaction energy ΔE_{IMI} is −88.7 kJ/mol, i.e., more than 44 kJ/mol per HB. Close values of this energy were obtained for p-n-propyloxycinnamic and p-n-propylbenzoic acids in [29] (see Note S1).

The significant value of the intermolecular interaction energy in carboxylic acid dimers is the reason for their noticeable presence in the gas phase.

3. Materials and Methods

Commercial (Aldrich) 4-n-propyloxybenzoic acid was used without additional purification.

3.1. Conditions of a Synchronous GED/MS Experiment

The gas phase electron diffraction patterns and mass spectra were recorded simultaneously using the technique described in ref. [30,31] at two nozzle-to-plate distances (338 and 598 mm). The POBA sample was evaporated at a temperature of 87(15) °C from an effusion cell made of X18H10T stainless steel. The cell nozzle size is 0.5 × 1.6 mm (diameter × length).

The conditions of the GED/MS experiments are shown in Table 4.

Table 4. Conditions of the synchronous GED/MS experiments.

Nozzle-to-Plate Distance, mm	338	598
Number of recorded plates	6	6
Electron beam current, μA	1.35	0.65
Wavelength of electrons, Å	0.03969(3)	0.03981(4)
Temperature of the effusion cell, K	360(15)	360(15)
Average exposure time, s	105	85
Residual gas pressure in diffraction chamber, Torr	$1.3 \cdot 10^{-6}$	$1.8 \cdot 10^{-6}$
s-range/Δs, $Å^{-1}$	2.6–27.0/0.1	1.3–15.0/0.1
Ionization voltage, V	50	50

For each distance nozzle-to-plate, 6 electron diffraction patterns of the studied substance and 2 diffraction patterns of the ZnO polycrystalline standard were obtained. The latter were taken before and after recording electron diffraction patterns of POBA vapors to control and accurately determine the electron wavelength.

3.2. The Separate MS Experiment

MS experiment was carried out on a serial MI-1201 mass spectrometer equipped for thermodynamic studies [22]. In the study of saturated vapors over POBA, a molybdenum effusion cell was used with the ratio "evaporation area/effusion orifice area" equal to 1000. The ionizing voltage in the ion source was 50 V. The temperature of the effusion cell was measured using a tungsten-rhenium thermocouple WRe-5/20.

3.3. Computational Details

The QC calculations were carried out using the Gaussian 09 software [32], within the density functional theory (DFT) using B97D [33] functionals with 6-311++G ** basis sets [34]. The use of the functional, which considers dispersion interaction, is important in theoretical studies of the considered objects capable of forming H-bonded complexes. The results of the QC calculations were visualized with the ChemCraft program [35]. The electron density distribution was analyzed using the NBO program [36] from the Gaussian 09 software.

The results of QC calculations were used for a comparative analysis of the geometric parameters, electronic, and vibrational characteristics of the monomeric and dimeric forms of acetic, benzoic, and POBA acids, in the calculation of thermodynamic functions at different temperatures.

For each molecular form, the geometry was optimized without any restrictions, the vibration frequencies and the force field were calculated in the harmonic approximation. The latter were used to calculate the amplitudes of vibrations and corrections to internuclear distances necessary for the structural analysis of GED data. In this case, a modified VibModule program [18] was used (for details of the modification, see note S2), which makes it possible to calculate the vibrational parameters for supermolecules containing intermolecular hydrogen bonds.

4. Conclusions

Performed study of the saturated vapor of p-n-propyloxybenzoic acid showed that an efficient method for detecting hydrogen-bonded carboxylic acid dimers is the electron diffraction method.

For recording gas IR spectra, a higher vapor pressure is required than in electron diffraction experiment, hence high temperatures, which lead to the dissociation of dimers. In the method of electron impact mass spectrometry, deep dissociative ionization of dimers is observed, and fragment ions with a mass exceeding the mass of the monomer have a very low intensity compared to monomeric ions, based on which a false conclusion can be made about the composition of the vapor.

One of the advantages of the GED method is the greater scattering power of dimers compared to monomers. This makes it possible to detect dimeric forms even with a significant predominance of monomers in the vapor.

It is shown that the geometrical parameters of the cyclic fragment of a dimer with two hydrogen bonds can be determined by the GED method. However, this method is not sensitive to the determination of conformations of OC_3H_7 substituents.

The decisive role in the successful registration of dimeric forms in aromatic carboxylic acid vapors by the GED method is the choice of experimental conditions, which depend on the design of the inlet system used.

The QC modeling is a necessary tool that supplements experimental studies in solving such problems. It is shown that due to the presence of two strong hydrogen bonds, not all cyclic dimers of the considered acids dissociate into monomers during sublimation.

Supplementary Materials: The following supporting information can be downloaded at https://www.mdpi.com/article/10.3390/ijms232315079/s1.

Author Contributions: Conceptualization, G.V.G. and N.I.G.; methodology, G.V.G., N.I.G.; investigation: N.I.G., K.E.B., A.V.K. and G.V.G.; formal analysis, N.I.G. and K.E.B., resources, N.I.G., G.V.G.; writing—original draft preparation, N.I.G.; writing—review and editing, G.V.G.; visualization, N.I.G., K.E.B.; supervision, G.V.G.; project administration, G.V.G.; funding acquisition, G.V.G. All authors have read and agreed to the published version of the manuscript.

Funding: This work was supported by the Russian Science Foundation (grant No 20-13-00359).

Institutional Review Board Statement: Not applicable.

Informed Consent Statement: Not applicable.

Data Availability Statement: Not applicable.

Acknowledgments: N.I.G. thanks Russian Ministry of Science and High Education (grant No FZZM-2020-0006) for support the structural analysis of electron diffraction data. The gas-phase electron diffraction/mass-spectrometric experiments were carried out using the GED/MS equipment (https://www.isuct.ru/department/ckp/structure/ged-ms) of the resources of the Center for Shared Use of Scientific Equipment of the ISUCT (with the support of the Ministry of Science and Higher Education of Russia, grant No. 075-15-2021-671). We thank A.V. Belyakov for his help in conducting the calculations.

Conflicts of Interest: The authors declare no conflict of interest.

References

1. Abdy, M.; Murdoch, A.; Martínez-Felipe, A. New insights into the role of hydrogen bonding on the liquid crystal behavior of 4-alkoxybenzoic acids: A detailed IR spectroscopy study. *Liq. Cryst.* **2016**, *43*, 2191–2207. [CrossRef]
2. Giricheva, N.; Fedorov, M.; Shpilevaya, K.; Syrbu, S.; Ditsina, O. Characteristics of the hydrogen bond and the structure of H-complexes of *p*-n-propyloxybenzoic acid and *p*-n-propyloxy-*p*′-cyanobiphenyl. *J. Struct. Chem.* **2017**, *58*, 9–16. [CrossRef]
3. Painter, P.; Cleveland, C.; Coleman, M. An Infrared Spectroscopic Study of p-n-Alkoxybenzoic Acids. Molecular Crystals and Liquid Crystals Science and Technology. *Mol. Cryst. Liq. Cryst. Sci. Technol. Sect. A. Mol. Cryst. Liq. Cryst.* **2000**, *348*, 269–293. [CrossRef]
4. Kato, T.; Jin, C.; Kaneuchi, F.; Uryu, T. Effect of the Molecular Orientation on the Stability of Hydrogen-Bonded Benzoic Acid Dimers. Infrared Study of Liquid-Crystalline 4-Alkylbenzoic Acids. *Bull. Chem. Soc. Jpn.* **1993**, *66*, 3581–3584. [CrossRef]
5. Demus, D.; Zaschke, H. *Flussiqe Kristalle in Tabellen II*; VEB Deutscher Verlag für Grundstoffindustrie: Leipzig, Germany, 1984; 468p. [CrossRef]
6. Giricheva, N.; Syrbu, S.; Bubnova, K.; Fedorov, M.; Kiselev, M.; Girichev, G. H-complexes in the "4-n-alkoxybenzoic acid: 4-pyridyl 4′-n-alkoxybenzoate" system. IR spectroscopy and quantum chemical calculations. *J. Mol. Liq.* **2019**, *277*, 833–842. [CrossRef]
7. Yaremko, A.; Ratajczak, H.; Barnes, A.; Baran, J.; Durlak, P.; Latajka, Z. Fermi resonance and strong anharmonic effects in the absorption spectra of the m-OH (m-OD) vibration of solid H- and D-benzoic acid. *Chem. Phys.* **2009**, *364*, 51–63. [CrossRef]
8. Paterson, D.; Martinez-Felipe, A.; Jansze, S.; Marcelis, A.; Storey, J.; Imrie, C. New insights into the liquid crystal behavior of hydrogen-bonded mixtures provided by temperature-dependent FTIR spectroscopy. *Liq. Cryst.* **2015**, *5–6*, 928–939. [CrossRef]
9. Meganathan, C.; Sebastian, S.; Kurt, M.; Lee, K.; Sundaraganesan, N. Molecular structure, spectroscopic (FTIR, FTIR gas phase, FT-Raman) first-order hyperpolarizability and HOMO–LUMO analysis of 4-methoxy-2-methyl benzoic acid. *J. Raman Spectrosc.* **2010**, *41*, 1369–1378. [CrossRef]
10. Bakker, J.; Aleese, L.; von Helden, G.; Meijer, G. The infrared absorption spectrum of the gas phase neutral benzoic acid monomer and dimer. *J.Chem. Phys.* **2003**, *119*, 11180. [CrossRef]
11. Reva, I.; Stepanian, S. An infrared study on matrix-isolated benzoic acid. *J. Mol. Struct.* **1995**, *349*, 337–340. [CrossRef]
12. Stepanian, S.; Reva, I.; Radchenko, E.; Sheina, G. Infrared spectra of benzoic acid monomers and dimers in argon matrix. *Vib. Spectrosc.* **1996**, *11*, 123–133. [CrossRef]
13. Florio, G.; Silbert III, E.; Zwier, T. Fluorescence-dip IR spectra of jet-cooled benzoic acid dimer in its ground and First excited singlet states. *Faraday Discuss.* **2001**, *118*, 315–330. [CrossRef] [PubMed]
14. Vishnevskiy, Y.; Blomeyer, S.; Reuter, C. Low Pressure Gas Electron Diffraction: An Experimental Setup and Case Studies. *Rev. Sci. Instrum.* **2020**, *91*, 074104. [CrossRef] [PubMed]
15. Aarset, K.; Page, E.; Rice, D. Molecular Structures of Benzoic Acid and 2-Hydroxybenzoic Acid, Obtained by Gas-Phase Electron Diffraction and Theoretical Calculations. *J. Phys. Chem. A.* **2006**, *110*, 9014–9019. [CrossRef] [PubMed]
16. Vishnevskiy, Y.; Blomeyer, S.; Reuter, C.; Pimenov, O.; Shlykov, S. Combined gas electron diffraction and mass spectrometric experimental setup at Bielefeld University. *Rev Sci Instrum.* **2020**, *91*, 73103. [CrossRef]
17. Andersen, B.; Seip, H.; Strand, T.; Stolevik, R. Procedure and Computer Programs for the Structure Determination of Gaseous Molecules from Electron Diffraction Data. *Acta Chem. Scand.* **1969**, *23*, 3224–3234. [CrossRef]
18. Vishnevskiy, Y.; Zhabanov, Y. New implementation of the first-order perturbation theory for calculation of interatomic vibrational amplitudes and corrections in gas electron diffraction. *J. Phys. Conf. Ser.* **2015**, *633*, 012076. [CrossRef]
19. Giricheva, N.; Fedorov, M.; Shpilevaya, K.; Syrbu, S.; Girichev, G. Intramolecular vibrations and structural nonrigidity of hydrogen-bonded complexes. *Liq. Cryst. Their Appl.* **2018**, *18*, 14–25. [CrossRef]
20. Zhabanov, Y.; Zakharov, A.; Giricheva, N.; Shlykov, S.; Koifman, O.; Girichev, G. To the limit of gas-phase electron diffraction: Molecular structure of magnesium octa(m-trifluoromethylphenyl)porphyrazine. *J. Mol. Struct.* **2015**, *1092*, 104–112. [CrossRef]
21. Hamilton, W. Significance tests on the crystallographic R factor. *Acta Crystallogr.* **1965**, *18*, 502–510. [CrossRef]
22. Pelipets, O.V. *Investigation of the Thermodynamics of ErCl3, EuBr2 and EuCl2 Evaporation and the Structure of Molecular Forms According to High-Temperature Mass Spectrometry and Gas Electron Diffraction Data*; ISU: Ivanovo, Russia, 2000; 146p. (In Russian)
23. Lebedev, A. *Mass Spectrometry in Organic Chemistry*; BINOM: Moscow, Russia, 2003; 493p. (In Russian)

24. Mori, Y.; Kitagawa, T.; Yamamoto, T.; Yanada, K.; Nagahara, S. Mass Spectrometric Studies of Some Carboxylic Acid Dimers Produced by a Nozzle Beam: HCO_2H, CH_3CO_2H, $C_2H_5CO_2H$, and n-C3H7CO2H. *Bull. Chem. Soc. Jpn.* **1980**, *53*, 3492–3495. [CrossRef]
25. Karle, J.; Brockway, L. An Electron Diffraction Investigation of the Monomers and Dimers of Formic, Acetic and Trifluoroacetic Acids and the Dimer of Deuterium Acetate. *J.Am. Chem. Soc.* **1944**, *66*, 574–584. [CrossRef]
26. Derissen, J. A reinvestigation of the molecular structure of acetic acid monomer and dimer by gas electron diffraction. *J. Mol. Struct.* **1971**, *7*, 67–80. [CrossRef]
27. Giricheva, N.; Fedorov, M.; Syrbu, S.; Shpilevaya, K.; Fedoseev, O. Geometric and electronic characteristics of H-complexes based on aromatic carboxylic acids with N···H–O type hydrogen bonds. In *Organic and Hybrid Nanomaterials: Preparation and Prospects of Application: Monograph*; Razumova, V., Klyuev, M., Eds.; Ivanovo State University: Ivanovo, Russia, 2017; pp. 149–174. (In Russian)
28. Bryan, R.F.; Hartley, P. An X-Ray Study of the *p*-n-Alkoxybenzoic Acids. Part V. Crystal Structures of the Nematogenic Acids Having Three and Five Alkyl-Chain Carbon Atoms. *Mol. Cryst. Liq. Cryst.* **1980**, *62*, 259–279. [CrossRef]
29. Giricheva, N.; Bubnova, K.; Zhabanov, Y.; Fedorov, M.; Girichev, G. Structural and dynamic non-rigidity of hydrogen-bonded complexes of A···A and A···B···A types and odd-even effect. *J. Mol. Liq.* **2022**, *350*, 118521. [CrossRef]
30. Girichev, G.; Utkin, A.; Revichev, Y. Modernization of electron diffraction recorder EMR-100 for gas research. *Prib. Tekh. Eksp.* **1984**, *27*, 187–190. (In Russian)
31. Girichev, G.; Shlykov, S.; Revichev, Y. Equipment for studying the structure of molecules of valence-unsaturated compounds. *Prib. Tekh. Eksp.* **1986**, *4*, 167. (In Russian)
32. Frisch, M.J.; Trucks, G.W.; Schlegel, H.B.; Scuseria, G.E.; Robb, M.A.; Cheeseman, J.R.; Scalmani, G.; Barone, V.; Mennucci, B.; Petersson, G.A.; et al. *Gaussian 09*; Revision D.02.; Gaussian, Inc.: Wallingford, CT, USA, 2009.
33. Grimme, S. Semiempirical GGA-type density functional constructed with a long-range dispersion correction. *J. Compt. Chem.* **2006**, *27*, 1787–1799. [CrossRef]
34. McLean, A.; Chandler, G. Contracted Gaussian basis sets for molecular calculations. I. Second row atoms, Z=11–18. *J. Chem. Phys.* **1980**, *72*, 5639–5648. [CrossRef]
35. Zhurko, G.; Zhurko, D. Chemcraft—Graphical Software for Visualization of Quantum Chemistry Computations. Available online: https://www.chemcraftprog.com (accessed on 21 March 2022).
36. Glendening, E.; Reed, A.; Carpenter, J.; Weinhold, F. *NBO*; Version 3.1; Gaussian Inc.: Pittsburgh, PA, USA, 2003.

International Journal of
Molecular Sciences

MDPI

Article

Nature of Beryllium, Magnesium, and Zinc Bonds in Carbene···MX_2 (M = Be, Mg, Zn; X = H, Br) Dimers Revealed by the IQA, ETS-NOCV and LED Methods

Filip Sagan [1], Mariusz Mitoraj [1] and Mirosław Jabłoński [2,*]

[1] Faculty of Chemistry, Jagiellonian University, Gronostajowa 2, 30-387 Cracow, Poland
[2] Faculty of Chemistry, Nicolaus Copernicus University in Toruń, Gagarina 7, 87-100 Toruń, Poland
* Correspondence: teojab@chem.umk.pl; Tel.: +48-056-611-4695

Abstract: The nature of beryllium–, magnesium– and zinc–carbene bonds in the cyclopropenylidene···MX_2 (M = Be, Mg, Zn; X = H, Br) and imidazol-2-ylidene···MBr_2 dimers is investigated by the joint use of the topological QTAIM-based IQA decomposition scheme, the molecular orbital-based ETS-NOCV charge and energy decomposition method, and the LED energy decomposition approach based on the state-of-the-art DLPNO-CCSD(T) method. All these methods show that the C···M bond strengthens according to the following order: Zn < Mg << Be. Electrostatics is proved to be the dominant bond component, whereas the orbital component is far less important. It is shown that QTAIM/IQA underestimates electrostatic contribution for zinc bonds with respect to both ETS-NOCV and LED schemes. The σ carbene→MX_2 donation appears to be much more important than the MX_2 →carbene back-donation of π symmetry. The substitution of hydrogen atoms by bromine (X in MX_2) strengthens the metal–carbene bond in all cases. The physical origin of rotational barriers has been unveiled by the ETS-NOCV approach.

Keywords: carbene; imidazol-2-ylidene; cyclopropenylidene; beryllium bond; magnesium bond; zinc bond; interaction; IQA; ETS-NOCV; LED

1. Introduction

Carbenes are undoubtedly one of the most important groups of organic compounds [1–18]. They are characterized by having only two valence carbon atoms, thanks to which it generally forms two single covalent bonds to the substituents, CR_1R_2. As a consequence, the carbon atom retains two unbound electrons, which in a singlet state occupy the same p orbital, creating a lone electron pair, which is the cause of nucleophilic properties of carbenes. In other words, singlet carbenes are good Lewis bases. It is therefore not surprising that carbenes participate in various kinds of intermolecular interactions, such as hydrogen bonds [19–27], lithium bonds [16,18,28–30], beryllium bonds [16,18,31–35], magnesium bonds [16,18,34–38], triel bonds [18,39–42], tetrel bonds [18,43–45], pnictogen bonds [18,46–48], chalcogen bonds [18,49], halogen bonds [18,50–53] (in particular to the iodine atom [50,51]), and aerogen bonds [54]. However, carbenes in particular are known for their great ease of bonding with transition metal atoms [8,12,16,36,55–61] to form various adducts, which is the goal of numerous studies in organometallic chemistry. In this case, the N-heterocyclic carbenes (NHCs), especially those derived from imidazol-2-ylidene, are of particular importance [4,8,9,11,12,15–18,27]. In turn, within the group of the NHC–M complexes, those involving zinc are of increasing practical importance [35,36,38,62–78], due in part to the easy availability and relatively low cost of precursors of such complexes. Some practical applications of NHC–Zn complexes in organometallic catalysis have recently been extensively described [62–65]. It is worth mentioning, however, that apart from the lone electron pair which gives singlet carbenes nucleophilic properties, singlet carbenes also feature electrophilic properties [7,79–84] resulting from the presence of an empty p orbital perpendicular to the plane formed by CR_1R_2.

Very recently, one of us (M. J.) has investigated [35] the beryllium [85–93], magnesium [85,92–96] and zinc (spodium) [78,97–100] bonds in a large group of dimers formed by carbenes ($(NH_2)_2C$, imidazol-2-ylidene, imidazolidin-2-ylidene, tetrahydropyrymid-2-ylidene, and cyclopropenylidene) or carbodiphosphoranes ($(PH_3)_2C$ and $(NH_3)_2C$) and MX_2 (M = Be, Mg, Zn and X = H, F, Cl, Br, Me) molecules. This was the first study of beryllium, magnesium, and zinc bonds of the C$\cdots$M type. Moreover, in the vast majority of cases, in earlier systems with the beryllium or magnesium bond, the Lewis base was a small molecule involving either atom being good electron donors [85–87,90,92,93,95] or π-electron bonds [88]. On the other hand, theoretical research on zinc bonds is rather sporadic [35,78,97,100].

Among many important findings was the demonstration that carbene$\cdots MX_2$ dimers are characterized by a very high charge transfer from carbene to the MX_2 molecule [35]. Namely, it was from two to almost four times greater than in the case of HOH$\cdots OH_2$ and HOH$\cdots NH_3$ dimers. This result showed that the carbene$\cdots MX_2$ (M = Be, Mg, Zn; X = H, F, Cl, Br, Me) dimers are undoubtedly systems in which the charge transfer effect plays a large role, especially in the presence of very polarizable bromine atoms as X. This, in turn, proves the strong interorbital interactions between the carbene molecule and MX_2. Importantly, it has been shown that in terms of the electron density and the total electronic energy density values computed at the bond critical point [101] of C$\cdots$M the C$\cdots$Zn zinc bond is similar to the C$\cdots$Be beryllium bond. Treating the value of electron density as a measure of bond strength [101], this result has shown that the zinc bonds should be of similar strength to the beryllium bonds, while the magnesium bonds should be significantly weaker than them [35]. Moreover, together with positive values of the Laplacian of the electron density, negative values of the total electronic energy density obtained for the beryllium and zinc bonds have indicated a significant degree of electron sharing, in turn reflecting a high degree of covalence [102].

A somewhat surprisingly different picture of the similarity of these three types of bonding has been obtained on the basis of the delocalization index, δ(C,M), which describes the exchange of the electrons in the basins of atoms C and M [103–106]. Namely, the δ(C,M) values for the zinc bond were large, while those for the beryllium and magnesium bond were much lower and similar to each other [35]. Thus, taking into account the relationship of δ(A,B) to the exchange energy [107] and therefore treating δ(A,B) as a measure of the covalent character of an A–B bond [108], this result has shown that the zinc bond is much more covalent than the beryllium and magnesium bond. On the other hand, the NCI-based (the abbreviation NCI is derived from the Non-Covalent Interactions index [109,110]) analysis has shown [35] that the zinc bond is the strongest; however, the beryllium bond is only slightly weaker.

In light of the previously obtained results [35] briefly recalled here, it is necessary to investigate comparatively the physical nature of the beryllium, magnesium and TM-type zinc bonds, not only qualitatively but also quantitatively. This is the main goal of this article, and this nature will be analyzed using complementary energy decomposition methods, IQA (i.e., the Interacting Quantum Atoms approach [111,112]), which provides a local picture of bonding and ETS-NOCV (i.e., the combination of the Extended Transition State (ETS) method [113] with the Natural Orbitals for Chemical Valence (NOCV) method [114–118]), which is more suited for the description of bonding between molecular fragments. Results will also be compared with the state-of-the-art DLPNO-CCSD(T) (i.e., the Domain-based Localized Pair-Natural Orbital Singles and Doubles Coupled Cluster with perturbative Triples) [119–123] calculations and the Local Energy Decomposition (LED) scheme [124,125].

2. Results and Discussion

As already mentioned in the Introduction section, one of us (M. J.) has recently investigated [35] beryllium, magnesium and zinc (spodium) bonds in a large group of dimers formed by either a carbene ($(NH_2)_2C$, imidazol-2-ylidene, imidazolidin-2-ylidene, tetrahydropyrymid-2-ylidene, and cyclopropenylidene) or a carbodiphosphorane ($(PH_3)_2C$ and $(NH_3)_2C$) and the MX_2 (M = Be, Mg, Zn and X = H, F, Cl, Br, Me) species. While the C$\cdots$M contact has been shown to be certainly dominant in all of these dimers, some of them contain additional intermolecular interactions (either hydrogen bond or dihydrogen bond). Since the purpose of this article is to quantify the physical nature of the C$\cdots$M bond, such dimers are not considered here. In other words, in this article, only those dimers in which the described interaction C$\cdots$M is undoubtedly dominant will be considered. The structural characteristics of these dimers will be briefly discussed in the first subsection. Then, in the second and third subsections, the description of the C$\cdots$M interaction obtained by the IQA and ETS-NOCV methods, respectively, will be presented. The correspondence of the results with the outcomes originating from the DLPNO-CCSD(T) method, and based on it, LED energy decomposition will be shown in the fourth subsection. Finally, in the fifth subsection, we will analyze the energy changes of dimers during the rotation of carbene with respect to the $ZnBr_2$ subunit.

2.1. Structural Characteristics of the Considered Dimers

To meet the aforementioned condition, in this article only the cyclopropenylidene$\cdots MX_2$ (M = Be, Mg, Zn; X = H, Br) and, additionally, imidazol-2-ylidene$\cdots MBr_2$ dimers will be studied (see Figures S1 and S2). It is worth noting that in the former case, the lack of additional intermolecular interactions (i.e., apart from the described C$\cdots$M) is ensured by the perpendicular orientation of the cyclopropenylidene and MX_2 molecules (see the representative cyclopropenylidene$\cdots ZnBr_2$ dimer in Figure 1).

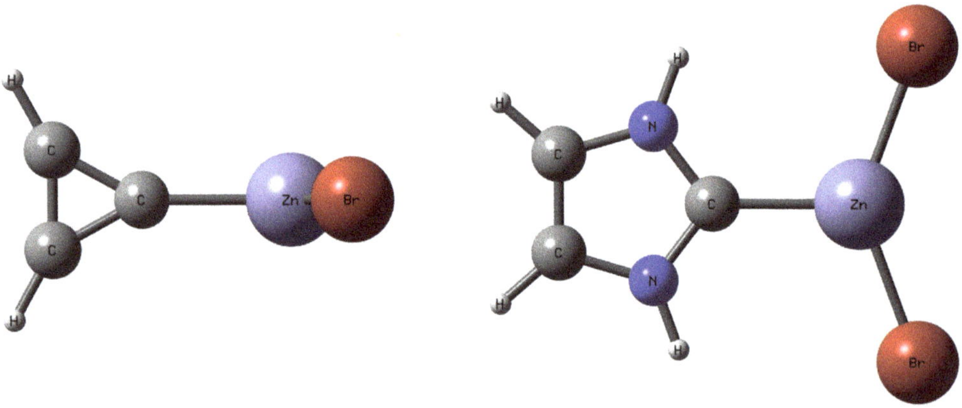

Figure 1. Fully optimized cyclopropenylidene$\cdots ZnBr_2$ (**left**) and imidazol-2-ylidene$\cdots ZnBr_2$ (**right**) dimer structures.

In the latter case, however, although the flatness of the entire dimer admits the possibility of the presence of N–H$\cdots$Br hydrogen bonds, these interactions should be relatively weaker than C$\cdots$M due to the large distance H$\cdots$Br and the almost parallel orientation of the N–H and Zn–Br bonds relative to each other (see the representative imidazol-2-ylidene$\cdots ZnBr_2$ dimer in Figure 1). A thorough analysis of these intermolecular N-H$\cdots$Br hydrogen bonds will also be performed. The values of some fundamental parameters characterizing the considered carbene$\cdots MX_2$ dimers are presented in Table 1.

Table 1. Some fundamental data characterizing cyclopropenylidene$\cdots MX_2$ (X = H, Br) and imidazol-2-ylidene$\cdots MBr_2$ dimers (M = Be, Mg, Zn): C$\cdots$M and M-X distance, change of the MX bond length (in Å), XMX, CMX, LCL angles (in degrees), dissociation energy (in kcal/mol), charge transfer (in a.u.).

MX_2	$d_{C\cdots M}$	d_{MX}	Δd_{MX}	α_{XMX}	α_{CMX}	α_{LCL}	$\Delta\alpha_{LCL}$	D_0	CT [a]
				cyclopropenylidene					
BeH_2	1.743	1.374	0.039	135.1	112.4	56.6	0.8	29.1	−0.317
MgH_2	2.268	1.740	0.035	148.3	105.9	56.7	0.9	20.9	−0.273
ZnH_2	2.121	1.575	0.036	144.6	107.7	57.3	0.9	15.2	−0.279
$BeBr_2$	1.764	2.043	0.093	134.5	112.8	57.3	1.5	35.4	−0.414
$MgBr_2$	2.206	2.376	0.052	145.4	107.3	57.0	1.3	32.1	−0.331
$ZnBr_2$	2.061	2.298	0.083	137.1	111.5	57.3	1.5	28.2	−0.380
				imidazol-2-ylidene					
$BeBr_2$	1.765	2.059	0.109	132.9	113.6	104.2	3.5	48.6	−0.464
$MgBr_2$	2.173	2.393 [b]	0.069 [c]	147.9	106.1 [d]	103.5	2.8	43.7	−0.363
$ZnBr_2$	2.037	2.311	0.097	139.2	110.4	104.2	3.4	41.2	−0.432

[a] The charge transfer was calculated based on Hirshfeld's atomic charges [126,127]. [b] The average value determined from the two values 2.386 Å and 2.400 Å. [c] The average value determined from the two values 0.062 Å and 0.076 Å. [d] The average value determined from the two values 102.9° and 109.2°.

As can be seen from Figure 1 and Table 1 (see the α_{XMX} angle), the characteristic feature of the considered dimers is the clearly bent structure of the MX_2 molecule. It is worth noting that this bend is similar in the case of MX_2 molecules with beryllium or zinc, while it is clearly smaller in the case of the magnesium atom, which forms more ionic and thus more resistant M–X bonds. Thus, the value of the angle α_{XMX} suggests the similarity between the Be and Zn atoms. It should be noted that the bending of the MX_2 molecule is a characteristic effect of the presence of a beryllium or magnesium bond in dimers with different Lewis bases [85,86,88,92]. Moreover, Martín-Sómer et al. [86] have shown that in the case of $H_3N\cdots BeH_{2-n}X_n$ (X = F, Cl, Br; $n \leq 2$) dimers, the nonlinearity of the $BeH_{2-n}X_n$ molecule is due to the decrease in LUMO energy. It is also worth noting that for the bond length M–X the relation Be<Zn<Mg applies so that for a given X the magnesium atom forms the longest M–X bond. It should also be emphasized that the formation of the C$\cdots$M bond leads to significant elongation of the M–X bond (see Δd_{MX} in Table 1) and a greater opening of the L–C–L angle in the carbene molecule (see $\Delta\alpha_{LCL}$ in Table 1). For the same MX_2 molecule, both of these effects are clearly greater in imidazol-2-ylidene than in cyclopropenylidene, suggesting that the C$\cdots$M interaction should be stronger in the dimers of imidazol-2-ylidene. This is confirmed by the computed dissociation energies [35]. For imidazol-2-ylidene, these energies are in the 41.2–48.6 kcal/mol range, while for cyclopropenylidene, only 28.2–35.4 kcal/mol (for the same set of MX_2 molecules, i.e., MBr_2), similar outcomes are valid when considering other XC and DLPNO-CCSD(T) method, Table S1. The greatest dissociation energy value characterizes the dimers with Be, while the smallest D_0 are noted in the dimers containing Zn, which can only partially be explained by the shortest distance C$\cdots$M in the case when M = Be. This is because in the case of Mg and Zn, the distance C$\cdots$Mg is longer than that of C$\cdots$Zn, but the magnesium bond is stronger than the zinc bond. For example, for the cyclopropenylidene$\cdots MH_2$ dimers, the C$\cdots$M lengths for Be, Mg, and Zn, respectively, are 1.743, 2.268, and 2.121 Å, while the dissociation energies are 29.1, 20.9, and 15.2 kcal/mol. This result shows that the dissociation energy is not completely dependent on the distance C$\cdots$M. Such a trend has already been observed in literature [35].

It has already been mentioned in Introduction that the carbene$\cdots MX_2$ dimers considered here are characterized by a very high charge transfer effect from the Lewis base, i.e., the carbene molecule, to the Lewis acid, i.e., the MX_2 molecule, and this effect is particularly large in the presence of highly polarizable bromine atoms. Indeed, in the case of the imidazol-2-ylidene$\cdots$ $BeBr_2$ dimer, the value of the charge transfer (based on chemically reliable [128–130] Hirshfeld atomic charges [126,127]) is as high as −0.464 au. It is enough to mention that the charge transfer determined on the same level of theory in the case of the water dimer is only −0.098 au. This comparison suggests the presence of strong interorbital interactions between the carbene molecule and the MX_2 unit. Their significance, however, can be quantified by ETS-NOCV and LED methods.

2.2. IQA-Based Results

As already mentioned in the Methods section, the IQA approach is a unique method that enables the determination of the interaction energy of any two atoms and the decomposition of this energy into individual terms, according to Equations (1) and (2). The results of such decomposition carried out for the contact C$\cdots$M are shown in Table 2.

Table 2. The IQA-based energy (in a.u.) terms (see Methods) computed for the C$\cdots$M interaction in the carbene$\cdots$ MX_2 dimers.

MX_2	E_{neen}	E_{nn}	E_{ee}	$E_{ee,C}$	$E_{ee,xc}$	$\%E_{ee,xc}$	E_{elst}	E_{int}
				cyclopropenylidene				
BeH_2	−12.2839	7.2846	4.6505	4.6963	−0.0458	13.1	−0.3030	−0.3488
MgH_2	−32.4999	16.8017	15.5098	15.5405	−0.0307	16.3	−0.1577	−0.1884
ZnH_2	−90.1277	44.8987	45.0634	45.1726	−0.1091	65.9	−0.0564	−0.1655
$BeBr_2$	−12.1017	7.1990	4.5383	4.5810	−0.0428	11.7	−0.3217	−0.3644
$MgBr_2$	−33.3934	17.2713	15.9040	15.9383	−0.0343	15.7	−0.1838	−0.2181
$ZnBr_2$	−92.5816	46.2217	46.1682	46.2894	−0.1212	63.2	−0.0705	−0.1917
				imidazol-2-ylidene				
$BeBr_2$	−11.3609	7.1976	4.1093	4.1554	−0.0461	85.5	−0.0079	−0.0539
$MgBr_2$	−32.0341	17.5351	14.5370	14.5767	−0.0397	−104.7	0.0777	0.0379
$ZnBr_2$	−88.3086	46.7541	41.5418	41.6735	−0.1317	1037.0	0.1190	−0.0127

A certain inconvenience of IQA (which is also a characteristic feature of energy decomposition methods) is that the components of the interaction energy are generally much larger than the interaction energy itself. This is obviously due to the close cancellation of the negative and largest E_{neen} and the positive sum of E_{nn} and $E_{ee,C}$. It is obvious that the rapid growth of the relevant components with the change Be→Mg→Zn results from both the increase in the atomic number of M, i.e., the charge of the atomic nucleus, and the increase in the number of electrons. However, it is much more important that according to IQA, the C$\cdots$M (M = Be, Mg, Zn) interaction in all the dimers with the participation of cyclopropenylidene is stabilizing, which is indicated by negative values of the obtained interaction energies. For the same X, the binding effect increases in the Zn→Mg→Be direction and is slightly stronger when X = Br, so that the strongest C$\cdots$M bond is for $BeBr_2$ (−0.3644 a.u.).

It is interesting to look at the values of the exchange-correlation energy ($E_{ee,xc}$) as well as the percentage share of this component in the total interaction energy ($\%E_{ee,xc}$). Of course, independently of X, $E_{ee,xc}$ increases (i.e., becomes more negative) in the direction Mg→Be→Zn, with the corresponding values being slightly greater for Br than for H. Importantly, the values for Zn are significantly greater than those for Be or Mg (e.g., $E_{ee,xc}$ is −0.1091 a.u. in cyclopropenylidene$\cdots$ ZnH_2 and only −0.0458 a.u. and −0.0307 a.u. in cyclopropenylidene$\cdots$ BeH_2 and cyclopropenylidene$\cdots$ MgH_2, respectively). The contribution of negative electrostatic energy (E_{elst}) in the case of the cyclopropenylidene dimers quickly decreases in the Be→Mg→Zn series, as a consequence of which the percentage of the exchange-correlation energy quickly increases from ca. 12–13% for the beryllium bond, to ca. 16% for the magnesium bond and up to ca. 63–66% for the zinc bond. This result shows that the IQA method suggests that the zinc bond is highly covalent, whereas the beryllium and magnesium bonds are electrostatic in nature. Of course, such a result was to be expected, because, as already mentioned in the Introduction section, the same trend has already been obtained earlier for the delocalization index δ(C,M) [35], which, as is known [107,108], correlates very well with the exchange-correlation energy.

Interestingly, IQA has given quite small interaction energies for the imidazol-2-ylidene$\cdots MBr_2$ dimers, which do not match the significant dissociation energies for these complexes; Table 1. Even more surprisingly, a positive interaction energy value (0.0379 a.u.) has been obtained for the imidazol-2-ylidene$\cdots MgBr_2$ dimer. This destabilizing effect of C$\cdots$Mg (0.0379 a.u.) is due to a significant destabilizing electrostatic contribution (0.0777 a.u.), which exceeds the negative exchange-correlation energy (−0.0397, i.e., ca. 51% only). In the case of imidazol-2-ylidene$\cdots ZnBr_2$, E_{elst} is also positive (0.1190 a.u.; and therefore the purely Coulombic contribution gives no bonding), but unlike for imidazol-2-ylidene$\cdots MgBr_2$, this contribution is excessively overshadowed (over 110%) by $E_{ee,xc}$ (−0.1317 a.u.), which gives stabilizing C$\cdots$Zn bond, which could be considered as anti-electrostatic and covalent.

These diatomic IQA data do not explain the stability of the imidazol-2-ylidene$\cdots MgBr_2$ species. To this end, we have also summed the interaction energies corresponding to the pairs of atoms in the groups representing the {carbene}$\cdots${MX_2} fragmentation (i.e., all the combinations where one of the atoms is in the {carbene} moiety and the other one in {MX_2}). In such treatment, which in fact ignores many-body effects, all the E_{int} values are negative, as expected. Furthermore, the IQA fragment-based data (see Table 3) indicate domination of E_{elst} over E_{xc} for the Be- and Mg-based systems (ca. 51–58% of E_{int}), with the Zn-based species as the only exceptions and the share of electrostatics in their total stabilization reaching ca. 16–21% only.

Table 3. Sums of the IQA-based energy terms (in kcal/mol) computed (MP2/6-311+G(d,p)) for groups of atoms reflecting the ETS-NOCV fragmentation, that is {carbene}$\cdots${MX_2}.

MX_2	E_{int}	E_{elst}	$\%E_{elst}$	$E_{ee,xc}$	$\%E_{ee,xc}$
		cyclopropenylidene			
BeH_2	−143.28	−73.68	51.4	−69.60	48.6
MgH_2	−71.81	−38.55	53.7	−33.25	46.3
ZnH_2	−91.45	−14.86	16.2	−76.59	83.8
$BeBr_2$	−162.74	−90.98	55.9	−71.76	44.1
$MgBr_2$	−88.25	−49.41	56.0	−38.85	44.0
$ZnBr_2$	−105.00	−18.61	17.7	−86.38	82.3
		imidazol-2-ylidene			
$BeBr_2$	−190.24	−110.00	57.8	−80.27	42.2
$MgBr_2$	−113.04	−63.87	56.5	−49.17	43.5
$ZnBr_2$	−127.25	−26.19	20.6	−101.05	79.4

For the imidazol-2-ylidene complexes, among all the {carbene}$\cdots${MX_2} atomic pairs, the most stabilizing according to IQA are N$\cdots$M interactions, providing from ca. −150 kcal/mol in the case of Zn via. ca. −220 kcal/mol (Mg) to ca. −250 kcal/mol (Be), followed by C$\cdots$Br (from ca. −43 kcal/mol to ca. −39 kcal/mol) and H$\cdots$Br (from ca. −48 kcal/mol to ca. −32 kcal/mol), Table S2. Surprisingly, the C$\cdots$M interactions are not among the most stabilizing ones, ranging from +20 kcal/mol (Mg) to −40 kcal/mol (Be). In the case of cyclopropenylidene complexes, the C$\cdots$M interactions are definitely the most important, ranging from ca. −100 to over −200 kcal/mol, Tables S3 and S4.

In the fragment-based treatment, a quite different picture of the C$\cdots$M bond nature has been obtained. Namely, for cyclopropenylidene$\cdots MX_2$, the exchange-correlation energy component became more prominent (ca. 45% for Be and Mg, and up to 80%

in the case of Zn; Table 3). Concurrently, in the fragment-based approach, all the MX_2 complexes with imidazol-2-ylidene have negative E_{int} values, as well as their electrostatic and exchange-correlation components; Table 3. Moreover, the character of these interactions changed in the case of Be and Mg as the electrostatic energy became dominant.

Clearly, the fragment-based approach seems to be much more credible for the description of the carbene$\cdots$metal bonds in both cyclopropenylidene and imidazol-2-ylidene complexes. However, the lack of the C$\cdots$M interaction among the most important contributions is alarming, as much as the counterintuitive positive E_{int} value for the imidazol-2-ylidene$\cdots MgBr_2$ dimer (Table 2). The most probable explanation is a quite significant positive QTAIM charge on the metal atoms (e.g., the QTAIM-based atomic charge of Mg is as much as 1.687 a.u. compared to 0.616 a.u. only obtained from the Hirshfeld method) [35]. This, coupled with the incorrect sign of charges on the carbene carbon atom (already stressed in the first work concerning these systems [35]), puts the reliability of IQA electrostatics in question. Since the IQA method clearly depends on the charge distribution between QTAIM atomic basins and therefore provides variable outcomes for different charge density partitionings, we will switch to ETS-NOCV, which estimates E_{elst} in a more reliable way through the quasi-classical approach based on self-consistently converged fragment electron densities; see Methods.

2.3. ETS-NOCV-Based Results

In this section, we will present the results obtained by means of the molecular orbital-based ETS-NOCV method. We have used singlet {carbene} and {MX_2} fragments which have appeared to be more suitable than triplets since they deliver the smallest orbital interaction contribution. The obtained values of the ETS-NOCV energy terms (see Equation (5)) are listed in Table 4.

Table 4. The ETS-NOCV-based energy (in kcal/mol) contributions to carbene$\cdots MX_2$ bonding.

MX_2	ΔE_{int}	ΔE_{elst}	ΔE_{Pauli}	ΔE_{disp}	ΔE_{orb}	ΔE^{σ}_{orb}	ΔE^{π}_{orb}	$\Delta E^{NH\cdots Br}_{orb}$
				cyclopropenylidene				
BeH_2	−45.69	−69.29 (56%)	77.03	−2.56 (2%)	−50.87 (41%)	−40.33	−10.11	n/a
MgH_2	−26.20	−45.85 (70%)	39.55	−2.58 (4%)	−17.32 (26%)	−14.16	−2.88	n/a
ZnH_2	−23.63	−75.38 (68%)	87.86	−2.87 (3%)	−33.24 (30%)	−27.31	−5.19	n/a
$BeBr_2$	−52.23	−82.67 (58%)	90.98	−4.90 (3%)	−55.64 (39%)	−47.43	−6.34	n/a
$MgBr_2$	−38.57	−57.23 (69%)	44.25	−4.14 (5%)	−21.44 (26%)	−17.67	−2.56	n/a
$ZnBr_2$	−37.74	−95.45 (67%)	104.58	−4.52 (3%)	−42.34 (30%)	−36.96	−3.53	n/a
				imidazol-2-ylidene				
$BeBr_2$	−66.86	−103.25 (61%)	102.10	−6.61 (4%)	−59.10 (35%)	−47.92	−5.24	−1.71
$MgBr_2$	−50.49	−77.00 (71%)	58.24	−6.01 (5%)	−25.72 (24%)	−18.57	−3.13	−1.10
$ZnBr_2$	−50.92	−122.82 (68%)	128.22	−6.45 (4%)	−49.87 (28%)	−42.15	−3.70	−1.11

Contrary to the atomic resolution-based IQA results, all ETS-NOCV interaction energies (ΔE_{int}) are negative and correspond better to the bond dissociation energies (D_0 values in Table 1). Bonds formed by beryllium are the strongest, followed by energetically similar magnesium and zinc bonds. Systems with Br atoms are bonded noticeably stronger than with H atoms, and complexes with imidazol-2-ylidene are stabilized more than their cyclopropenylidene counterparts. The carbene$\cdots$metal bonds are dominated by electrostatics (ΔE_{elst}, Table 4), with its percentage contribution to the total stabilization being 67–71% in the case of Mg and Zn dimers and 56–61% in systems with Be. This is similar to other reports of bonding in transition metal complexes of imidazol-2-ylidene-based NHCs [12,56–59,61] and other model carbenes [55,58,61]. For example, it turned out that the electrostatic contribution to C$\cdots$M bonds in imidazol-2-ylidene complexes with Cu, Ag and Au halides accounts for more than 65% of the binding energy [56]. This contribution was also very similar (ca. 2/3 of the interaction energy) in the various NHC–TM complexes studied by Tonner et al. [58]. It is worth mentioning at this point the excellent review on the NHC–M bond by Jacobsen et al. [12], in which it was found that NHC–M bonds are mainly electrostatic in origin with a percentage contribution from 60% to 80%.

Orbital interaction is the second strongest energy term, reaching up to 40% of total stabilization in beryllium systems and 25–30% in the case of Mg and Zn systems, Table 4. The carbene→metal donation (the LP(C)→ σ^*(MX) transfer in MO language) dominates by far (72–87%) over the metal→carbene back-donation in π symmetry, with 16–20% contribution in the case of systems with X = H and 7–12% for X = Br. The clearly lower back-donation effect is visible in the latter case, resulting from the electron-withdrawing properties of Br atoms. It should be pointed out that in many transition metal complexes, the π-back-bonding may be far more pronounced and can reach even 40% of the overall orbital interaction energy [60]. Finally, it should be commented that the geometry distortion term appeared to be the most pronounced for Be-species, Table S5. Also noteworthy, weak (up to -1.7 kcal/mol, i.e., ca. 3% of the orbital interaction energy) side N–H$\cdots$Br hydrogen bonds have been identified in the imidazol-2-ylidene$\cdots$$MBr_2$ complexes, Figure 2.

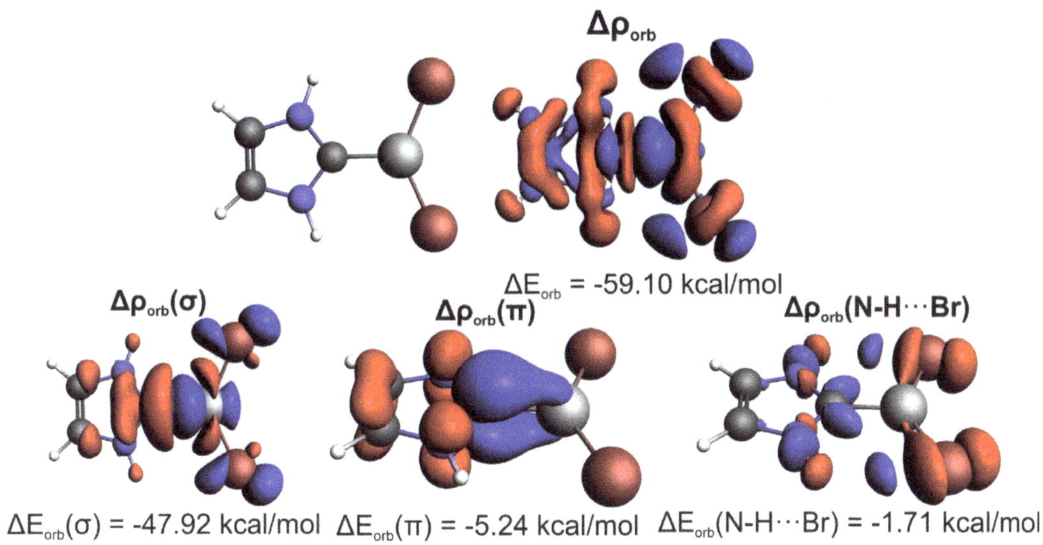

Figure 2. Dominant NOCV-based deformation density contributions determined for imidazol-2-ylidene$\cdots$$BeBr_2$ dimer.

Such a negligible share of N–H$\cdots$Br hydrogen bonds really confirms the correct choice of systems with the dominant C$\cdots$M interaction. As ETS-NOCV is only capable of separating the ΔE_{orb} contribution, it is necessary to support it with complementary IQA results, which indicate a strong electrostatic component ($E_{elst}^{H\cdots Br}$ represents 95% of $E_{int}^{H\cdots Br}$ amounting to -45.5 kcal/mol). Such pronounced interaction energies are common for a two-atomic approach. For example, for the hydrogen bond in water, the $\Delta E_{int}^{O\cdots H}$ value amounts to ca. -100 kcal/mol [111].

2.4. LED-Based Results

Local Energy Decomposition [124,125] allows for partitioning of the binding energy at the DLPNO-CCSD(T) [120–122] level of theory, often considered as the 'gold standard' for calculating interaction energies. The same fragments as before, i.e., {carbene} and {MX_2}, have been chosen. The binding energies obtained at the DLPNO-CCSD(T) level of theory are consistent with the DFT-based energies, that is, the beryllium bonds are the strongest, followed by magnesium and zinc (spodium) bonds; compare Tables 1 and 5.

Table 5. DPLNO-CCSD(T)/ccpv-tz Local Energy Decomposition analysis of the {carbene}···{MX_2} bonding (carbene names shortened for clarity: cpy = cyclopropenylidene and imi = imidazol-2-ylidene). Energies in kcal/mol.

	cpy–$BeBr_2$	cpy–$MgBr_2$	cpy–$ZnBr_2$	imi–$BeBr_2$	imi–$MgBr_2$	imi–$ZnBr_2$
E_{int}	−39.60	−34.28	−29.08	−54.28	−46.37	−42.63
E^{ref}_{elst}	−218.44	−152.51	−327.12	−251.84	−190.31	−393.52
E^{ref}_{exch}	−20.12	−11.84	−33.73	−20.84	14.59	−39.32
$E^{ref}_{geo-prep}$	14.88	6.23	10.76	17.58	6.73	11.24
$E^{ref}_{el-prep}$	190.33	126.75	325.62	210.63	158.28	387.67
$E^{C}_{non-disp}$	-0.69	0.12	1.91	−2.74	−1.83	−0.17
E^{C}_{disp}	−5.35	−2.93	−6.14	−6.43	−4.10	−7.67
E^{C}_{T}	−0.21	−0.10	−0.38	−0.63	−0.55	−0.85
CT 1→2	−16.96	−9.52	−13.44	−19.17	−11.11	−16.32
CT 2→1	−3.33	−2.43	−7.58	−3.56	−2.53	−8.14

Local Energy Decomposition partitioning confirms the importance of electrostatics in metal–carbene bonding, with $E^{ref}_{elst} \approx 90\%$ of total stabilization for all compounds, Table 5. Dispersion contribution of ca. 2% of the total interaction energy proved to be very small. The charge-transfer contributions, bottom of the Table 5, confirm qualitatively the picture obtained by ETS-NOCV, i.e., the carbene → MX_2 donation being far more important than MX_2 → carbene back–donation.

2.5. Flat vs. Perpendicular Structure of the Carben···MX_2 Dimer

Looking at Figure 1, showing the cyclopropenylidene ···$ZnBr_2$ and imidazol-2-ylidene···$ZnBr_2$ dimers, the fundamental difference between them in the orientation of the $ZnBr_2$ molecule relative to the carbene plane is clearly visible. Namely, the $ZnBr_2$ and carbene units are perpendicular to each other in cyclopropenylidene···$ZnBr_2$, while imidazol-2-ylidene···$ZnBr_2$ is planar. Taking into account the fact that the cause of the twisting of the metal-unit in relation to the carbene plane was described in the literature only very sporadically [78], and taking advantage of the relatively high structural simplicity of the studied systems, it was tempting to investigate the cause of the structural differences regarding the twisting of the MX_2 and carbene planes in relation to each other. For this purpose, we have performed additional calculations for the twisted structures of both dimers, i.e., cyclopropenylidene···$ZnBr_2$ and imidazol-2-ylidene···$ZnBr_2$, in which the twist angle (denoted by θ) L-C-Zn-Br (L is either C or N) was changed every 10° from 0° to 90°. Figure 3 (left) shows the energy profile related to the change in θ from 0° to 90°.

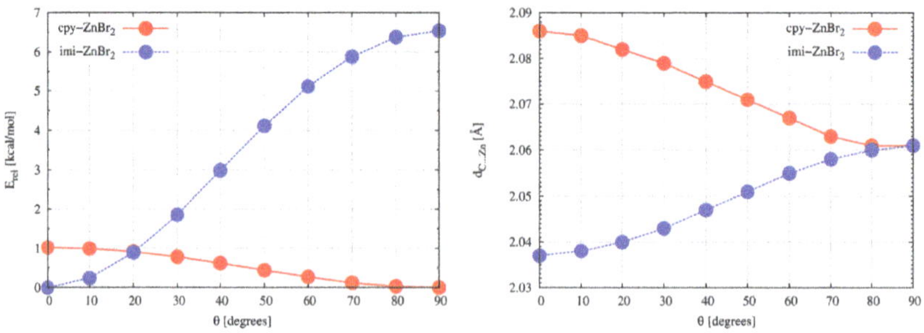

Figure 3. Relation between the relative energy (i.e., determined in relation to the minimum for a given dimer) (left) or the distance $d_{C\cdots Zn}$ (right) and the twist angle of the $ZnBr_2$ unit plane with respect to the carbene plane for dimers cyclopropenylidene···$ZnBr_2$ (cpy-$ZnBr_2$) and imidazol-2-ylidene···$ZnBr_2$ (imi-$ZnBr_2$).

It is clearly visible that although the equilibrium geometry of the cyclopropenylidene $\cdots ZnBr_2$ dimer corresponds to the perpendicular structure (i.e., with $\theta = 90°$), the planar structure ($\theta = 0°$) is energetically less stable by only 1 kcal/mol so that the rotation of the $ZnBr_2$ unit around the C$\cdots$Zn bond axis is practically free, Figure 3. Dimer imidazol-2-ylidene$\cdots ZnBr_2$ presents a completely different situation, namely the flat structure of this dimer is more stable than the perpendicular structure by about 6.5 kcal/mol which proves hindered rotation of $ZnBr_2$ due to the presence of adjacent N–H bonds, Figure 1. Planarization clearly favors the formation of residual hydrogen bonds N–H$\cdots$Br giving rise to Zn$\cdots$C bond contraction by ca. 0.023 Å; Figure 3, right.

According to the Walsh rule [131], a molecule adopts the conformation that most ensures stabilization of HOMO. A Walsh diagram showing the dependence of the orbital energies of the three highest occupied orbitals, i.e., HOMO, HOMO-1 and HOMO-2, on the twist angle θ is shown in Figure 4.

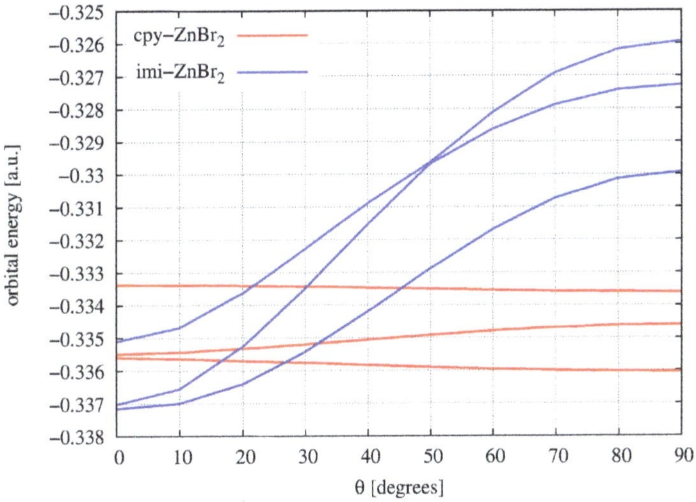

Figure 4. Dependence of the orbital energy of the three highest occupied orbitals (i.e., HOMO, HOMO-1 and HOMO-2) on the twist angle of the $ZnBr_2$ unit plane with respect to the carbene plane for dimers cyclopropenylidene$\cdots ZnBr_2$ (cpy-$ZnBr_2$) and imidazol-2-ylidene$\cdots ZnBr_2$ (imi-$ZnBr_2$).

It is clearly visible that in the case of imi-$ZnBr_2$, i.e., the imidazol-2-yliden$\cdots ZnBr_2$ complex, the $90° \rightarrow 0°$ rotation leads to a greater stabilization of all the orbitals considered, and thus also of HOMO. Thus, based on Walsh's rule, the planar structure should be more stable for this complex than the perpendicular structure, which is indeed true (see Figures 1 and 3). By the way, it is worth noting that both structures are characterized by reversed positions of HOMO and HOMO-1. In the case of the cyclopropenylidene$\cdots ZnBr_2$ dimer, the $0° \rightarrow 90°$ rotation leads to the stabilization of HOMO, but this stabilization is so minimal that it is doubtful to see this stabilization as the cause of the perpendicular structure of this dimer (Figure 1). Nevertheless, small changes in orbital energies may perhaps explain a small change in total energy (Figure 3) and almost unhindered rotation around the C$\cdots$Zn axis.

However, in order to make a more reliable assessment of structural changes, ETS-NOCV contributions along rotation angle have been determined and depicted in Figure 5.

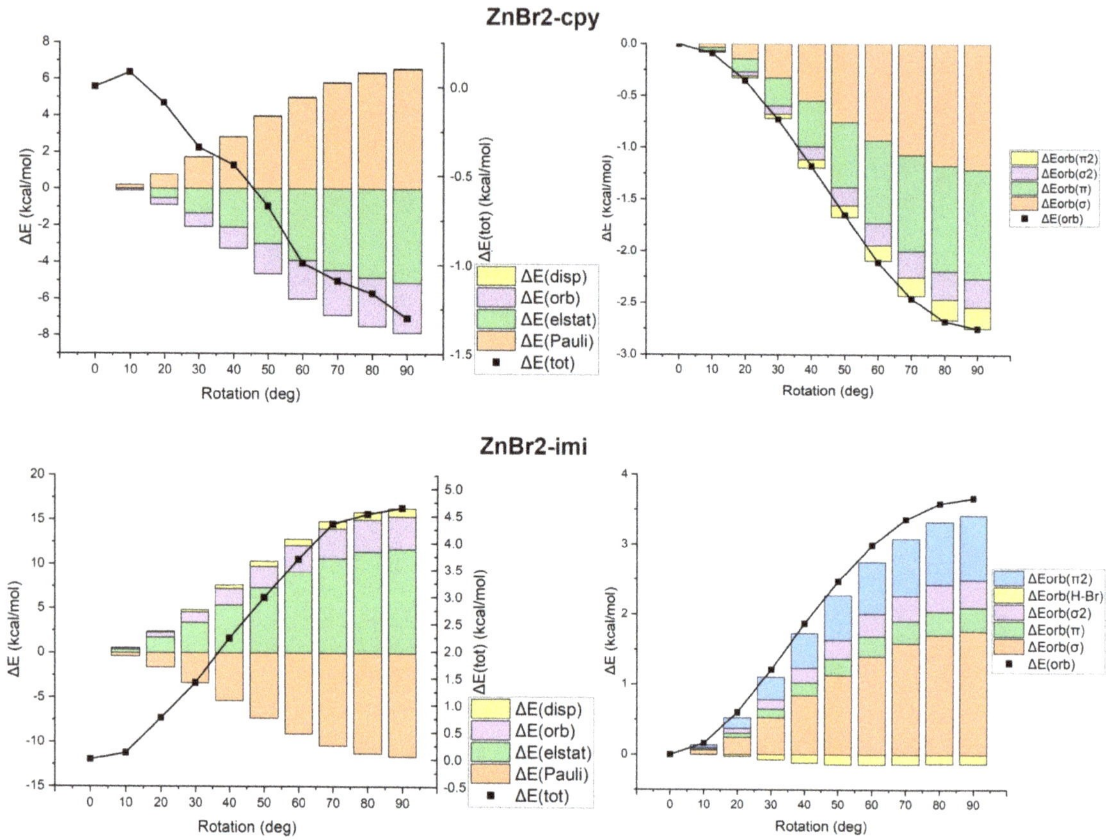

Figure 5. Changes in the values of the ETS-NOCV-based interaction energy components (left) or the orbital energy components (right) with the change of the rotation angle of the $ZnBr_2$ unit plane with respect to the carbene plane for dimers cyclopropenylidene$\cdots$$ZnBr_2$ (changes relative to θ = 0°; $ZnBr_2$-cpy) and imidazol-2-ylidene$\cdots$$ZnBr_2$ (changes relative to θ = 90°; $ZnBr_2$-imi).

It can be seen that the significant shortening (0.024 Å) of the C$\cdots$Zn distance upon planarization of the imidazol-2-ylidene$\cdots$$ZnBr_2$ dimer (see Figure 3, right) is due to an increase in the orbital interaction term and, to a larger extent, the electrostatic contributions, which causes a drop of ΔE_{int} by ca. 4.7 kcal/mol. The former, i.e., the orbital interactions, results mainly from the amplification of the σ(Zn–C) component and, to a lesser extent, the π(Zn–C) component, Figure 5. Interestingly, the effect of twisting (0° $\rightarrow$ 90°) the cyclopropenylidene$\cdots$$ZnBr_2$ dimer also causes a similar shortening (0.025 Å) of the distance C$\cdots$Zn (Figure 3, right). Nevertheless, the overall C$\cdots$Zn interaction energy (ΔE_{int}) changes only slightly by ca. 1.3 kcal/mol, which is consistent with a facile rotation. Such small variations in C$\cdots$Zn bond interaction energies are caused by modifications of both orbital interaction and electrostatic contributions, Figure 5. Finally, it is worth mentioning that rotation around the C$\cdots$Zn axis in the cyclopropenylidene$\cdots$$ZnBr_2$ dimer has practically no effect on the Zn–Br bond length, while in the imidazol-2-ylidene$\cdots$$ZnBr_2$ dimer the transition from perpendicular to planar structure slightly lengthens the bond by 0.006 Å. This effect can be attributed to a larger LP(C) $\rightarrow$ σ^*(Zn–Br) donation.

3. Methods and Materials

Geometries of monomers and dimers were fully optimized on the ωB97X-D/6-311++G (2df,2p) level of theory, that is utilizing the ωB97X-D range-separated hybrid GGA exchange-correlation functional [132] of Density Functional Theory (DFT) [133–136] and the 6-311++G(2df,2p) basis set [137–143], which includes both polarization and diffuse functions. Mardirossian and Head-Gordon have shown [136] that ωB97X-D is one of the best exchange-correlation functionals among 200 for general purposes including intermolecular interactions. To increase the accuracy of the optimization procedure and numerical integration, cutoffs on forces and step size that are used to determine convergence were additionally tightened (0.000015 and 0.000010 for maximum force and its root mean square, respectively, and 0.000060 and 0.000040 for maximum displacement and its root mean square, respectively) and integration grid was increased to the (99, 590) one (UltraFine) having 99 radial shells and 590 angular points per shell. Vibration analysis showed that all the considered systems correspond to real minima on potential energy surfaces. Both geometry optimization and vibration analysis were performed by means of Gaussian 09 [144].

The Interacting Quantum Atoms (IQA) approach [111,112], which is based on Bader's QTAIM (i.e., Quantum Theory of Atoms in Molecules) [101], allows the total energy of a system to be divided into mono- and polyatomic contributions. Of the many useful terms defined by the IQA method, this article will utilize the interatomic interaction energy:

$$E_{\mathrm{int}}^{\mathrm{E_1E_2}} = E_{\mathrm{nn}}^{\mathrm{E_1E_2}} + E_{\mathrm{ne}}^{\mathrm{E_1E_2}} + E_{\mathrm{en}}^{\mathrm{E_1E_2}} + E_{\mathrm{ee}}^{\mathrm{E_1E_2}} \quad (E_1 \neq E_2) \tag{1}$$

where $E_{\mathrm{nn}}^{\mathrm{E_1E_2}}$ is the repulsion energy between nuclei of atoms $\mathrm{E_1}$ and $\mathrm{E_2}$, $V_{\mathrm{ne}}^{\mathrm{E_1E_2}}$ is the attraction energy between the nucleus of the atom $\mathrm{E_1}$ and the electrons of the atom $\mathrm{E_2}$, $E_{\mathrm{en}}^{\mathrm{E_1E_2}}$ is the attraction energy between electrons of the atom $\mathrm{E_1}$ and the nucleus of the atom $\mathrm{E_2}$ and $E_{\mathrm{ee}}^{\mathrm{E_1E_2}}$ is the interatomic two-electron repulsion energy. The sum of the middle two terms gives the energy of the interatomic nucleus–electron attraction ($E_{\mathrm{neen}}^{\mathrm{E_1E_2}}$). Importantly, the interelectron repulsion energy can be further divided into a sum of the purely classical (Coulombic) contribution and the exchange-correlation (i.e., the non-classical term) energy:

$$E_{\mathrm{ee}}^{\mathrm{E_1E_2}} = E_{\mathrm{ee,C}}^{\mathrm{E_1E_2}} + E_{\mathrm{ee,xc}}^{\mathrm{E_1E_2}} \tag{2}$$

Moreover, the sum of the first three terms in Equation (1) and the classical interelectron energy contribution gives the electrostatic energy, such that:

$$E_{\mathrm{int}}^{\mathrm{E_1E_2}} = E_{\mathrm{elst}}^{\mathrm{E_1E_2}} + E_{\mathrm{ee,xc}}^{\mathrm{E_1E_2}} \tag{3}$$

As one can see, all the terms in the IQA equations have strictly defined physical meanings. It is also worth emphasizing that an extremely valuable feature of the IQA decomposition method is the fact that it does not require any reference system and that the interatomic interaction energy can be determined for any pair of $\mathrm{E_1}$ and $\mathrm{E_2}$ atoms, e.g., not necessarily connected with each other by a bond path. In this article, the interatomic interaction energy and its components will be determined for the C$\cdots$M (M = Be, Mg, Zn) interaction. Moreover, for an estimation of the total interaction energy between carbene and metal moieties, pair atomic energies between all atoms in both fragments were summed, such as for moieties $\mathcal{G}$ and $\mathcal{H}$:

$$E_{\mathrm{int}}^{\mathcal{GH}} = \sum_{\mathrm{A}\in\mathcal{G}} \sum_{\mathrm{B}\in\mathcal{H}} E_{\mathrm{int}}^{\mathrm{AB}} \tag{4}$$

The AIMAll program [145] was utilized for this purpose. MP2/6-311+G(d,p)-based [146,147] wave functions generated in Gaussian09 were used based on the aforementioned geometries.

The ETS-NOCV method [118] is a combination of the ETS Ziegler–Rauk method [113] with the Natural Orbitals for Chemical Valence (NOCV) method [114–117]. In the ETS method, the total interaction energy between fragments A and B (atoms, fragments and molecules) is decomposed according to the following equation:

$$\Delta E_{\text{total}} = \Delta E_{\text{dist}} + \Delta E_{\text{elst}} + \Delta E_{\text{Pauli}} + \Delta E_{\text{disp}} + \Delta E_{\text{orb}} = \Delta E_{\text{dist}} + \Delta E_{\text{int}} \tag{5}$$

where ΔE_{dist} is the distortion energy that is needed to transform the structures of the isolated fragments A and B into the geometries they have in the created A–B system, ΔE_{elst} is the energy of the electrostatic interaction of fragments A and B (in their final positions in the molecule, but frozen electron densities), ΔE_{Pauli} is the Pauli repulsion energy resulting from the repulsion between electrons with same spins of fragments A and B, ΔE_{disp} is the dispersion energy, and ΔE_{orb} is the orbital interaction energy associated with the formation of the A–B system (e.g., an A–B bond) and results from the interaction between the occupied orbitals of the A subsystem with the virtual orbitals of the B subsystem and vice versa. This term also contains the polarization energy related to the electron density changes in each of the A and B subsystems and the interfragment charge transfer. The last four terms of Equation (5) form the interaction energy, ΔE_{int}. On the other hand, the NOCV orbitals are eigenvectors of the differential charge and bond orders matrix, $\mathbf{\Delta P} = \mathbf{P} - \mathbf{P^0}$, where $\mathbf{P}$ is the density matrix of a molecule (e.g., A–B), and $\mathbf{P^0}$ is the sum of the density matrices of the orthogonalized fragments A and B, $\mathbf{P^0_A} + \mathbf{P^0_B}$. The overriding feature of the NOCV orbitals representation is the ability to decompose the differential electron density, $\Delta\rho_{\text{orb}} = \rho_{\text{AB}} - \rho^0_{\text{A}} - \rho^0_{\text{B}}$, into chemically meaningful $\Delta\rho^k_{\text{orb}}$ diagonal contributions:

$$\Delta\rho_{\text{orb}} = \sum_{k=1}^{M/2} \nu_k[-\psi^2_{-k} + \psi^2_k] = \sum_{k=1}^{M/2} \Delta\rho^k_{\text{orb}} \tag{6}$$

where ν_k are the eigenvalues obtained by diagonalizing the matrix $\mathbf{\Delta P}$, and M is the number of basis set functions. Visualization of contributions $\Delta\rho^k_{\text{orb}}$ enables the identification and characterization of the individual components (σ, π, δ, etc.) of a bond. Importantly, under the ETS-NOCV method, it is possible to quantify the energies (ΔE^k_{orb}) corresponding to the individual $\Delta\rho^k_{\text{orb}}$ components:

$$\Delta E_{\text{orb}} = \sum_{k=1}^{M/2} \nu_k[-F^{\text{TS}}_{-k,-k} + F^{\text{TS}}_{k,k}] = \sum_{k=1}^{M/2} \Delta E^k_{\text{orb}} \tag{7}$$

Calculations under the ETS-NOCV method were performed on the BLYP-D3(BJ) [148–150]/TZP-ZORA [151,152] level of theory using the ADF program [153]. Such a computational protocol proved to be adequate for bonding analysis both in our and other previous studies [154,155]. In this work, we will focus on the instantaneous interaction energy ΔE_{int}, since total energy has already been studied in Ref. [35].

The Domain-based Localized Pair-Natural Orbital Singles and Doubles Coupled Cluster with perturbative Triples [120–123] (DLPNO-CCSD(T)) approach is a relatively new approximation of the Coupled Clusters method, which in turn is often referred to as a 'golden standard' for quantum-chemical calculations [142]. DLPNO-CCSD(T) itself recovers >95% of triples contribution and 99.8% of correlation energy, and thanks to its scaling being close to DFT, it is the first CCSD(T) approximation capable of calculating the entire protein [122]. It is performed by localizing the orbitals and generating the Projected Atomic Orbitals from occupied localized orbitals, followed by creation of Pair Natural Orbitals from selected PAOs [119,120].

Local Energy Decomposition [124,125] was created for the DLPNO-CCSD(T) approach in order to harness its accuracy in determining interaction energies and provide an energy decomposition scheme on the CCSD(T) level of theory. It decomposes the binding energy into the following quantities:

$$\Delta E = \Delta E_{\text{geo-prep}} + \Delta E_{\text{int}} \tag{8}$$

The first component is the energy needed to distort fragments A and B from their optimal geometries to their final state in the AB adduct. On the other hand, ΔE_{int} is the interaction energy of said fragments, which can be further decomposed into reference (HF) energy, $\Delta E_{\text{int}}^{\text{HF}}$, and a correlation contribution ($\Delta E_{\text{int}}^{\text{C}}$):

$$\Delta E_{\text{int}} = \Delta E_{\text{int}}^{\text{HF}} + \Delta E_{\text{int}}^{\text{C}} \quad (9)$$

The Hartree–Fock part can in turn be expanded into:

$$\Delta E_{\text{int}}^{\text{HF}} = \Delta E_{\text{el-prep}}^{\text{HF}} + \Delta E_{\text{elstat}} + \Delta E_{\text{exch}} \quad (10)$$

where $\Delta E_{\text{el-prep}}^{\text{HF}}$ is the energy needed to distort the electronic structure of fragments into the state optimal for fragment interaction. Electrostatic interaction between monomers is denoted by ΔE_{elstat}, and the HF exchange by ΔE_{exch}. Correlation energy is decomposed into:

$$\Delta E_{\text{int}}^{\text{C}} = \Delta E_{\text{no-disp}}^{\text{C}} + \Delta E_{\text{disp}}^{\text{C}} + \Delta E_{\text{int}}^{\text{C-(T)}} \quad (11)$$

Terms $\Delta E_{\text{disp}}^{\text{C}}$ and $\Delta E_{\text{no-disp}}^{\text{C}}$ are correlation energy contributions associated with dispersion and non-dispersive corrections, respectively. The term denoted by $\Delta E_{\text{int}}^{\text{C-(T)}}$ is the triples correction arising from CCSD(T).

4. Conclusions

This article presents a comparative study of a series of metal–carbene bonds between main- and transition group metals: beryllium, magnesium and zinc, and model carbenes: cyclopropenylidene (a regular carbene), and imidazol-2-ylidene (N-heterocyclic carbene, NHC). The physical nature of these carbene$\cdots$$MX_2$ (X = H, Br) interactions has been described for the first time by the joint use of topological QTAIM-based IQA decomposition scheme, molecular orbital-based ETS-NOCV charge and energy decomposition method, as well as LED energy decomposition basing on the state-of-the-art DLPNO-CCSD(T) method.

All methods agree on the increasing bond strength in a series: Zn < Mg << Be. For beryllium and magnesium bonds, electrostatics proved to be the most important bond component (55–90% of total stabilization), followed by stabilization due to electron sharing (10–45% of total stabilization), while dispersion interactions proved to be marginal (2–5%). QTAIM/IQA underestimates electrostatic contribution for zinc bonds with respect to both ETS-NOCV and LED schemes which, similarly to Be and Mg species, identify the following importance of Zn–carbene bond constituents: $E_{\text{elstat}} > E_{\text{orb}} > E_{\text{dispersion}}$. Both ETS-NOCV and LED also provide a consistent picture of these carbene bonds on an orbital level, with the σ carbene→MX_2 donation strongly dominating over the MX_2 → carbene back-donation of π symmetry. Substitution of hydrogen atoms by bromines (X in MX_2) strengthens the metal–carbene bond in all cases due to the amplification of electrostatic forces as well as donation and back-donation effects. The former is likely due to making metal atoms more electrophilic. Finally, the conformational preferences (planar vs twisted) are controlled by the nature of C–Zn bonding as unveiled by ETS-NOCV analyses performed along rotation coordinate pathways.

Supplementary Materials: The following are available at https://www.mdpi.com/article/10.3390/ijms232314668/s1.

Author Contributions: Conceptualization, M.M. and M.J.; methodology, F.S., M.M. and M.J.; software, F.S., M.M. and M.J.; validation, F.S., M.M. and M.J.; formal analysis, F.S., M.M. and M.J.; investigation, F.S., M.M. and M.J.; data curation, F.S., M.M. and M.J.; writing–original draft preparation, F.S., M.M. and M.J.; writing–review and editing, F.S., M.M. and M.J.; visualization, F.S. and M.J.; project administration, M.J. All authors have read and agreed to the published version of the manuscript.

Funding: This research received no external funding.

Data Availability Statement: Data available from the authors on reasonable request.

Acknowledgments: This research was supported in part by PLGrid Infrastructure.

Conflicts of Interest: The authors declare no conflict of interest.

Abbreviations

The following abbreviations are used in this manuscript:

QTAIM	Quantum Theory of Atoms in Molecules
IQA	the Interacting Quantum Atoms approach
ETS	the Extended Transition State method
NOCV	the Natural Orbitals for Chemical Valence method
DLPNO	Domain-based Localized Pair-Natural Orbital
CCSD(T)	the coupled cluster (CC) single-double-triple method
LED	the Local Energy Decomposition scheme

References

1. Kirmse, W. *Carbene Chemistry*; Academic Press: Cambridge, MA, USA, 1964.
2. Hubert, A.J. *Catalysis in C_1 Chemistry*; Springer: Dordrecht, The Netherlands, 1983.
3. Schubert, U. *Advances in Metal Carbene Chemistry*; Springer: Dordrecht, The Netherlands, 1989.
4. Herrmann, W.A.; Köcher, C. N-Heterocyclic Carbenes. *Angew. Chem. Int. Ed. Engl.* **1997**, *36*, 2162–2187. [CrossRef]
5. Bourissou, D.; Guerret, O.; Gabbaï, F.P.; Bertrand, G. Stable Carbenes. *Chem. Rev.* **2000**, *100*, 39–91. [CrossRef] [PubMed]
6. Bertrande, G. *Carbene Chemistry: From Fleeting Intermediates to Powerful Reagents*; FontisMedia S.A.: Lausanne, Switzerland; Marcel Dekker, Inc.: New York, NY, USA, 2002.
7. Moss, R.A.; Platz, M.S.; Jones, M., Jr. (Eds.) *Reactive Intermediate Chemistry*; John Wiley & Sons, Inc.: Hoboken, NJ, USA, 2005.
8. Scott, N.M.; Nolan, S.P. Stabilization of Organometallic Species Achieved by the Use of N-Heterocyclic Carbene (NHC) Ligands. *Eur. J. Inorg. Chem.* **2005**, *2005*, 1815–1828. [CrossRef]
9. Nolan, S.P. *N-Heterocyclic Carbenes in Synthesis*; Wiley-VCH: Weinheim, Germany, 2006.
10. Carey, F.A.; Sundberg, R.J. *Carbenes, Part B: Reactions and Synthesis. Advanced Organic Chemistry*; Springer: New York, NY, USA, 2007.
11. Díez-González, S.; Nolan, S.P. Stereoelectronic parameters associated with N-heterocyclic carbene (NHC) ligands: A quest for understanding. *Coord. Chem. Rev.* **2007**, *251*, 874–883. [CrossRef]
12. Jacobsen, H.; Correa, A.; Poater, A.; Costabile, C.; Cavallo, L. Understanding the M–(NHC) (NHC = N-heterocyclic carbene) bond. *Coord. Chem. Rev.* **2009**, *253*, 687–703. [CrossRef]
13. de Frémont, P.; Marion, N.; Nolan, S.P. Carbenes: Synthesis, properties, and organometallic chemistry. *Coord. Chem. Rev.* **2009**, *253*, 862–892. [CrossRef]
14. Moss, R.A.; Doyle, M.P. *Contemporary Carbene Chemistry*; John Wiley & Sons, Inc.: Hoboken, NJ, USA, 2013.
15. Nelson, D.J.; Nolan, S.P. Quantifying and understanding the electronic properties of N-heterocyclic carbenes. *Chem. Soc. Rev.* **2013**, *42*, 6723–6753. [CrossRef]
16. Bellemin-Laponnaz, S.; Dagorne, S. Group 1 and 2 and Early Transition Metal Complexes Bearing N-Heterocyclic Carbene Ligands: Coordination Chemistry, Reactivity, and Applications. *Chem. Rev.* **2014**, *114*, 8747–8774. [CrossRef]
17. Hopkinson, M.N.; Richter, C.; Schedler, M.; Glorius, F. An overview of N-heterocyclic carbenes. *Nature* **2014**, *510*, 485–496. [CrossRef]
18. Nesterov, V.; Reiter, D.; Bag, P.; Frisch, P.; Holzner, R.; Porzelt, A.; Inoue, S. NHCs in Main Group Chemistry. *Chem. Rev.* **2018**, *118*, 9678–9842. [CrossRef]
19. Pople, J.A.; Raghavachari, K.; Frisch, M.J.; Binkley, J.S.; Schleyer, P.v.R. Comprehensive Theoretical Study of Isomers and Rearrangement Barriers of Even Electron Polyatomic Molecules H_mABH_n (A, B = C, N, O, and F). *J. Am. Chem. Soc.* **1983**, *105*, 6389–6398. [CrossRef]
20. Pople, J.A. A theoretical search for the methylenefluoronium ylide. *Chem. Phys. Lett.* **1986**, *132*, 144–146. [CrossRef]
21. Arduengo, A.J., III; Gamper, S.F.; Tamm, M.; Calabrese, J.C.; Davidson, F.; Craig, H.A. A Bis(carbene)–Proton Complex: Structure of a C–H–C Hydrogen Bond. *J. Am. Chem. Soc.* **1995**, *117*, 572–573. [CrossRef]
22. Alkorta, I.; Elguero, J. Carbenes and Silylenes as Hydrogen Bond Acceptors. *J. Phys. Chem.* **1996**, *100*, 19367–19370. [CrossRef]
23. Jabłoński, M.; Palusiak, M. Divalent carbon atom as the proton acceptor in hydrogen bonding. *Phys. Chem. Chem. Phys.* **2009**, *11*, 5711–5719. [CrossRef]
24. Giffin, N.A.; Makramalla, M.; Hendsbee, A.D.; Robertson, K.N.; Sherren, C.; Pye, C.C.; Masuda, J.D.; Clyburne, J.A.C. Anhydrous TEMPO-H: Reactions of a good hydrogen atom donor with low-valent carbon centres. *Org. Biomol. Chem.* **2011**, *9*, 3672–3680. [CrossRef]
25. Gerbig, D.; Ley, D. Computational methods for contemporary carbene chemistry. *WIREs Comput. Mol. Sci.* **2013**, *3*, 242–272. [CrossRef]

26. Samanta, R.C.; De Sarkar, S.; Fröhlich, R.; Grimme, S.; Studer, A. N-Heterocyclic carbene (NHC) catalyzed chemoselective acylation of alcohols in the presence of amines with various acylating reagents. *Chem. Sci.* **2013**, *4*, 2177–2184. [CrossRef]
27. Jabłoński, M. Coexistence of the Carbene$\cdots$H-D Hydrogen Bond and Other Accompanying Interactions in Forty Dimers of N-Heterocyclic-Carbenes (I, IMe_2 , I^iPr_2, I^tBu_2, $IMes_2$, $IDipp_2$, IAd_2; I = imidazol-2-ylidene) and Some Fundamental Proton Donors (HF, HCN, H_2O, MeOH, NH_3). *Molecules* **2022**, *27*, 5712. [CrossRef]
28. Li, Q.; Wang, H.; Liu, Z.; Li, W.; Cheng, J.; Gong, B.; Sun, J. Ab Initio Study of Lithium-Bonded Complexes with Carbene as an Electron Donor. *J. Phys. Chem. A* **2009**, *113*, 14156–14160. [CrossRef]
29. Wang, Y.; Xie, Y.; Abraham, M.Y.; Wei, P.; Schaefer, H.F., III; Schleyer, P.v.R.; Robinson, G.H. A Viable Anionic N-Heterocyclic Dicarbene. *J. Am. Chem. Soc.* **2010**, *132*, 14370–14372. [CrossRef] [PubMed]
30. Zhi-Feng, L.; Sheng, Y.; Hui-Xue, L. Theoretical prediction characters of unconventional weak bond with carbene as electron donors and Li–Y (Y = OH, H, F, NC and CN) as electron acceptors. *J. Mol. Struct. THEOCHEM* **2010**, *952*, 56–60. [CrossRef]
31. Herrmann, W.A.; Runte, O.; Artus, G. Synthesis and structure of an ionic beryllium—"Carbene" complex. *J. Organomet. Chem.* **1995**, *501*, C1–C4. [CrossRef]
32. Gilliard, R.J., Jr.; Abraham, M.Y.; Wang, Y.; Wei, P.; Xie, Y.; Quillian, B.; Schaefer, H.F., III; Schleyer, P.v.R.; Robinson, G.H. Carbene-Stabilized Beryllium Borohydride. *J. Am. Chem. Soc.* **2012**, *134*, 9953–9955. [CrossRef] [PubMed]
33. Arrowsmith, M.; Hill, M.S.; Kociok-Köhn, G.; MacDougall, D.J.; Mahon, M.F. Beryllium-Induced C–N Bond Activation and Ring Opening of an N-Heterocyclic Carbene. *Angew. Chem. Int. Ed.* **2012**, *51*, 2098–2100. [CrossRef]
34. Walley, J.E.; Wong, Y.-O.; Freeman, L.A.; Dickie, D.A.; Gilliard, R.J., Jr. N-Heterocyclic Carbene-Supported Aryl- and Alk- oxides of Beryllium and Magnesium. *Catalysts* **2019**, *9*, 934. [CrossRef]
35. Jabłoński, M. Study of Beryllium, Magnesium, and Spodium Bonds to Carbenes and Carbodiphosphoranes. *Molecules* **2021**, *26*, 2275. [CrossRef]
36. Arduengo, A.J., III; Dias, H.V.R.; Davidson, F.; Harlow, R.L. Carbene adducts of magnesium and zinc. *J. Organomet. Chem.* **1993**, *462*, 13–18. [CrossRef]
37. Arrowsmith, M.; Hill, M.S.; MacDougall, D.J.; Mahon, M.F. A Hydride-Rich Magnesium Cluster. *Angew. Chem. Int. Ed.* **2009**, *48*, 4013–4016. [CrossRef]
38. Arnold, P.L.; Casely, I.J.; Turner, Z.R.; Bellabarba, R.; Tooze, R.B. Magnesium and zinc complexes of functionalised, saturated N-heterocyclic carbene ligands: Carbene lability and functionalisation, and lactide polymerisation catalysis. *Dalton Trans.* **2009**, 7236–7247. [CrossRef]
39. Arduengo, A.J., III; Dias, H.V.R.; Calabrese, J.C.; Davidson, F. A Stable Carbene-Alane Adduct. *J. Am. Chem. Soc.* **1992**, *114*, 9724–9725. [CrossRef]
40. Li, X.-W.; Su, J.; Robinson, G.H. Syntheses and molecular structure of organo-group 13 metal carbene complexes. *Chem. Commun.* **1996**, *23*, 2683–2684. [CrossRef]
41. Hibbs, D.E.; Hursthouse, M.B.; Jones, C.; Smithies, N.A. Synthesis, crystal and molecular structure of the first indium trihydride complex, $[InH_3\{CN(Pr^i)C_2Me_2N(Pr^i)\}]$. *Chem. Commun.* **1998**, *8*, 869–870. [CrossRef]
42. Merceron, N.; Miqueu, K.; Baceiredo, A.; Bertrand, G. Stable (Amino)(phosphino)carbenes: Difunctional Molecules. *J. Am. Chem. Soc.* **2002**, *124*, 6806–6807. [CrossRef]
43. Wang, Y.; Robinson, G.H. Unique homonuclear multiple bonding in main group compounds. *Chem. Commun.* **2009**, 5201–5213. [CrossRef]
44. Del Bene, J.E.; Alkorta, I.; Elguero, J. Carbon–Carbon Bonding between Nitrogen Heterocyclic Carbenes and CO_2. *J. Phys. Chem. A* **2017**, *121*, 8136–8146. [CrossRef]
45. Liu, M.; Li, Q.; Li, W.; Cheng, J. Carbene tetrel-bonded complexes. *Struct. Chem.* **2017**, *28*, 823–831. [CrossRef]
46. Wang, Y.; Xie, Y.; Abraham, M.Y.; Gilliard, R.J., Jr.; Wei, P.; Schaefer, H.F., III; Schleyer, P.v.R.; Robinson, G.H. Carbene-Stabilized Parent Phosphinidene. *Organometallics* **2010**, *29*, 4778–4780. [CrossRef]
47. Abraham, M.Y.; Wang, Y.; Xie, Y.; Wei, P.; Schaefer, H.F., III; Schleyer, P.v.R.; Robinson, G.H. Carbene Stabilization of Diarsenic: From Hypervalency to Allotropy. *Chem. Eur. J.* **2010**, *16*, 432–435. [CrossRef]
48. Patel, D.S.; Bharatam, P.V. Divalent N(I) Compounds with Two Lone Pairs on Nitrogen. *J. Phys. Chem. A* **2011**, *115*, 7645–7655. [CrossRef]
49. Zhao, Q.; Feng, D.; Sun, Y.; Hao, J.; Cai, Z. Theoretical Investigations on the Weak Nonbonded C=S$\cdots CH_2$ Interactions: Chalcogen-Bonded Complexes With Singlet Carbene as an Electron Donor. *Int. J. Quant. Chem.* **2011**, *111*, 3881–3887. [CrossRef]
50. Arduengo, A.J., III; Kline, M.; Calabrese, J.C.; Davidson, F. Synthesis of a Reverse Ylide from a Nucleophilic Carbene. *J. Am. Chem. Soc.* **1991**, *113*, 9704–9705. [CrossRef]
51. Kuhn, N.; Kratz, T.; Henkel, G. A Stable Carbene Iodine Adduct: Secondary Bonding in 1,3-Diethyl-2-iodo-4,5-dimethylimidazolium Iodide. *J. Chem. Soc. Chem. Commun.* **1993**, 1778–1779. [CrossRef]
52. Li, Q.; Wang, Y.; Liu, Z.; Li, W.; Cheng, J.; Gong, B.; Sun, J. An unconventional halogen bond with carbene as an electron donor: An ab initio study. *Chem. Phys. Lett.* **2009**, *469*, 48–51. [CrossRef]
53. Esrafili, M.D.; Mohammadirad, N. Insights into the strength and nature of carbene$\cdots$halogen bond interactions: A theoretical perspective. *J. Mol. Model.* **2013**, *19*, 2559–2566. [CrossRef] [PubMed]
54. Esrafili, M.D.; Sabouri, A. Carbene–aerogen bonds: An Ab Initio Study. *Mol. Phys.* **2017**, *115*, 971–980. [CrossRef]

55. Cases, M.; Frenking, G.; Duran, M.; Solà, M. Molecular Structure and Bond Characterization of the Fischer-Type Chromium–Carbene Complexes $(CO)_5Cr{=}C(X)R$ (X = H, OH, OCH_3, NH_2, $NHCH_3$ and R = H, CH_3, $CH{=}CH_2$, Ph, $C{\equiv}CH$). *Organometallics* **2002**, *21*, 4182–4191. [CrossRef]
56. Nemcsok, D.; Wichmann, K.; Frenking, G. The Significance of π Interactions in Group 11 Complexes with N-Heterocyclic Carbenes. *Organometallics* **2004**, *23*, 3640–3646. [CrossRef]
57. Frenking, G.; Solà, M.; Vyboishchikov, S.F. Chemical bonding in transition metal carbene complexes. *J. Organomet. Chem.* **2005**, *690*, 6178–6204. [CrossRef]
58. Tonner, R.; Heydenrych, G.; Frenking, G. Bonding Analysis of N-Heterocyclic Carbene Tautomers and Phosphine Ligands in Transition-Metal Complexes: A Theoretical Study. *Chem. Asian J.* **2007**, *2*, 1555–1567. [CrossRef]
59. Radius, U.; Bickelhaupt, F.M. Bonding of Imidazol-2-ylidene Ligands in Nickel Complexes. *Organometallics* **2008**, *27*, 3410–3414. [CrossRef]
60. Srebro, M.; Michalak, A. Theoretical Analysis of Bonding in N-Heterocyclic Carbene–Rhodium Complexes. *Inorg. Chem.* **2009**, *48*, 5361–5369. [CrossRef]
61. Andrada, D.M.; Holzmann, N.; Hamadi, T.; Frenking, G.; Beilstein, J. Direct estimate of the internal π-donation to the carbene centre within N-heterocyclic carbenes and related molecules. *Org. Chem.* **2015**, *11*, 2727–2736. [CrossRef]
62. Santoro, O.; Nahra, F.; Cordes, D.B.; Slawin, A.M.Z.; Nolan, S.P.; Cazin, C.S.J. Synthesis, characterization and catalytic activity of stable $[(NHC)H][ZnXY_2]$ (NHC = *N*-Heterocyclic carbene, X, Y = Cl, Br) species. *J. Mol. Catal.* **2016**, *423*, 85–91. [CrossRef]
63. Dagorne, S. Recent Developments on N-Heterocyclic Carbene Supported Zinc Complexes: Synthesis and Use in Catalysis. *Synthesis* **2018**, *50*, 3662–3670. [CrossRef]
64. Procter, R.J.; Uzelac, M.; Cid, J.; Rushworth, P.J.; Ingleson, M.J. Low-Coordinate NHC–Zinc Hydride Complexes Catalyze Alkyne C–H Borylation and Hydroboration Using Pinacolborane. *ACS Catal.* **2019**, *9*, 5760–5771. [CrossRef]
65. Specklin, D.; Fliedel, C.; Dagorne, S. Recent Representative Advances on the Synthesis and Reactivity of N-Heterocyclic-Carbene-Supported Zinc Complexes. *Chem. Rec.* **2021**, *21*, 1130–1143. [CrossRef]
66. Arduengo, A.J., III; Davidson, F.; Krafczyk, R.; Marshall, W.J.; Tamm, M. Adducts of Carbenes with Group II and XII Metallocenes. *Organometallics* **1998**, *17*, 3375–3382. [CrossRef]
67. Abernethy, C.D.; Baker, R.J.; Cole, M.L.; Davies, A.J.; Jones, C. Reactions of a carbene stabilised indium trihydride complex, $[InH_3\{CN(Mes)\text{-}C_2H_2N(Mes)\}]$ Mes = mesityl, with transition metal complexes. *Trans. Met. Chem.* **2003**, *28*, 296–299. [CrossRef]
68. Wang, D.; Wurst, K.; Buchmeiser, M.R. N-heterocyclic carbene complexes of Zn(II): Synthesis, X-ray structures and reactivity. *J. Organomet. Chem.* **2004**, *689*, 2123–2130. [CrossRef]
69. Jensen, T.R.; Breyfogle, L.E.; Hillmyer, M.A.; Tolman, W.B. Stereoelective polymerization of D,L-lactide using N-heterocyclic carbene based compounds. *Chem. Commun.* **2004**, 2504–2505. [CrossRef] [PubMed]
70. Jensen, T.R.; Schaller, C.P.; Hillmyer, M.A.; Tolman, W.B. Zinc N-heterocyclic carbene complexes and their polymerization of D,L-lactide. *J. Organomet. Chem.* **2005**, *690*, 5881–5891. [CrossRef]
71. Anantharaman, G.; Elango, K. Synthesis and Characterization of NHC-Stabilized Zinc Aryloxide and Zinc Hydroxyaryloxide. *Organometallics* **2007**, *26*, 1089–1092. [CrossRef]
72. Budagumpi, S.; Endud, S. Group XII Metal–N-Heterocyclic Carbene Complexes: Synthesis, Structural Diversity, Intramolecular Interactions, and Applications. *Organometallics* **2013**, *32*, 1537–1562. [CrossRef]
73. Schnee, G.; Fliedel, C.; Avilés, T.; Dagorne, S. Neutral and Cationic N-Heterocyclic Carbene Zinc Adducts and the $BnOH/Zn(C_6F_5)_2$ Binary Mixture–Characterization and Use in the Ring-Opening Polymerization of β-Butyrolactone, Lactide, and Trimethylene Carbonate. *Eur. J. Inorg. Chem.* **2013**, *2013*, 3699–3709. [CrossRef]
74. Fliedel, C.; Vila-Viçosa, D.; Calhorda, M.J.; Dagorne, S.; Avilés, T. Dinuclear Zinc–N-Heterocyclic Carbene Complexes for Either the Controlled Ring-Opening Polymerization of Lactide or the Controlled Degradation of Polylactide Under Mild Conditions. *ChemCatChem* **2014**, *6*, 1357–1367. [CrossRef]
75. Collins, L.R.; Moffat, L.A.; Mahon, M.F.; Jones, M.D.; Whittlesey, M.K. Lactide polymerisation by ring-expanded NHC complexes of zinc. *Polyhedron* **2016**, *103*, 121–125. [CrossRef]
76. Tian, J.; Chen, Y.; Vayer, M.; Djurovic, A.; Guillot, R.; Guermazi, R.; Dagorne, S.; Bour, C.; Gandon, V. Exploring the Limits of π-Acid Catalysis Using Strongly Electrophilic Main Group Metal Complexes: The Case of Zinc and Aluminium. *Chem. Eur. J.* **2020**, *26*, 12831–12838. [CrossRef]
77. Roy, M.M.D.; Baird, S.R.; Ferguson, M.J.; Rivard, E. Toward N-heterocyclic carbene stabilized zinc sulfides. *Mendeleev Commun.* **2021**, *31*, 173–175. [CrossRef]
78. Jabłoński, M. Theoretical Study of N-Heterocyclic-Carbene–ZnX_2 (X = H, Me, Et) Complexes. *Materials* **2021**, *14*, 6147. [CrossRef]
79. Schoeller, W.W. Electrophilicity and nucleophilicity in singlet carbenes. II. Electrophilic selectivity. *Tetrahedron Lett.* **1980**, *21*, 1509–1510. [CrossRef]
80. Goumri-Magnet, S.; Polishchuk, O.; Gornitzka, H.; Marsden, C.J.; Baceiredo, A.; Bertrand, G. The Electrophilic Behavior of Stable Phosphanylcarbenes Towards Phosphorus Lone Pairs. *Angew. Chem. Int. Ed.* **1999**, *38*, 3727–3729. [CrossRef]
81. Moss, R.A.; Wang, L.; Cang, H.; Krogh-Jespersen, K. Extremely reactive carbenes: Electrophiles and nucleophiles. *J. Phys. Org. Chem.* **2017**, *30*, e3555. [CrossRef]
82. Jabłoński, M. The first theoretical proof of the existence of a hydride-carbene bond. *Chem. Phys. Lett.* **2018**, *710*, 78–83. [CrossRef]
83. Jabłoński, M. In search for a hydride-carbene bond. *J. Phys. Org. Chem.* **2019**, *32*, e3949. [CrossRef]

84. Yourdkhani, S.; Jabłoński, M. Physical nature of silane··· carbene dimers revealed by state-of-the-art ab initio calculations. *J. Comput. Chem.* **2019**, *40*, 2643–2652. [CrossRef]
85. Yáñez, M.; Sanz, P.; Mó, O.; Alkorta, I.; Elguero, J. Beryllium Bonds, Do They Exists? *J. Chem. Theory Comput.* **2009**, *5*, 2763–2771. [CrossRef]
86. Martín-Sómer, A.; Lamsabhi, A.M.; Mó, O.; Yáñez, M. The importance of deformation on the strength of beryllium bonds. *Comput. Theor. Chem.* **2012**, *998*, 74–79. [CrossRef]
87. Eskandari, K. Characteristics of beryllium bonds: A QTAIM study. *J. Mol. Model.* **2012**, *18*, 3481–3487. [CrossRef]
88. Villanueva, E.F.; Mó, O.; Yáñez, M. On the existence and characteristics of π-beryllium bonds. *Phys. Chem. Chem. Phys.* **2014**, *16*, 17531–17536. [CrossRef]
89. Zhong, A.; Chen, D.; Li, R. Revisiting the beryllium bonding interactions from energetic and wavefunction perspectives. *Chem. Phys. Lett.* **2015**, *633*, 265–272. [CrossRef]
90. Eskandari, K. Nature of beryllium bonds in view of interacting quantum atoms and natural energy decomposition analysis. *Comput. Theor. Chem.* **2016**, *1090*, 74–79. [CrossRef]
91. Montero-Campillo, M.M.; Mó, O.; Yáñez, M.; Alkorta, I.; Elguero, J. The beryllium bond. *Adv. Inorg. Chem.* **2019**, *73*, 73–121.
92. Alkorta, I.; Legon, A.C. Non-Covalent Interactions Involving Alkaline-Earth Atoms and Lewis Bases B: An ab Initio Investigation of Beryllium and Magnesium Bonds, B··· MR_2 (M = Be or Mg, and R = H, F or CH_3). *Inorganics* **2019**, *7*, 35. [CrossRef]
93. Jabłoński, M. On the Uselessness of Bond Paths Linking Distant Atoms and on the Violation of the Concept of Privileged Exchange Channels. *Chem. Open* **2019**, *8*, 497–507. [CrossRef]
94. Yang, X.; Li, Q.; Cheng, J.; Li, W. A new interaction mechanism of $LiNH_2$ with MgH_2: Magnesium bond. *J. Mol. Model.* **2013**, *19*, 247–253. [CrossRef]
95. Xu, H.-L.; Li, Q.-Z.; Scheiner, S. Effect of magnesium bond on the competition between hydrogen and halogen bonds and the induction of proton and halogen transfer. *ChemPhysChem* **2018**, *19*, 1456–1464. [CrossRef]
96. Sanz, P.; Montero-Campillo, M.M.; Mó, O.; Yáñez, M.; Alkorta, I.; Elguero, J. Intramolecular magnesium bonds in malonaldehyde-like systems: A critical view of the resonance-assisted phenomena. *Theor. Chem. Acc.* **2018**, *137*, 97. [CrossRef]
97. Lupinetti, A.J.; Jonas, V.; Thiel, W.; Strauss, S.H.; Frenking, G. Trends in Molecular Geometries and Bond Strengths of the Homoleptic d^{10} Metal Carbonyl Cations $[M(CO)_n]^{x+}$ (M^{x+} = Cu^+, Ag^+, Au^+, Zn^{2+}, Cd^{2+}, Hg^{2+}; n = 1–6): A Theoretical Study. *Chem. Eur. J.* **1999**, *5*, 2573–2583. [CrossRef]
98. Joy, J.; Jemmis, E.D. Contrasting Behavior of the Z Bonds in X–Z··· Y Weak Interactions: Z = Main Group Elements Versus the Transition Metals. *Inorg. Chem.* **2017**, *56*, 1132–1143. [CrossRef]
99. Wang, S.R.; Arrowsmith, M.; Braunschweig, H.; Dewhurst, R.D.; Dömling, M.; Mattock, J.D.; Pranckevicius, C.; Vargas, A. Monomeric 16-Electron π-Diborene Complexes of Zn(II) and Cd(II). *J. Am. Chem. Soc.* **2017**, *139*, 10661–10664. [CrossRef]
100. Kalhor, P.; Wang, Y.; Yu, Z. The Structures of $ZnCl_2$-Ethanol Mixtures, a Spectroscopic and Quantum Chemical Calculation Study. *Molecules* **2021**, *26*, 2498. [CrossRef]
101. Bader, R.F.W. *Atoms in Molecules: A Quantum Theory*; Oxford University Press: New York, NY, USA, 1990.
102. Cremer, D.; Kraka, E. Chemical Bonds without Bonding Electron Density–Does the Difference Electron-Density Analysis Suffice for a Description of the Chemical Bond? *Angew. Chem. Int. Ed. Engl.* **1984**, *23*, 627–628. [CrossRef]
103. Bader, R.F.W.; Stephens, M.E. Spatial Localization of the Electronic Pair and Number Distributions in Molecules. *J. Am. Chem. Soc.* **1975**, *97*, 7391–7399. [CrossRef]
104. Fradera, X.; Austen, M.A.; Bader, R.F.W. The Lewis Model and Beyond. *J. Phys. Chem. A* **1999**, *103*, 304–314. [CrossRef]
105. Fradera, X.; Poater, J.; Simon, S.; Duran, M.; Solà, M. Electron-pairing analysis from localization and delocalization indices in the framework of the atoms-in-molecules theory. *Theor. Chem. Acc.* **2002**, *108*, 214–224. [CrossRef]
106. Firme, C.L.; Antunes, O.A.C.; Esteves, P.M. Relation between bond order and delocalization index of QTAIM. *Chem. Phys. Lett.* **2009**, *468*, 129–133. [CrossRef]
107. Rafat, M.; Popelier, P.L.A. *The Quantum Theory of Atoms in Molecules: From Solid State to DNA and Drug Design*; Matta, C.F., Boyed, R.J., Eds.; Wiley-VCH: Weinheim, Germany, 2007; pp. 121–140.
108. García-Revilla, M.; Francisco, E.; Popelier, P.L.A.; Pendás, A.M. Domain-Averaged Exchange-Correlation Energies as a Physical Underpinning for Chemical Graphs. *ChemPhysChem* **2013**, *14*, 1211–1218. [CrossRef]
109. Johnson, E.R.; Keinan, S.; Mori-Sánchez, P.; Contreras-García, J.; Cohen, A.J.; Yang, W. Revealing Noncovalent Interactions. *J. Am. Chem. Soc.* **2010**, *132*, 6498–6506. [CrossRef]
110. Contreras-García, J.; Johnson, E.R.; Keinan, S.; Chaudret, R.; Piquemal, J.-P.; Beratan, D.N.; Yang, W. NCIPLOT: A Program for Plotting Noncovalent Interaction Regions. *J. Chem. Theory Comput.* **2011**, *7*, 625–632. [CrossRef]
111. Blanco, M.A.; Pendás, A.M.; Francisco, E. Interacting Quantum Atoms: A Correlated Energy Decomposition Scheme Based on the Quantum Theory of Atoms in Molecules. *J. Chem. Theory Comput.* **2005**, *1*, 1096–1109. [CrossRef] [PubMed]
112. Guevara-Vela, J.M.; Francisco, E.; Rocha-Rinza, T.; Pendás, A.M. Interacting Quantum Atoms–A Review. *Molecules* **2020**, *25*, 4028. [CrossRef]
113. Ziegler, T. ; Rauk, A. On the calculation of bonding energies by the Hartree Fock Slater method. *Theoret. Chim. Acta (Berl.)* **1977**, *46*, 1–10.
114. Mitoraj, M.; Michalak, A. Donor–Acceptor Properties of Ligands from the Natural Orbitals for Chemical Valence. *Organometallics* **2007**, *26*, 6576–6580. [CrossRef]

115. Mitoraj, M.; Michalak, A. Natural orbitals for chemical valence as descriptors of chemical bonding in transition metal complexes. *J. Mol. Model.* **2007**, *13*, 347–355. [CrossRef] [PubMed]
116. Michalak, A.; Mitoraj, M.; Ziegler, T. Bond Orbitals from Chemical Valence Theory. *J. Phys. Chem. A* **2008**, *112*, 1933–1939. [CrossRef]
117. Mitoraj, M.P.; Michalak, A.; Ziegler, T. A Combined Charge and Energy Decomposition Scheme for Bond Analysis. *J. Chem. Theory Comput.* **2009**, *5*, 962–975. [CrossRef]
118. Mitoraj, M.P.; Michalak, A.; Ziegler, T. On the Nature of the Agostic Bond between Metal Centers and β-Hydrogen Atoms in Alkyl Complexes. An Analysis Based on the Extended Transition State Method and the Natural Orbitals for Chemical Valence Scheme (ETS-NOCV). *Organometallics* **2009**, *28*, 3727–3733. [CrossRef]
119. Saebø, S.; Pulay, P. Local Treatment of Electron Correlation, *Annu. Rev. Phys. Chem* **1993**, *44*, 213–236. [CrossRef]
120. Neese, F.; Wennmohs, F.; Hansen, A. Efficient and accurate local approximations to coupled-electron pair approaches: An attempt to revive the pair natural orbital method. *J. Chem. Phys.* **2009**, *130*, 114108. [CrossRef]
121. Neese, F.; Hansen, A.; Liakos, D.G. Efficient and accurate approximations to the local coupled cluster singles doubles method using a truncated pair natural orbital basis. *J. Chem. Phys.* **2009**, *131*, 064103. [CrossRef]
122. Riplinger, C.; Sandhoefer, B.; Hansen, A.; Neese, F. Natural triple excitations in local coupled cluster calculations. *J. Chem. Phys.* **2013**, *139*, 134101. [CrossRef]
123. Guo, Y.; Riplinger, C.; Becker, U.; Liakos, D.G.; Minenkov, Y.; Cavallo, L.; Neese, F. Communication: An improved linear scaling perturbative triples correction for the domain based local pair-natural orbital based singles and doubles coupled cluster method [DLPNO-CCSD(T)]. *J. Chem. Phys.* **2018**, *148*, 011101. [CrossRef]
124. Altun, A.; Saitow, M.; Neese, F.; Bistoni, G. Local Energy Decomposition of Open-Shell Molecular Systems in the Domain-Based Local Pair Natural Orbital Coupled Cluster Framework. *J. Chem. Theory Comput.* **2019**, *15*, 1616–1632. [CrossRef]
125. Altun, A.; Izsák, R.; Bistoni, G. Local energy decomposition of coupled-cluster interaction energies: Interpretation, benchmarks, and comparison with symmetry-adapted perturbation theory. *Int. J. Quantum Chem.* **2021**, *121*, e26339. [CrossRef]
126. Hirshfeld, F.L. Bonded-Atom Fragments for Describing Molecular Charge Densities. *Theor. Chim. Acta* **1977**, *44*, 129–138. [CrossRef]
127. Ritchie, J.P.; Bachrach, S.M. Some Methods and Applications of Electron Density Distribution Analysis. *J. Comput. Chem.* **1987**, *8*, 499–509. [CrossRef]
128. Ritchie, J.P. Electron Density Distribution Analysis for Nitromethane, Nitromethide, and Nitramide. *J. Am. Chem. Soc.* **1985**, *107*, 1829–1837. [CrossRef]
129. Wiberg, K.B.; Rablen, P.R. Atomic Charges. *J. Org. Chem.* **2018**, *83*, 15463–15469. [CrossRef]
130. Jabłoński, M.; Krygowski, T.M. Study of the influence of intermolecular interaction on classical and reverse substituent effects in para-substituted phenylboranes. *New J. Chem.* **2020**, *44*, 9656–9670. [CrossRef]
131. Walsh, A.D. The Electronic Orbitals, Shapes, and Spectra of Polyatomic Molecules. Part I. AH_2 Molecules. *J. Chem. Soc.* **1953**, 2260–2266. [CrossRef]
132. Chai, J.-D.; Head-Gordon, M. Long-range corrected hybrid density functionals with damped atom–atom dispersion corrections. *Phys. Chem. Chem. Phys.* **2008**, *10*, 6615–6620. [CrossRef] [PubMed]
133. Hohenberg, P.; Kohn, W. Inhomogeneous Electron Gas. *Phys. Rev.* **1964**, *136*, B864–B871. [CrossRef]
134. Kohn, W.; Sham, L.J. Self-Consistent Equations Including Exchange and Correlation Effects. *Phys. Rev.* **1965**, *140*, A1133–A1138. [CrossRef]
135. Parr, R.G.; Yang, W. *Density-Functional Theory of Atoms and Molecules*; Oxford University Press: New York, NY, USA, 1989.
136. Mardirossian, N.; Head-Gordon, M. Thirty years of density functional theory in computational chemistry: An overview and extensive assessment of 200 density functionals. *Mol. Phys.* **2017**, *115*, 2315–2372. [CrossRef]
137. Krishnan, R.; Binkley, J.S.; Seeger, R.; Pople, J.A. Self-consistent molecular orbital methods. XX. A basis set for correlated wave functions. *J. Chem. Phys.* **1980**, *72*, 650–654. [CrossRef]
138. McLean, A.D.; Chandler, G.S. Contracted Gaussian basis sets for molecular calculations. I. second row atoms, Z = 11–18. *J. Chem. Phys.* **1980**, *72*, 5639–5648. [CrossRef]
139. Clark, T.; Chandrasekhar, J.; Spitznagel, G.W.; Schleyer, P.v.R. Efficient Diffuse Function-Augmented Basis Sets for Anion Calculations. III. The 3-21+G Basis Set for First-Row Elements, Li–F. *J. Comput. Chem.* **1983**, *4*, 294–301. [CrossRef]
140. Frisch, M.J.; Pople, J.A.; Binkley, J.S. Self-consistent molecular orbital methods 25. Supplementary functions for Gaussian basis sets. *J. Chem. Phys.* **1984**, *80*, 3265–3269. [CrossRef]
141. Curtiss, L.A.; McGrath, M.P.; Blandeau, J.-P.; Davis, N.E.; Binning, R.C., Jr.; Radom, L. Extension of Gaussian-2 theory to molecules containing third-row atoms Ga–Kr. *J. Chem. Phys.* **1995**, *103*, 6104–6113. [CrossRef]
142. Jensen, F. *Introduction to Computational Chemistry*; John Wiley & Sons Ltd.: Chichester, UK, 2007.
143. Pritchard, B.P.; Altarawy, D.; Didier, B.; Gibson, T.D.; Windus, T.L. New Basis Set Exchange: An Open, Up-to-Date Resource for the Molecular Sciences Community. *J. Chem. Inf. Model.* **2019**, *59*, 4814–4820. [CrossRef]
144. Frisch, M.J.; Trucks, G.W.; Schlegel, H.B.; Scuseria, G.E.; Robb, M.A.; Cheeseman, J.R.; Scalmani, G.; Barone, V.; Mennucci, B.; Petersson, G.A.; et al. *Gaussian 09*; Gaussian, Inc.: Wallingford, CT, USA, 2013.
145. Keith, T.A. *AIMAll* (Version 15.05.18); TK Gristmill Software: Overland Park, KS, USA, 2015. Available online: Aim.tkgristmill.com (accessed on 17 September 2022)

146. Møller, C.; Plesset, M.S. Note on an Approximation Treatment for Many-Electron Systems. *Phys. Rev.* **1934**, *46*, 618–622. [CrossRef]
147. Head-Gordon, M.; Pople, J.A.; Frisch, M.J. MP2 Energy Evaluation by Direct Methods. *Chem. Phys. Lett.* **1988**, *153*, 503–506. [CrossRef]
148. Becke, A.D. Density-fnnctional exchange-energy approximation with correct asymptotic behavior. *Phys. Rev. A* **1988**, *38*, 3098–3100. [CrossRef]
149. Lee, C.; Yang, W.; Parr, R.G. Development of the Colle-Salvetti correlation-energy formula into a functional of the electron density. *Phys. Rev. B* **1988**, *37*, 785–789. [CrossRef]
150. Grimme, S.; Ehrlich, S.; Goerigk, L. Effect of the Damping Function in Dispersion Corrected Density Functional Theory. *J. Comput. Chem.* **2011**, *32*, 1456–1465. [CrossRef]
151. van Lenthe, E.; Baerends, E.J.; Snijder, J.G. Relativistic regular two-component Hamiltonians. *J. Chem. Phys.* **1993**, *99*, 4597–4610. [CrossRef]
152. Snijder, J.G.; Sadlej, A.J. Perturbation versus variation treatment of regular relativistic Hamiltonians. *Chem. Phys. Lett.* **1996**, *252*, 51–61. [CrossRef]
153. Baerends, E.J.; Ziegler, T.; Atkins, A.J.; Autschbach, J.; Baseggio, O.; Bashford, D.; Bééces, A.; Bickelhaupt, F.M.; Bo, C.; Boerrigter, P.M.; et al. *ADF 2022.1*; SCM, Theoretical Chemistry; Vrije Universiteit: Amsterdam, The Netherlands, 2022.
154. Sagan, F.; Mitoraj, M.P. Non-covalent Interactions in Selected Transition Metal Complexes. In *Transition Metals in Coordination Environments: Computational Chemistry and Catalysis Viewpoints*; Broclawik, E., Borowski, T., Radoń, M., Eds.; Springer International Publishing: Cham, Switzerland, 2019; pp. 65–89.
155. Stasyuk, O.A.; Sedlak, R.; Fonseca-Guerra, C.; Hobza, P. Comparison of the DFT-SAPT and Canonical EDA Schemes for the Energy Decomposition of Various Types of Noncovalent Interactions. *J. Chem. Theory Comput.* **2018**, *14*, 3440–3450. [CrossRef]

International Journal of
Molecular Sciences

MDPI

Article

Molecular Structure, Vibrational Spectrum and Conformational Properties of 4-(4-Tritylphenoxy)phthalonitrile-Precursor for Synthesis of Phthalocyanines with Bulky Substituent

Natalia V. Tverdova [1], Nina I. Giricheva [2], Vladimir E. Maizlish [3], Nikolay E. Galanin [3] and Georgiy V. Girichev [1,*]

[1] Department of Physics, Ivanovo State University of Chemistry and Technology, Sheremetevsky Avenue 7, 153000 Ivanovo, Russia
[2] Nanomaterial Research Institute, Ivanovo State University, Ermak Street 39, 153025 Ivanovo, Russia
[3] Department of Fine Organic Chemistry and Technology, Ivanovo State University of Chemistry and Technology, Sheremetevsky Avenue 7, 153000 Ivanovo, Russia
[*] Correspondence: g.v.girichev@mail.ru or girichev@isuct.ru; Tel.: +7-4932-35-9874

Abstract: By DFT method with B3LYP, PBE, CAM-B3LYP, and B97D functionals, it was found that the molecule 4-(4-tritylphenoxy)phthalonitrile (TPPN) has four conformers. The geometric structure, vibrational frequencies, electronic characteristics, and thermodynamic functions of conformers, as well as the structure and energy of transition states, were determined. IR spectrum of TPPN film contains vibrational bands belonging to different conformers. The assignment of bands was performed basing the distribution of normal vibration energy on internal coordinates. A synchronous electron diffraction/mass spectrometric experiment was performed to determine the structure of conformers in a saturated TPPN vapor. The elemental composition of the ions recorded in the mass spectrum indicates the thermal stability of TPPN at least up to T = 200 °C. The difference in the structure of tetrasubstituted metal phthalocyanines, which can be synthesized from different TPPN conformers, has been shown.

Keywords: gas electron diffraction; mass spectrometry; DFT calculations; IR spectrum; molecular structure; 4-(4-tritylphenoxy)phthalonitrile; conformers

1. Introduction

Currently, considerable attention is paid to phthalocyanines containing bulk trityl, due to their manifestation of liquid crystal properties [1–4], as well as the prospects for their use in thin-film electronics [5]. In addition, phthalocyanine compounds with high solubility are currently needed [6,7]. An increase in solubility can be achieved by introducing substituents of a certain nature [7,8]. In [5], novel tetra (4-tritylphenoxy) substituted copper, zinc, and cobalt phthalocyanines were obtained by reaction of 4-(4-tritylphenoxy)phthalonitrile (TPPN) (Scheme 1) with $CuCl_2$, $ZnCl_2$, $CoCl_2$ metal salts (scheme in two dimensions of metal substituted phthalocyanine molecules is given in Supplementary Materials, Section S1). It has been shown in Ref. [5] that the presence of bulky 4-tritylphenoxy substituents on the peripheral positions prevents the aggregation in organic solvents, such as $CHCl_3$, CH_2Cl_2, THF, DMF, DMSO, and adds interesting properties to phthalocyanines, such as high solubility.

It is well known that tetrasubstituted phthalocyanines are usually obtained as a mixture of four constitutional isomers [9]. This can be considered a positive or negative factor, depending on the purpose of applied phthalocyanines. For using these compounds as materials in thin films, it is even better to have an isomeric mixture that will not crystallize very easily [2]. The presence of bulky 4-tritylphenoxy substituents in the β-position can lead to regioselectivity of substituted phthalocyanines.

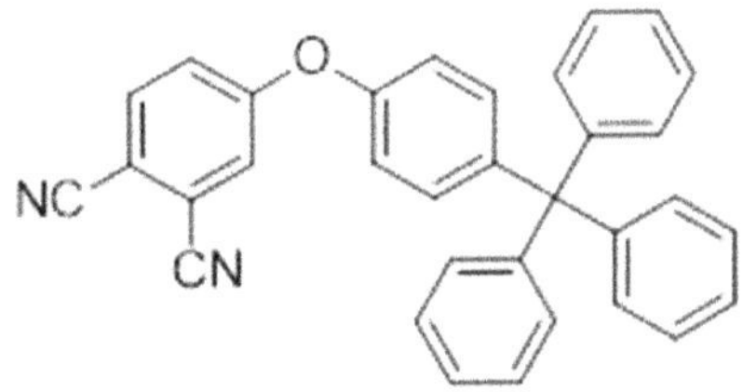

Scheme 1. Molecule of 4-(4-tritylphenoxy)phthalonitrile.

However, 4-tritylphenoxy substituents have a structural feature that is usually overlooked and which can affect the thermodynamic, spectral, and electro-optical properties of metal phthalocyanines. This feature of 4-tritylphenoxy substituents is related to the conformational properties of TPPN, which is used as the precursor for the synthesis of phthalocyanines with trityl groups as peripheral substituents.

Obtaining reliable experimental data for the 4-(4-tritylphenoxy)phthalonitrile and their subsequent study is an actual problem due to the lack of any experimental work on the structure of this compound.

This paper describes the first experimental study (gas electron diffraction in combination with mass spectrometry, as well as IR spectroscopy) of the molecular structure and conformational properties of 4-(4-tritylphenoxy)phthalonitrile supplemented by quantum chemical calculations.

2. Results and Discussion

2.1. DFT Calculations

2.1.1. Conformational Analysis

The TPPN molecule has six torsion coordinates, which can give rise to several conformers that differ in the mutual orientation of trityl, phenoxy (PhO), and phthalonitrile (PhN) fragments. The atom numbering is shown in Figure 1.

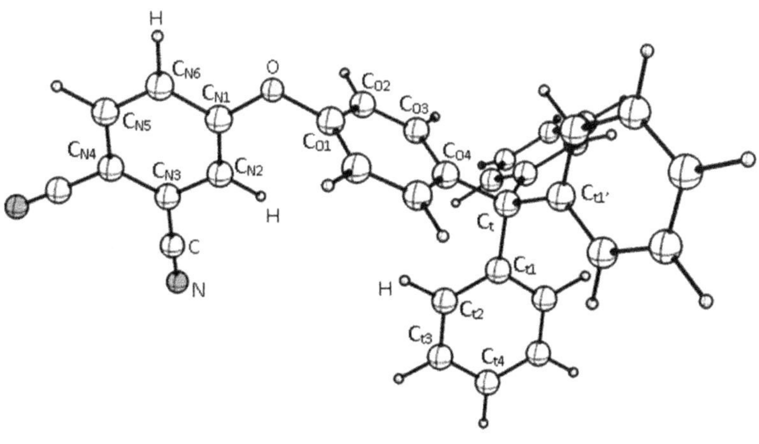

Figure 1. Atom numbering in the TPPN molecule.

To determine the structure of the found conformers (Figure 2) and the barriers between them, the potential functions of internal rotation were calculated (B3LYP/cc-pVTZ): Around the C_{O1}-O (Figure 3), C_t-C_{O4} (Figure 4), C_{N1}-O (Figure 5) bonds, and around C_t-C_{t1} bond in the trityl group. The $\phi[C_{O2}C_{O1}OC_{N1}]$, $\phi[C_{O3}C_{O4}C_tC_{t1}]$, and $\phi[C_{O1}OC_{N1}C_{N2}]$ dihedral angles were scanned with a step of 2° or 10°.

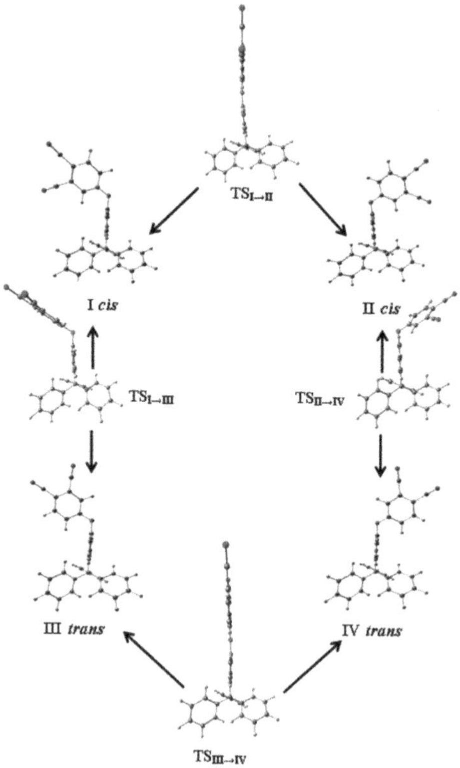

Figure 2. Conformers and transition states of the TPPN molecule. Projections with the same orientation of the trityl fragment are shown.

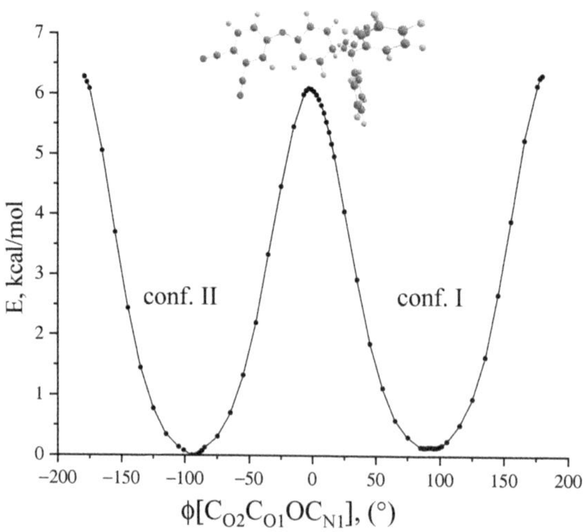

Figure 3. The potential function of internal rotation of O-PhN fragment around the C_{O1}-O bond. Structure $TS_{I\rightarrow II}$ Structures of conformers I and II are shown in Figure 2.

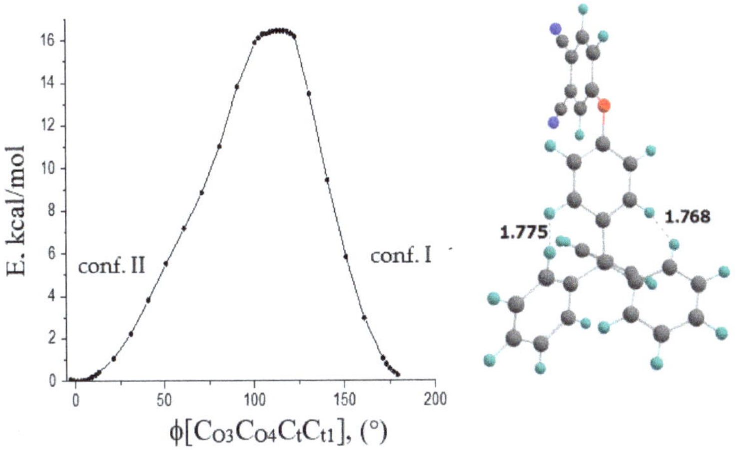

Figure 4. The potential function of internal rotation of PhO-PhN group around the C_{O4}-C_t bond in *cis*-conformers I and II. On the right: TS structure.

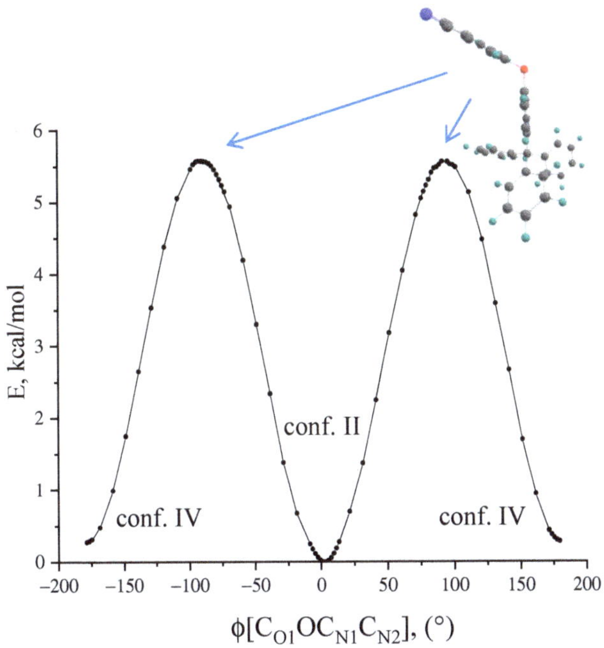

Figure 5. The potential function of internal rotation of PhN fragment around the C_{N1}-O bond. Structures of conformers II and IV are shown in Figure 2.

The potential function of the internal rotation of the O-PhN group around the C_{O1}-O bond has two deep minima (Figure 3) corresponding to *cis*-conformers I and II with $\phi[C_{O1}OC_{N1}\ C_{N2}] \approx 0°$ (Table 1). The potential function of the internal rotation of the O-PhN group for *trans*-conformers III→IV with torsion angle $\phi[C_{O1}OC_{N1}C_{N2}] \approx 180°$ (Table 1) has a similar form. In *cis*-conformers, the C_{O1}–O bond eclipses the C_{N1}-C_{N2} bond, whereas in *trans*-conformers, the dihedral angle $C_{O1}OC_{N1}C_{N2}$ is approximately 180°. As a result, *cis*-conformers I and II differ from the corresponding *trans*-conformers III and IV by the position of the C≡N substituents in the phthalonitrile moiety (Figure 2).

Table 1. Selected characteristics of 4-(4-tritylphenoxy)phthalonitrile conformers according to DFT/B3LYP/cc-pVTZ.

Characteristics	*Cis*-Forms		*Trans*-Forms	
	I	II	III	IV
ΔE_{total}, kcal/mol	0.10	0	0.34	0.27
E_{HOMO}, eV	−6.8	−6.8	−6.9	−6.8
E_{LUMO}, eV	−2.4	−2.4	−2.3	−2.3
E_{LUMO}-E_{HOMO}, eV	4.4	4.4	4.6	4.5
μ, D	8.2	8.1	10.1	10.0
Dipole moment direction in I and III conformers	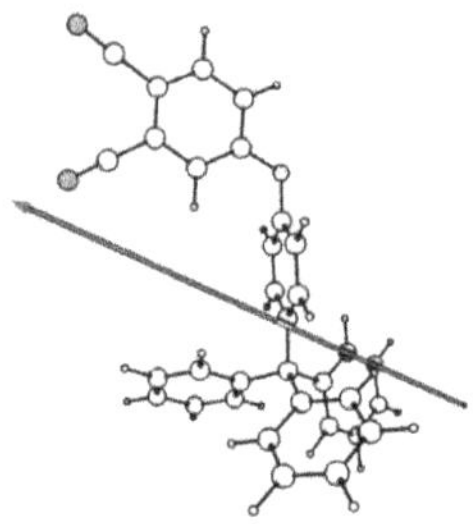		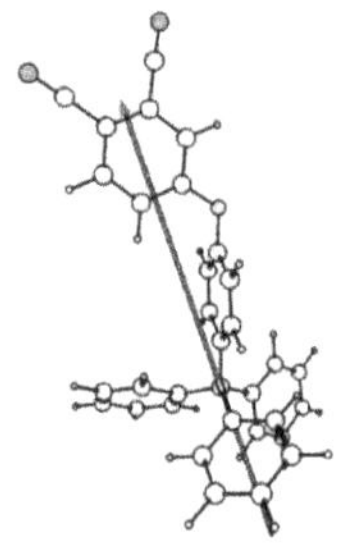	
$C_{O1}OC_{N1}C_{N2}$	−4.2	1.4	−176.1	−178.0
$C_{O4}C_tC_{t1}C_{t2}$	−2.3	−1.4	−6.2	−1.1
$C_{O3}C_{O4}C_tC_{t1}$	67.6	−116.4	69.8	−115.6
$C_{N1}OC_{O1}C_{O2}$	99.3	−94.8	99.7	−96.5
min $r(H_{PhN}\cdots H_{trt})$	3.74	4.06	3.79	4.15
ΔG°_{298}, kcal/mol	0.12	0.40	0.63	0.0
ΔG°_{478}, kcal/mol	0.28	0.76	0.95	0.0
χ_i, mol% T = 298 K	30.4	19.2	13.0	37.4
χ_i, mol% T = 478 K	29.1	17.6	14.3	39.0
S°_{298} cal/mol K	197.3	196.2	196.4	198.2
C_{v298} cal/mol K	114.3	114.2	114.3	114.3

The barrier between *cis*-conformers I and II and, accordingly, between *trans*-conformers III and IV is practically the same and amounts to 6.3 kcal/mol. In transition states $TS_{I\rightarrow II}$ and $TS_{III\rightarrow IV}$ the dihedral angle $\phi[C_{O2}C_{O1}OC_{N1}]$ between the planes of the PhO and PhN fragments is close to 0° (Figures 2 and 3).

The second way of the transition between conformers I→II and III→IV can be realized by rotating the PhO-PhN fragment around the C_{O4}-C_t bond. In this case, in the $TS_{I\rightarrow II}$ and $TS_{III\rightarrow IV}$ transition states, strong steric interaction occurs between PhO and trityl $C(Ph)_3$ fragments, which leads to a high energy barrier > 16 kcal/mol (Figure 4).

The difference in the two *cis*-conformers I and II, as well as in the two *trans*-conformers III and IV, lies in the different orientation of the PhN group relative to trityl fragment, namely, in conformers I and III, the dihedral angle $C_{O2}C_{O1}OC_{N1}$ should be defined as $\varphi = +99°$, and in conformers II and IV $\varphi = -95°$ (Figure 3, Table 1).

Figure 5 demonstrates that *cis*→*trans* transition can be realized by rotation of the PhN fragment around the O-C_{N1} bond. The energy barrier between the *cis*-conformer II and *trans*-conformer IV is 5.6 kcal/mol and corresponds to $TS_{II\rightarrow IV}$ in which the bond angle C_{O1}-O-C_{N1} between the PhO and PhN fragments is 118°, and dihedral angle $\phi[C_{O1}OC_{N1}C_{N2}]$ is 90°.

Additional calculations were carried out to estimate the barrier of internal rotation of the phenyl ring of the trityl group around the C_t-C_{t1} bond. It has been shown that an independent rotation by ~40° of one of the phenyl fragments leads to a strong steric repulsion between the Ph fragments in trityl. As a result, the phenyl rings cannot rotate but perform synchronous torsional vibrations. The amplitude of these vibrations $\Delta\phi$ depends on the thermal energy RT and is estimated as ±16° at T = 298 K and ±18° at T = 478 K.

2.1.2. The General Structural Motives and Electronic Characteristics of Conformers

Theoretical calculations performed using four DFT functionals lead to consistent results. All variants of the DFT method predict the presence of four C_1 symmetry conformers for the TPPN molecule. For identical molecular forms, the differences in the corresponding internuclear distances do not exceed 0.002 Å, in the values of bond angles 1°, and less than 7° in the values of torsion angles (Table S1).

All conformers have a similar structure of $(Ph)_3C(PhO)$ fragment. The central site $C(C)_4$ is a distorted tetrahedron with identical C-C bonds, two C-C-C bond angles that are smaller than the tetrahedral one, and four C-C-C bond angles larger than the tetrahedral one (Table S1). The bond lengths and bond angles in this fragment are the same in all conformers. Differences are observed in the values of torsion angles of *cis*- and *trans*-conformers as well as bond lengths in the PhN fragment. In addition, in conformers I and III, the shortest distance H···H between the hydrogen atoms of the trityl group and the phthalonitrile PhN fragment is ~3.7 Å and in conformers II and IV ~4.1 Å.

Table 1 shows the selected characteristics of conformers of 4-(4-tritylphenoxy)phthalonitrile molecule according to DFT/B3LYP/cc-pVTZ (Table S2 shows the characteristics of the conformers calculated with the functionals B97D, CAM-B3LYP, PBE).

All conformers have close energies of frontier orbitals and large dipole moments (Table 1), however, the direction of dipole moments in *cis*-conformers differs significantly from their direction in *trans*-conformers.

Figure 6 shows the frontier MOs of conformer II. At the electron transition from HOMO to LUMO, the electron density is transferred from the trityl fragment to phthalonitrile PhN. The frontier orbitals of other conformers have a similar form.

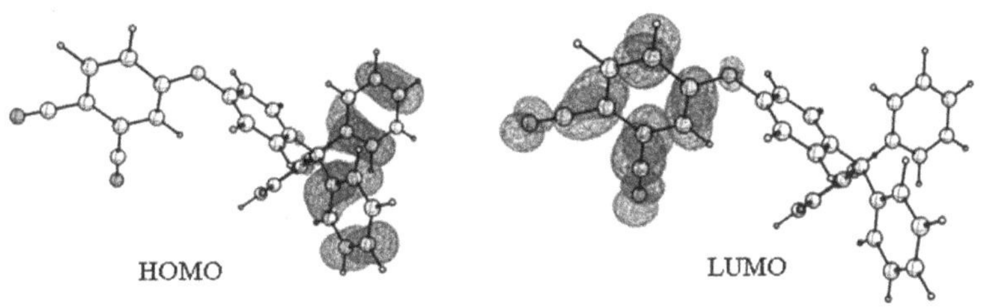

Figure 6. View of the frontier MOs of *cis*-conformer II.

Table 1 lists the relative energies of conformers ΔE_{total}, relative Gibbs free energies ΔG°_{T}, and mole fractions χ_i of conformers in saturated vapor calculated for T = 298 K and T = 478 K (the latter corresponds to the temperature of the GED experiment). The difference in the ratio of ΔE_{total} and ΔG°_{T} for the four conformers is due to the entropy factor.

The conformers have close electronic energies. The difference in total energy ΔE between conformers I-IV does not exceed 0.4 kcal/mol. This increases the probability that the 4-(4-tritylphenoxy)phthalonitrile may contain different conformations in the unit crystal cell and have a complex conformational composition of the gas phase (Tables 1 and S2). Since the barriers between conformers are quite high and exceed 5 kcal/mol, conformational transitions are unlikely even at the temperature of GED experiment (the thermal energy RT is less than 1 kcal/mol). Thus, conformational diversity may arise during the synthesis of a compound.

In this case, the mixture of conformers should be characteristic of both the condensed and gas phases. We tried to find the validity of this assumption when interpreting the IR spectra and GED data.

2.2. Vibrational Spectrum

The infrared spectrum for the films of TPPN is shown in Figure 7, along with the calculated spectrum for a mixture of *cis*- and *trans*-conformers.

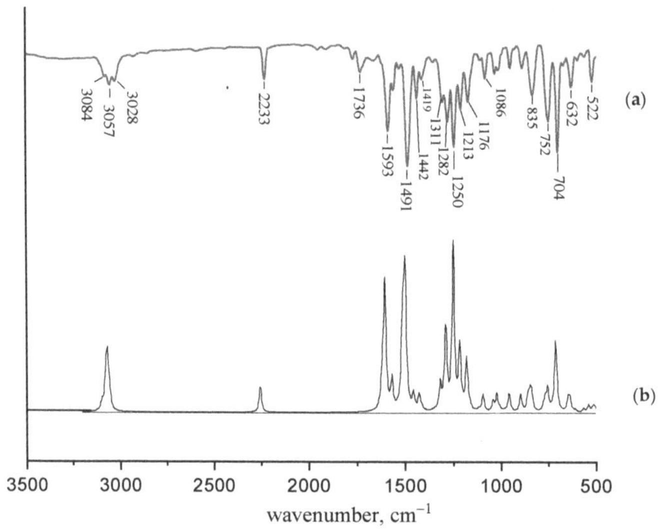

Figure 7. Experimental FTIR spectrum for the film of TPPN (**a**), and calculated spectrum for a mixture of *cis*- and *trans*-conformers (**b**).

The harmonic vibrational frequencies were calculated for the fully optimized geometries of I–IV conformers at B3LYP/cc-pVTZ level. Analysis of the calculated IR spectra showed that the spectra of the two conformers I and II (*cis*) and two conformers III and IV (*trans*) almost coincide (Figure S1).

In Table 2, the experimental vibrational frequencies ν_{exp} of TPPN recorded in this work and in [5] are compared with the calculated frequencies ω_{theor}, which have the highest intensity in the corresponding spectral region for *cis*- and *trans*-conformers (Figure 7).

The values of the experimental frequencies ν_{exp} in the two works are in good agreement with each other. However, in Ref. [5], low-intensity bands at 1419, 1311, and 1033 cm^{-1} were not noted, and an erroneous assignment of the frequency at 1253 cm^{-1} to the C-O-C bending vibration was given.

The correctness of the assignment of frequencies was checked on the basis of the dependencies ω_{theor} = f(ν_{exp}), which had a linear character with a correlation coefficient almost equal to 1 (Figure 8).

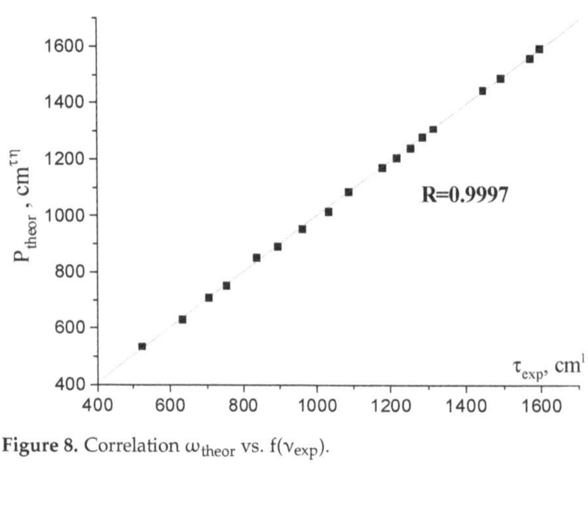

Figure 8. Correlation ω_{theor} vs. f(ν_{exp}).

Table 2. Experimental (ν_{exp}) and theoretical (ω_{theor}) vibrational frequencies for *cis*- and *trans*-conformers of 4-(4-tritylphenoxy)phthalonitrile [1,2,3].

[4]	ν_{exp}, cm^{-1}	ω_{theor} [2], cm^{-1} *cis*	ω_{theor}, cm^{-1} *trans*	Distribution Energy of Normal Vibrations over Internal Coordinates
	3084 w [4]	3101	3101	$\nu(C_t$-H)
3051 $(C\text{-}H)_{Ar}$	3057 m	3073	3073	$\nu(C\text{-}H)_t$, $\nu(C\text{-}H)_{PhO}$
	3028 w	2986	2986	$\nu(C\text{-}H)_t$
2231 (C≡N)	2233 m	2256 m	2256 m	ν(C≡N)
1593 (C=C)	1593 s	1595 s	1597 s	$\nu(C\text{-}C)_{PhN}$, $\nu(C\text{-}C)_{PhO}$
1566	1566 w	1561 vw	1558 m	$\nu(C\text{-}C)_{PhN}$, $\delta(CCH)_{PhN}$
	1505 sh	1505 m	1505 m	$\nu(C\text{-}C)_{PhO}$, $\delta(CCH)_{PhO}$
1487	1491 s	1492 s	1488 s	$\nu(C\text{-}C)_{PhN}$, $\delta(CCH)_{PhN}$, ν(C-O)
1442	1442 m	1448 w		$\nu(C\text{-}C)_{PhN}$, $\delta(CCH)_{PhN}$
	1419 w		1420 m	$\nu(C\text{-}C)_{PhN}$, $\delta(CCH)_{PhN}$
	1311 w	1310 s		$\nu(C\text{-}C)_{PhN}$, ν(C-O), $\delta(CCC)_{PhN}$
1288	1282 m	1279 m	1283 s	ν(C-O), $\nu(C\text{-}C)_{PhN}$, $\delta(CCH)_{PhN}$
1253 (Ar-O-Ar)	1250 s	1243 s	1248 s	ν(C-O), $\nu(C\text{-}C)_{PhN}$, $\nu(C_N\text{-}C)$,$\delta(CCH)_{PhN}$
1213	1213 m	1208 m	1211 m	ν(C-O), $\nu(C\text{-}C)_{PhN}$, $\delta(CCC)_{PhO}$,$\delta(CCH)_t$
1174	1176 m	1172 m	1174 m	ν(C-O), ν(C-C), $\delta(CCH)_{PhO}$, $\delta(CCH)_{PhN}$
1082	1086 w	1087 w	1092 w	$\nu(C_N\text{-}C)$, δ(CCH) ν(C-O), $\nu(C\text{-}C)_{PhN}$, δ(CCC)
	1033 w		1036 vw	$\nu(C\text{-}C)_t$
1014	1014 w	1017 w	1017 w	$\nu(C\text{-}C)_{PhO}$, $\delta(CCC)_{PhO}$, $\delta(CCH)_{PhO}$
956	954 w	953 w	954 w	$\rho(C\text{-}H)_{PhO}$, δ(CCC)
885	890 w	892 w	892 w	$\rho(C\text{-}H)_{PhO}$, $\tau(CCCC)_{PhO}$
825	835 m	854 w	853 w	$\rho(C_N\text{-}C)$, $\rho(C\text{-}H)_{PhO}$, ρ(O-C), $\rho(C\text{-}H)_{PhN}$
750	752 s	752 w	752 w	$\rho(C\text{-}H)_t$, $\rho(C\text{-}C)_t$, $\tau(CCCC)_t$
700	704 s	709 m	708 m	$\rho(C\text{-}H)_t$, $\delta(CCC)_{PhN}$, $\rho(C\text{-}C)_t$, $\tau(CCCC)_{PhO}$
632	632 w	632 w	635 w	$\delta(CCC)_{C4}$, $\delta(CCC)_t$, $\rho(CC_N)$
	522 w	537 w	536 w	$\rho(C\text{-}C)_{PhN}$, $lb(C_{PhN}C_NN)$

[1] see Figure 1 for atom numbering. PhO—phenoxy fragment, PhN—phthalonitrile fragment, t—trityl fragment. [2] scaled values of ω_{theor} are given. Scale is found as the average value of the ratio ν_{exp}/ω_{calc}. The scale value is 0.965 for frequencies in the range of 3500–1650 cm^{-1} and 0.978 for frequencies in the range of 1650–500 cm^{-1}. [3] relative intensities: w—weak, vw—very weak, s—strong, m—medium. [4] internal coordinates: ν(X-Y)—bond stretching, δ(XYZ)—deformation of XYZ valence angle, ρ(XY)—out of plane motion of X-Y bond, lb(XYZ)—bending of XYZ linear fragment, τ(XYVZ)—torsion about Y-V bond.

For all frequencies of normal vibrations, the distribution of potential energy (DPE) over internal vibrational coordinates is calculated. For this, the VibModule [10] program was used. The last column of Table 2 shows the natural coordinates that make the maximum contributions to an individual normal vibration.

The relative intensities are indicated for the vibrational bands. The assignment for all 168 vibrational frequencies of the four conformers of the TPPN molecule is given in Table S3.

It should be noted that most bands in the IR spectrum possess a mixed nature (Table 2). For this reason, it is possible to discuss only the predominant contribution of various internal coordinates to the normal vibration. The exceptions are the stretching vibrations of C-H bonds, which are completely separated from other vibrations, and the stretching vibrations of C≡N bonds.

The bands in the range 1593–1491 cm^{-1} correspond to the vibrations of $\nu(C\text{-}C)_{PhN}$, $\delta(CCC)_{PhN}$, $\delta(CCH)_{PhN}$ in the phthalonitrile fragment. The bands at 1282, 1250, 1213, and 1176 cm^{-1} can be assigned to $\nu(C\text{-}O)_{PhN}$, $\nu(C\text{-}O)_{Ph}$ stretching vibrations in combination with ν(C-C), δ(CCH) modes. Note that in [1,5], the band at 1250 cm^{-1} is assigned to the Ph-O-Ph bending vibration instead of the C-O stretching vibration (although the force constant stretching of C-O bond (5.38 mdn/Å) is significantly higher than the force constant deformation of the C-O-C angle (1.20 mdn/Å)).

The frequencies associated with bending vibrations of the central site $C(C)_4$, phthalonitrile PhN, phenoxy PhO fragments, and the trityl group $C(Ph)_3$, according to calculations, are in the range below 1000 cm^{-1}. The low frequency region of the IR spectrum shows the presence of bands at 835, 752, 704, 632, 522 cm^{-1} associated with out-of-plane vibrations of the C-H, C_{PhN}-C_N bonds.

The main differences between the calculated IR spectra of the *cis*- and *trans*-conformers are in the intensity of the band at 1566 cm^{-1} as well as the presence of characteristic bands at 1311 cm^{-1}, 1442 cm^{-1} belonging to *cis*-conformers, and 1033 cm^{-1}, 1419 cm^{-1} belonging to *trans*-conformers (Table 2).

The bands at 1442 cm^{-1} (*cis*) and 1419 cm^{-1} (*trans*) correspond mainly to the stretching vibrations of the C-C bonds in the phthalonitrile fragment. It should be noted that the $r(C_{N3}\text{-}C_{N4})_{PhN}$ bond length in *cis*-conformers is shorter than in *trans*-conformers (Table 3), which is reflected in the difference in the $\nu(C\text{-}C)_{PhN}$ frequency.

An analysis of the recorded vibrational frequencies allows us to conclude that the IR spectrum indicates the presence of the *cis*- and *trans*-conformers of TPPN in the condensed state.

2.3. GED Structural Analysis

Geometry of each conformer of the TPPN molecule possessed C_1 symmetry and was described by 79 independent parameters. To reduce the correlation, the number of independent parameters was restricted to 14: C_t-C_{O4}, C_{O4}-C_{O3}, O-C_{O1}, C_{N3}-C, C-N, C_{t2}-H bond distances, $C_{O4}C_tC_{t1}$, $C_{O3}C_{O4}C_t$, $C_{O2}C_{O3}C_{O4}$, $C_{N1}OC_{O1}$, $C_{t3}C_{t2}H$ valence angles, and $C_{O3}C_{O4}C_tC_{t1}$, $C_{N1}OC_{O1}C_{O2}$, and $C_{t1}C_tC_{O4}C_{t1'}$ dihedral angles (Figure 1). The remaining 65 independent parameters were linked to the above-mentioned 14 parameters by the difference between non-equivalent structural parameters of the same type obtained from B3LYP/cc-pVTZ calculations.

VibModule program [10] was applied to calculate the vibrational corrections $\Delta r = r_{h1} - r_a$ and the starting values of root-mean-square vibrational amplitudes of each conformer at the temperature of the GED experiment using the harmonic approximation and also taking into account the non-linear interrelation between internal and Cartesian vibrational coordinates. Starting values of the geometrical parameters and vibrational amplitudes were taken from B3LYP/cc-pVTZ calculations. The dependent parameters were determined in terms of the geometrically consistent r_{h1}-structure. The amplitudes were refined in groups corresponding to the different peaks on the radial distribution curve. The analysis of the electron diffraction intensities was carried out using the modified KCED-35 program [11].

Theory predicts that despite the different mutual orientations of the phthalonitrile and trithyl groups, the corresponding bond lengths and most nonbonded distances of conformers I–IV are very close. This indicates that the electron diffraction intensities may be insensitive to the conformational composition of this compound.

The calculated Gibbs free energies at the temperature of the GED experiment 478 K indicate that the ratio between the mole fractions of the *cis*- and *trans*-conformers in the gas phase is close to 1 (Table 1). In addition, DFT calculations predict that the *cis* I and *trans* IV forms predominate in vapor over the *cis* II and *tarns* III forms.

Table 3. Experimental and theoretical parameters of *cis*- and *trans*-conformers.

Parameter [1,2]	*cis*-Conformer I		*trans*-Conformer IV	
	DFT	GED	DFT	GED
C_t-C_{O4}	1.554	1.542(6) [4]	1.553	1.542(6)
C_{O4}-C_{O3}	1.402	1.409(3)	1.402	1.409(3)
O-C_{O1}	1.396	1.392(14)	1.395	1.391(14)
C≡N	1.152	1.148(7)	1.152	1.148(7)
C_{N3}-C	1.429	1.424(3)	1.429	1.425(3)
C_{t2}-H	1.078	1.101(6)	1.078	1.102(6)
$C_{O4}C_tC_{t1}$	105.1	103.7(2.3) [4]	104.9	103.6(2.3)
$C_{O3}C_{O4}C_t$	118.7	118.3(0.9)	118.7	118.2(0.9)
$C_{O2}C_{O3}C_{O4}$	121.9	121.8(0.8)	121.9	121.8(0.8)
$C_{N1}OC_{O1}$	120.3	120.3(4.6)	120.3	120.4(4.6)
$C_{t3}C_{t2}H$	118.6	fixed	118.6	fixed
$C_{t1}C_tC_{O4}C_{t1'}$	121.2	120.5(2.2) [4]	121.1	120.5(2.2)
$C_{O3}C_{O4}C_tC_{t1}$	−53.5	−57.2(3.0)	−54.1	−56.6(3.0)
$C_{N1}OC_{O1}C_{O2}$	99.3	82.3(27.3)	−96.5	−83.9(27.3)
Distorted tetrahedron CC_4				
C-C [3]	1.554	1.543(6)	1.554	1.543(6)
C-C-C [3] (4)	111.6	111.3(2.5)	111.7	111.4(2.5)
C-C-C [3] (2)	105.3	105.8(3.3)	105.1	105.6(3.3)
Parameters of PhN fragment				
C_{N2}-C_{N3}	1.395	1.402(3)	1.389	1.396(3)
C_{N3}-C_{N4}	1.407	1.414(3)	1.411	1.418(3)
C_{N4}-C_{N5}	1.400	1.407(3)	1.395	1.402(3)
C_{N5}-C_{N6}	1.379	1.386(3)	1.385	1.392(3)
$C_{N1}C_{N2}C_{N3}$	119.8	119.8(0.8)	120.3	120.2(0.8)
$C_{N5}C_{N6}C_{N1}$	120.1	120.1(0.8)	119.6	119.6(0.8)
Parameters of Ph fragment of trityl group				
C-C [3]	1.391	1.397(3)	1.391	1.397(3)
C-C-C [3]	120.0	120.0(1.2)	120.0	120.0(1.2)

[1] See Figure 1 for atom numbering. [2] distances in Å, bond angles in deg, [3] averaged values [4] Error limit given in parentheses was estimated as $\sigma = ((0.002\ \sigma)^2 + (2.5\ \sigma_{LS})^2)^{1/2}$ for distances and $\sigma = 3\ \sigma_{LS}$ for angles, where σ_{LS} is the standard deviation from LS refinement.

To assess the ability of the GED method to identify TPPN conformers I-IV, the theoretical functions sM(s) were calculated separately for each conformer based on structural parameters (bond lengths, angles, vibrational amplitudes, and shrinkage corrections) obtained using B3LYP/cc-pVTZ level (Figure S2), as well as theoretical radial distribution functions f(r) (Figure S3). Only a small difference in the f(r) functions is observed in the range of 3–8 Å for the *cis*- and *trans*-conformers.

The least-squares analysis of GED experimental intensities was performed under the assumption that the vapor contains separate conformers, as well as under the assumption that the vapor contains *cis*- and *trans*-conformers in mole percent of 30/70, 50/50, and 70/30. In these versions of the least squares analysis, the value of the disagreement factor R_f varied in the range of 4.13–4.28%. In accordance with the Hamilton criterion at 0.05 significance level [12], GED data do not allow giving preference to any variant of the vapor composition.

Despite the lack of success in the experimental solution of the conformational problem of TPPN, the GED method still allows us to obtain reliable structural parameters of the molecule. Table 3 shows that the calculated geometric parameters of equilibrium configurations of *cis*- and *trans*-conformers agree satisfactorily with the experimental ones.

The comparison of the experimental functions sM(s) and f(r) with the theoretical curves for the mole percent of 50/50 *cis*/*trans* vapor model are shown in Figures 9 and 10, respectively.

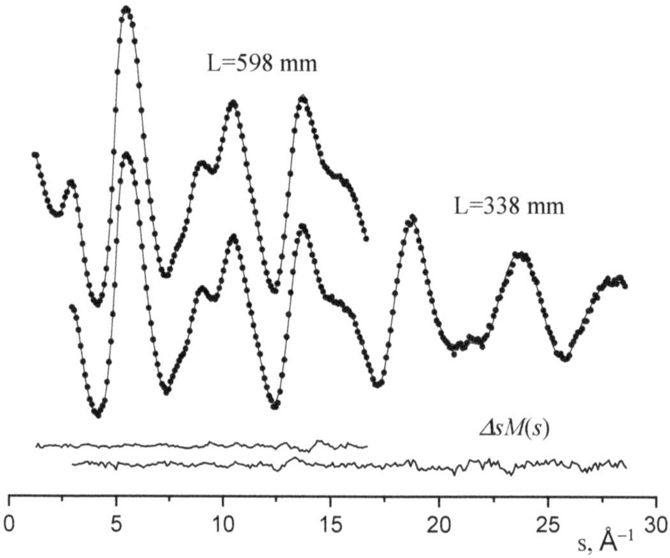

Figure 9. Experimental (dots) and theoretical (full line) molecular intensities sM(s) for mole percent of 50/50 cis/trans vapor model and residual curves for 4-(4-tritylphenoxy)phthalonitrile.

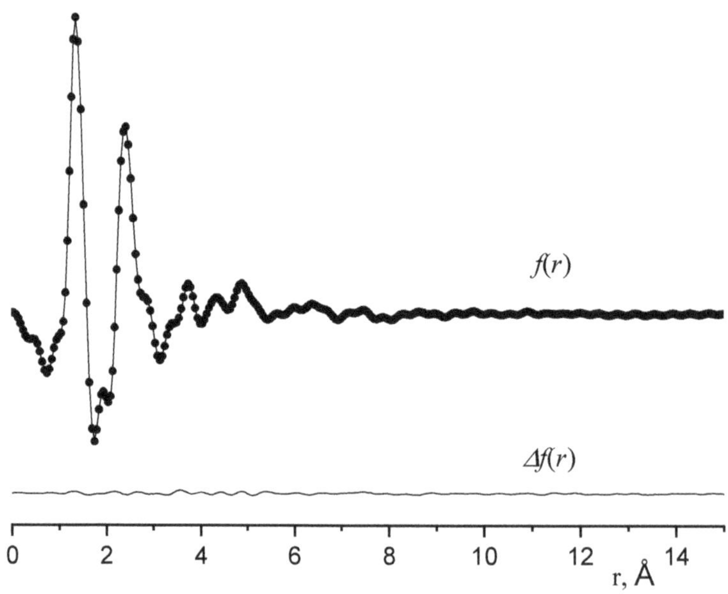

Figure 10. Experimental (dots) and theoretical (full line) radial distribution function f(r) for mole percent of 50/50 *cis/trans* vapor model and residual curve for 4-(4-tritylphenoxy)phthalonitrile.

The performed analysis of IR spectra, data, and the results of quantum chemical calculations indicates the existence of *cis*- and *trans*-conformers of TPPN, both in the condensed and in the gaseous state, and GED/MS data do not contradict this conclusion.

Figure 11 shows examples of positions of 4-tritylphenoxy substituents that can occur in tetrasubstituted metal phthalocyanines (MPc) during their synthesis by reaction of *cis*- (a) or *trans*-conformers (b) 4-(4-tritylphenoxy)phthalonitrile with metal salts. There are significant differences in the structure of the MPc conformers (a and b), which can lead to a difference in their physicochemical properties. Thus, for tetra-*cis*-substituted MPcs, there is a high probability of manifestation of regioselectivity, which reduces the steric repulsion between adjacent bulky substituents (Figure 11a). In the case of tetra-*trans*-substituted MPcs, regioisomers are possible. In addition, the possibility of the formation of tetrasubstituted MPcs with different *cis*- and *trans*-orientations of 4-tritylphenoxy substituents within the same metal complex cannot be ruled out.

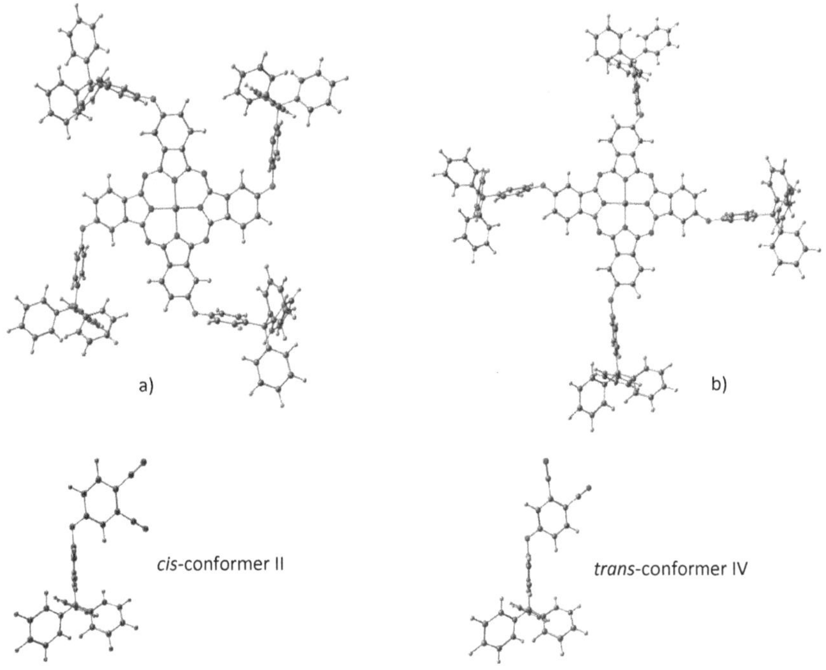

Figure 11. Tetra-*cis*-4-tritylphenoxy substituted metal phthalocyanines (**a**), tetra-*trans*-substituted metal phthalocyanines (**b**), bottom: *cis*- and *trans*-conformers of TPPN.

A series of copper (4-tritylphenoxy)phthalocyanines was synthesized in [4]. The authors note that the synthesized compounds exhibit mesomorphic properties and suggest that their mesomorphism originates from thermal fluctuations due to the free rotation of the bulky substituents. However, our data on barriers of internal rotations (Figures 3–5) show that the assertion of the authors of [4] about the free rotation of phenoxy and trityl groups is untenable. Nevertheless, it seems that the copper(II) phenoxy(phthalocyaninato) complex with trans-oriented 4-tritylphenoxy substituents has no serious steric obstacles to the formation of a columnar mesophase (Figure 11b).

3. Methods and Materials

3.1. Synthesis

Synthesis of TPPN was carried out according to the scheme used in [7,13], which is closely repeated in [4] and is described in detail in Supplementary Materials (Section S1).

The product was characterized by mass spectrometry (LDI–TOF, negative mode), ^{1}H NMR ($CDCl_3$), ^{13}C NMR ($CDCl_3$), FTIR spectrum, and elemental analysis methods. The results are given in Supplementary Materials Section S1.

3.2. The Infrared Spectrum

IR spectrum was recorded for the films deposited from a solution on the KRS-5 slides in the range of 400–4000 cm^{-1} using an Avatar 360 FTIR spectrometer. (Details in Section 2.2).

3.3. Synchronous Gas Electron Diffraction and Mass Spectrometric Experiment

The gas phase electron diffraction patterns and mass spectra were recorded simultaneously using the technique described in Refs. [14,15] at two nozzle-to-plate distances (338 and 598 mm). The conditions of the synchronous gas-phase electron diffraction/mass spectrometric (GED/MS) experiments are shown in Table S4.

The electron wavelength was obtained by polycrystalline ZnO. The optical densities were measured by a computer-controlled modified MD-100 microdensitometer [16] with a step of 0.1 mm along the diagonal of the plate. A 10 × 130 mm region was scanned. The number of equidistant scan lines was 33. The total intensity curves were obtained in the ranges s = 1.3–16.7 $Å^{-1}$ and s = 3.0–28.6 $Å^{-1}$.

Simultaneously with the registration of electron diffraction patterns, the mass spectra of TPPN vapors were recorded. According to the mass spectrometric data, only TPPN is present in the vapor, no volatile impurities are observed. Table 4 shows that the mass spectra of electron ionization (U_{ioniz} = 50 V) contain signals of the parent ion $[M]^+$ ($[C_{33}H_{22}N_2O]^+$) and lighter ions formed as a result of dissociative ionization of molecules: Ions formed upon breaking C_t-C_{t1} bonds (see Figure 1 for atom numbering), with the loss of one $[M\text{-}(C_6H_5)]^+$, two $[M\text{-}(C_6H_5)_2]^+$ and three $[[M\text{-}(C_6H_5)_3]^+$ phenyl groups of the trityl fragment, and additionally the loss of a substituent CN $[M\text{-}(C_6H_5)_3\text{-}CN]^+$, ions with loss of trityl fragment $[M\text{-}C(C_6H_5)_3]^+$ and group CN $[M\text{-}C(C_6H_5)_3\text{-}CN]^+$, trityl ion $[C(C_6H_5)_3]^+$, as well as the high intensive $[C(C_6H_5)_2]^+$ ion. In addition to ions associated with the detachment of individual groups, the mass spectrum contains ions related to the dissociative ionization of trityl phenyl fragments with the elimination of C_2H_n and C_3H_n groups ($[M\text{-}C_3H_3\text{-}C_2H_2]^+$ and $[(C_4H_4)_2CC_3H_3]^+$. The most intense is the ion with stoichiometry $[M\text{-}C_6H_5]^+$.

Table 4. Most intensive ions in mass spectrum of TPPN recorded during the GED experiment at 478 K (U_{ioniz} = 50 V).

Ion	M/E	Relative Abundance, %
$[M]^+ = [C_{33}H_{22}N_2O]^+$, parent ion	462	52
$[M\text{-}C_3H_3\text{-}C_2H_2]^+$	397	41
$[M\text{-}C_6H_5]^+$	385	100
$[C(C_6H_5)_3(C_6H_4)]^+$	319	40
$[M\text{-}(C_6H_5)_2]^+$	308	43
$[C(C_6H_5)_3]^+$	243	65
$[M\text{-}(C_6H_5)_3]^+$	231	44
$[M\text{-}(C(C_6H_5)_3)]^+$	219	40
$[M\text{-}(C_6H_5)_3\text{-}CN]^+$	205	38
$[M\text{-}(C(C_6H_5)_3)\text{-}CN]^+$	193	40
$[OC_6H_4CC_6H_5]^+$	181	35
$[C(C_6H_5)_2]^+$	166	63
$[(C_4H_4)_2CC_3H_3]^+$	155	42

3.4. Computational Details

Quantum-chemical modeling of the TPPN molecule was implemented using the Gaussian 09 program [17]. The calculations were performed using the density functional theory (DFT) method, variants of hybrid functionals: B3LYP [18–20], PBE [21], CAM-B3LYP [22], as well as the B97D functional [23], which considers dispersion interactions. The atoms of carbon, oxygen, nitrogen, and hydrogen were described using correlation-consistent three-exponential valence bases cc-pVTZ [24].

Using the DFT/B3LYP/cc-pVTZ method, the potential energy surface (PES) was scanned along torsion coordinates, and the presence of four conformers was established. For each conformer of the TPPN molecule, the geometry optimization was performed (see

Section 2.1) without symmetry restrictions. The calculation of the frequencies confirmed the correspondence of the optimized structures to the PES minima, and the calculated frequencies were used in the interpretation of the experimental IR spectrum.

4. Conclusions

The TPPN compound is thermally stable up to at least 200 °C, as evidenced by the data of the GED/MS experiment. This property of TPPN can be used in modeling CVD processes for obtaining thin films for various purposes.

The presence of bulky 4-tritylphenoxy substituents in the β-position can lead to regioselectivity of substituted phthalocyanines.

Using the DFT method with B3LYP, PBE, CAM-B3LYP, and B97D functionals, it was shown that the molecule TPPN has four conformers with the same structure of the -Ph-C(Ph)$_3$ fragment and different positions of the -O-PhN group. All conformers have a significant dipole moment and similar energies of frontier orbitals. At the HOMO → LUMO transition, the electron density is transferred from the trityl to the phthalonitrile fragment.

The structure of transition states between conformers is determined. The close values of the total energy of the conformers and high transition barriers suggest that the conformational diversity of TPPN is formed during the synthesis of the compound.

All vibrational bands in the experimental IR spectrum of TPPN have been assigned. A slight difference was noted in the theoretical IR spectra of the conformers, which, however, made it possible to determine that TPPN consists of a mixture of conformers in the condensed state.

The elemental composition of the ions registered in the mass spectrum of electron ionization has been established in saturated vapor TPPN.

GED study of saturated vapor TPPN was performed. The experimental geometrical parameters of the conformers are found, which agree with the calculated ones within the experimental error. Joint analysis of GED and QC data indicates that at T = 478 K, the gas phase of TPPN consists of a mixture of conformers.

It can be assumed that the use of TPPN for the synthesis of tetrasubstituted metal phthalocyanines will lead to the formation of conformers that differ significantly in the spatial structure of peripheral fragments and retain the conformational features of TPPN.

Moreover, within the same complex, 4-tritylphenoxy substituents can be in different conformations. This feature of such complexes can be used for practical purposes.

Supplementary Materials: The supporting information can be downloaded at https://www.mdpi.com/article/10.3390/ijms232213922/s1.

Author Contributions: Conceptualization, G.V.G.; methodology, G.V.G., N.I.G. and N.V.T.; investigation: N.V.T., N.I.G., V.E.M., N.E.G. and G.V.G.; formal analysis, N.V.T. and N.I.G., resources, G.V.G.; writing—original draft preparation, N.V.T. and N.I.G.; writing—review and editing, G.V.G.; visualization, N.V.T.; supervision, G.V.G.; project administration, G.V.G.; funding acquisition, G.V.G. All authors have read and agreed to the published version of the manuscript.

Funding: This work was supported by the Russian Science Foundation (grant No. 20-13-00359).

Data Availability Statement: Not applicable.

Acknowledgments: Vibrational analysis by N.I.G. was supported by Russian Ministry of Science and Higher Education (grant No FZZM-2020-0006). The synthesis and characterization of the sample by N.E.G. were supported by Russian Ministry of Science and Higher Education (grant No FZZM-2020-0010). The gas-phase electron diffraction/mass-spectrometric experiments were carried out using the GED/MS equipment (Available online: https://www.isuct.ru/department/ckp/structure/ged-ms (5 October 2022) of the resources of the Center for Shared Use of Scientific Equipment of the ISUCT (with the support of the Ministry of Science and Higher Education of Russia, grant No. 075-15-2021-671). We thank A.V. Belyakov for his help in conducting the calculations.

Conflicts of Interest: The authors declare no conflict of interest.

References

1. Usol'tseva, N.; Bykova, V.; Ananjeva, G.; Zharnikova, N.; Kudrik, E. Mesomorphism and Glass Formation of Phthalocyanine Metal Complexes with Bulky Substituents. *Mol. Cryst. Liq. Cryst.* **2004**, *411*, 329–336. [CrossRef]
2. Zharnikova, N.; Usol'tseva, N.; Kudrik, E.; Thelakkat, M. Synthesis, mesomorphism and electrochemical properties of tetrasubstituted zinc and copper phthalocyanines. *J. Mater. Chem.* **2009**, *19*, 3161–3167. [CrossRef]
3. Nakamura, H.; Sugiyama, K.; Ohta, K.; Yasutake, M. Phthalocyanine-based discotic liquid crystals switching from a molten alkyl chain type to a flying-seed-like type. *J. Mater. Chem. C* **2017**, *5*, 7297–7306. [CrossRef]
4. Takagi, Y.; Ohta, K.; Shimosugi, S.; Fujii, T.; Itoh, E. Flying-seed-like liquid crystals 2: Unprecedented guidelines to obtain liquid crystalline compounds. *J. Mater. Chem.* **2012**, *22*, 14418–14425. [CrossRef]
5. Solğun, D.G.; Horoz, S.; Ağirtaş, M.S. Synthesis of novel tetra (4-tritylphenoxy) substituted metallophthalocyanines and investigation of their aggregation, photovoltaic, solar cell properties. *Inorg. Nano-Met. Chem.* **2018**, *48*, 508–514. [CrossRef]
6. Yazıcı, A.; Sarıçiçek, B.; Altındal, A.; Salih, B.; Bekaroğlu, Ö. Synthesis and characterization, electrical and gas sensing properties of *N,N′*-bis(salicylidene)-1,2-phenylendiamine substituted novel mono and ball-type metallo phthalocyanines. *Inorg. Chim. Acta* **2015**, *428*, 83–92. [CrossRef]
7. Galanin, N.E.; Yakubov, L.A.; Kudrik, E.V.; Shaposhnikov, G.P. Synthesis and spectral properties of meso-phenyltetrabenzoazaporphyrins containing triphenylmethyl groups. *Russ. J. Gen. Chem.* **2008**, *78*, 1436–1440. [CrossRef]
8. Ağırtaş, M.S. Highly soluble phthalocyanines with hexadeca *tert*-butyl substituents. *Dyes Pigm.* **2008**, *79*, 247–251. [CrossRef]
9. Sommerauer, M.; Rager, C.; Hanack, M. Separation of 2(3),9(10),16(17),23(24)-Tetrasubstituted Phthalocyanines with Newly Developed HPLC Phases. *J. Am. Chem. Soc.* **1996**, *118*, 10085–10093. [CrossRef]
10. Vishnevskiy, Y.V.; Zhabanov, Y.A. New implementation of the first-order perturbation theory for calculation of interatomic vibrational amplitudes and corrections in gas electron diffraction. *J. Phys. Conf. Ser.* **2015**, *633*, 012076. [CrossRef]
11. Andersen, B.; Seip, H.M.; Strand, T.G.; Stolevik, R. Procedure and Computer Programs for the Structure Determination of Gaseous Molecules from Electron Diffraction Data. *Acta. Chem. Scand.* **1969**, *23*, 3224–3234. [CrossRef]
12. Hamilton, W.C. Significance tests on the crystallographic R factor. *Acta. Cryst.* **1965**, *18*, 502–510. [CrossRef]
13. Usol'tseva, N.V.; Bykova, V.V.; Anan'eva, G.A.; Elkin, I.A.; Kudrik, E.V.; Shaposhnikov, G.P. Search for new discotic glassy materials for use in optoelectronics. *Liq. Cryst. Appl.* **2001**, *1*, 74–83. (In Russian)
14. Girichev, G.V.; Utkin, A.N.; Revichev, Y.F. Modernization of electron diffraction recorder EMR-100 for gas research. *Prib. Tekh. Eksp.* **1984**, *27*, 187–190. (In Russian)
15. Girichev, G.V.; Shlykov, S.A.; Revichev, Y.F. Equipment for studying the structure of molecules of valence-unsaturated compounds. *Prib. Tekh. Eksp.* **1986**, *4*, 167. (In Russian)
16. Girichev, E.G.; Zakharov, A.V.; Girichev, G.V.; Bazanov, M.I. Automation of a physicochemical experiment: Photometry and voltammetry. *Izv. Vyssh. Uchebn. Zaved. Tekstil. Prom.* **2000**, *2*, 142–146. (In Russian)
17. Frisch, M.J.; Trucks, G.W.; Schlegel, H.B.; Scuseria, G.E.; Robb, M.A.; Cheeseman, J.R.; Scalmani, G.; Barone, V.; Mennucci, B.; Petersson, G.A.; et al. *Gaussian 09*; Revision D.02; Gaussian, Inc.: Wallingford, CT, USA, 2009.
18. Becke, A.D. Density-functional thermochemistry. III. The role of exact exchange. *J. Chem. Phys.* **1993**, *98*, 5648–5653. [CrossRef]
19. Becke, A.D. Density-functional exchange-energy approximation with correct asymptotic behavior. *Phys. Rev. A* **1988**, *38*, 3098–3101. [CrossRef]
20. Lee, C.; Yang, W.; Parr, R.G. Development of the Colle-Salvetti correlation-energy formula into a functional of the electron density. *Phys. Rev. B* **1988**, *37*, 785–789. [CrossRef]
21. Perdew, J.P.; Burke, K.; Ernzerhof, M. Generalized Gradient Approximation Made Simple. *Phys. Rev. Lett.* **1996**, *77*, 3865–3868, Erratum in *Phys. Rev. Lett.* **1997**, *78*, 1396. [CrossRef]
22. Yanai, T.; Tew, D.P.; Handy, N.C. A new hybrid exchange–correlation functional using the Coulomb-attenuating method (CAM-B3LYP). *Chem. Phys. Lett.* **2004**, *393*, 51–57. [CrossRef]
23. Grimme, S. Semiempirical GGA-type density functional constructed with a long-range dispersion correction. *J. Comp. Chem.* **2006**, *27*, 1787–1799. [CrossRef] [PubMed]
24. Dunning, T.H., Jr. Gaussian basis sets for use in correlated molecular calculations. I. The atoms boron through neon and hydrogen. *J. Chem. Phys.* **1989**, *90*, 1007–1024. [CrossRef]

International Journal of
Molecular Sciences

MDPI

Article

Spin-Crossover and Slow Magnetic Relaxation Behavior in Hexachlororhenate(IV) Salts of Mn(III) Complexes [Mn(5-Hal-sal$_2$323)]$_2$[ReCl$_6$] (Hal = Cl, Br)

Aleksandra V. Tiunova [1,2], Anna V. Kazakova [2], Denis V. Korchagin [2,*], Gennady V. Shilov [2], Sergey M. Aldoshin [2], Aleksei I. Dmitriev [2], Mikhail V. Zhidkov [2], Konstantin V. Zakharov [3] and Eduard B. Yagubskii [2,*]

1 Faculty of Fundamental Physical and Chemical Engineering, Lomonosov Moscow State University, 119991 Moscow, Russia
2 Federal Research Center of Problems of Chemical Physics and Medicinal Chemistry RAS, 142432 Chernogolovka, Russia
3 Department of Low Temperature Physics and Superconductivity, Lomonosov Moscow State University, 119991 Moscow, Russia
* Correspondence: korden@icp.ac.ru (D.V.K.); yagubski@gmail.com (E.B.Y.)

Abstract: The cationic complexes of Mn(III) with the 5-Hal-sal$_2$323 (Hal = Cl, Br) ligands and a paramagnetic doubly charged counterion [ReCl$_6$]$^{2-}$ have been synthesized: [Mn(5-Cl-sal$_2$323)]$_2$[ReCl$_6$] (**1**) and [Mn(5-Br-sal$_2$323)]$_2$[ReCl$_6$] (**2**). Their crystal structures and magnetic properties have been studied. These isostructural two-component ionic compounds show a thermally induced spin transition at high temperature associated with the cationic subsystem and a field-induced slow magnetic relaxation of magnetization at cryogenic temperature, associated with the anionic subsystem. The compounds are the first examples of the coexistence of spin crossover and field-induced slow magnetic relaxation in the family of known [Mn^{III}(sal$_2$323)] cationic complexes with various counterions.

Keywords: spin-crossover; Mn^{III}(sal$_2$323) complexes; hexahalorhenates; polyfunctional compounds; magnetic properties; single-ion magnet

1. Introduction

At present, the main class of Mn(III) spin-crossover compounds are cationic complexes of Mn(III) with the sal$_2$323 ligand and its derivatives [1] (Scheme 1), which were first obtained in 2003 [2], and the first spin transition in this class was discovered in 2006 [3]. Since that time, active investigations of these complexes started, aimed at studying the effects of substituents on the phenolate nucleus of the ligand, the nature of counterions and lattice solvents on the spin behavior of the complexes. The sal$_2$323 ligand is a condensation product of two molecules of salicylaldehyde with a flexible tetraamine 1,2-bis(3-aminopropylamino)ethane, Scheme 1. The use of this tetramine leads to a trans-arrangement of the oxygen atoms of the sal$_2$323 ligand around the manganese ion. As a result, the octahedral configuration of Mn(III) is strongly distorted and axial contraction of the Mn-O bonds occurs. Such a molecular structure of Mn(III) complexes with ligands of the sal$_2$323 family contributes to the appearance of reversible spin transitions from the S = 1 state to the S = 2 state upon heating (spin-crossover (SCO) behavior), accompanied by the population of $d_{x^2-y^2}$ orbital. During the transition, the equatorial Mn-N bonds lengthen, while the axial Mn-O bonds remain practically unchanged.

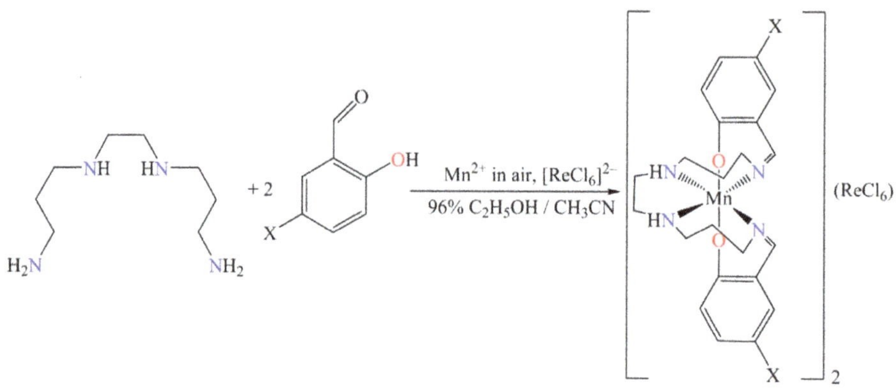

X= Cl (**1**)
X= Br (**2**)

Scheme 1. Synthesis of compounds **1** and **2**.

To date, several important results have been obtained in this class of Mn(III) spin-crossover complexes. Variation of substituents and counterions led to the synthesis of the complexes, which showed sharp spin transitions with significant hysteresis [4–13], including a conducting spin-crossover complex with an electroactive counterion TCNQ with a hysteresis width of 50 K, [Mn(5-Cl-$sal_2$323)]$TCNQ_{1.5}$·$2CH_3CN$ [7]. The Mn(III) complexes with a 3,5-dihalo-substituted $sal_2$323 ligand and tetraphenylborare as a counterion were recently synthesized and showed unusual structural and magnetic properties [8–12]. So, the dichlorine-containing complex displayed three structural phase transitions below 300 K and two spin crossovers coupling with structural transitions. One of the spin crossovers is characterized by the formation of an intermediate HS:LS phase from the initial high-spin state (HS), and the second is associated with the transition from the HS:LS phase to the low-spin state (LS) [11,12]. The giant magnetoelectric coupling was found at a Mn(III) spin crossover in [Mn(3,5-diBr-$sal_2$323]BPh_4 [14].

In this work, we used for the first time a doubly charged paramagnetic anion $[ReCl_6]^{2-}$ as a counterion in the Mn(III) cationic complexes with a monohalo-substituted ligand (5-Cl(Br)-$sal_2$323). To the best of our knowledge, all known complexes of the $[Mn(sal_2323)]^+$ family synthesized so far contain singly charged anions. An important argument in favor of choosing the $[ReCl_6]^{2-}$ as a counterion was the fact that this anion demonstrates a field-induced single-ion magnet behavior [15], like its analogs $[ReBr_6]^{2-}$, $[ReI_6]^{2-}$ and $[ReF_6]^{2-}$ [16–18]. Thus, one could expect the formation of complexes combining the Mn(III) spin crossover and $[ReCl_6]^{2-}$ single-ion magnetism in the same crystal lattice. Only a few compounds are known in the literature in which the spin transition coexists with single-molecule magnetism [19–27]. Two isostructural complexes $[Mn(5\text{-}Cl\text{-}sal_2323)]_2[ReCl_6]$ (**1**) and $[Mn(5\text{-}Br\text{-}sal_2323)]_2[ReCl_6]$ (**2**) were synthesized. Their crystal structures and magnetic properties have been studied. Herein, we present these results.

2. Results and Discussion

2.1. Synthesis

Complexes **1** and **2** were synthesized according to Scheme 1 in a solvent containing 96% ethanol and acetonitrile in a ratio of 1:1, see Section 3. Previously, the 5-Cl-$sal_2$323 and 5-Br-$sal_2$323 ligands were synthesized using the N,N′-bis(3-3-aminopropyl)ethylenediamine and the corresponding salicylaldehyde. Then, a solution containing $Mn(NO_3)_2$ $4H_2O$ and $K_2[ReCl_6]$ was added to the resulting solution of ligands. After mixing the two solutions, the color changed from yellow to dark red, which indicates the oxidation of Mn(II) to Mn(III). Black single crystals **1** and **2** were formed upon slow evaporation of the solvent.

2.2. Crystal Structures

The compounds **1** and **2** crystalize in the monoclinic system (centrosymmetric space group *C2/c*) and are isostructural. Both compounds do not undergo a temperature-induced phase transition in the studied range from 100 to 423 K. Electroneutral unit of **1** and **2** consists of two mononuclear cationic $[Mn(5\text{-}Cl(or\ Br)\text{-}sal_2323)]^+$ complexes and one $[ReCl_6]^{2-}$ counter anion (Figure 1). Asymmetric units are represented by halves of these moieties, the Mn and Re ions are on special positions. $ReCl_6{}^{2-}$ anions have a slightly distorted octahedral environment with bond lengths in a narrow range (2.3511(6)–2.3765(6) Å for **1** and 2.3505(8)–2.3760(7) Å for **2**). These values are in agreement with those in known compounds with $ReCl_6{}^{2-}$ anions [15,28].

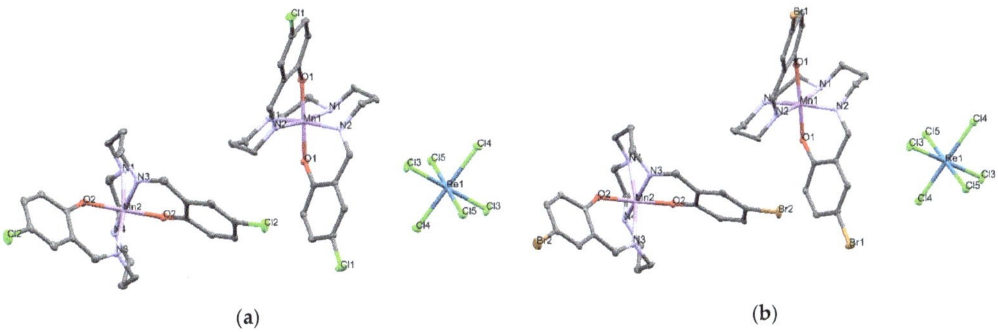

Figure 1. The general view of electroneutral units of compound **1** (**a**) and **2** (**b**). H atoms are omitted for clarity.

The $[Mn(5\text{-}Cl(or\ Br)\text{-}sal_2323)]^+$ complexes have a distorted hexacoordinated geometry with MnN_4O_2 coordination environment corresponding to the previously reported structures of cationic complexes $[Mn(R\text{-}sal_2323)]^+$ with similar hexadentate ligands [1]. The bond lengths Mn-N(or O) and octahedral distortion parameters (ζ, Σ and Θ) in coordination polyhedron $Mn^{III}N_4O_2$ of **1** and **2** (for different temperatures) are shown in Table 1. Two nonequivalent $[Mn(5\text{-}Cl(or\ Br)\text{-}sal_2323)]^+$ cations found in the structures of the compounds under consideration have significantly different Mn-N bond lengths and octahedral distortion parameters.

At 100 and 240 K, these nonequivalent Mn ions correspond to different spin states (LS and HS for Mn(1) and Mn(2), respectively). The average equatorial bond lengths $Mn–N_{im}$ and $Mn–N_{am}$ in complexes **1** and **2** fall in the range of that expected for LS and HS Mn(III) ions (Table 1) [1]. Upon increasing the temperature to 300K, the bond lengths Mn(1)–N show a slight increase in values that corresponds to the part of low-spin Mn(1) ions transit to a high-spin state (Table 1). A further gradual increase in temperature to 423 K reveals an increase in the bond lengths $Mn(1)–N_{im}$ and $Mn(1)–N_{am}$, which corresponds to an increase in the fraction of HS Mn(1) ions in the crystals, while these values do not reach values corresponding to the HS Mn(III) ion. For second nonequivalent ion Mn(2), the bond lengths and angles in coordination polyhedron correspond to the Mn(III) ion in HS state (Table 1) and are retained in all temperature ranges under consideration. The axial Mn–O bond lengths do not change significantly with increasing of temperature for both nonequivalent Mn ions.

Table 1. Selected bond lengths and octahedral distortion parameters [1] in **1** and **2**.

Compound	1											
Temp.	100 K		240 K		300 K		353 K		403 K		423 K	
Parameter	Mn1	Mn2	Mn1	Mn2	Mn1	Mn2	Mn1	Mn2	Mn1	Mn2	Mn1	Mn2
Mn-O, Å	1.874(2)	1.866(2)	1.871(2)	1.866(2)	1.868(2)	1.865(2)	1.866(2)	1.864(2)	1.865(2)	1.860(2)	1.864(2)	1.860(2)
Mn-N_{im}, Å	1.981(2)	2.091(2)	1.983(2)	2.098(2)	2.004(2)	2.101(2)	2.039(2)	2.099(2)	2.067(2)	2.097(3)	2.071(3)	2.098(3)
Mn-N_{am}, Å	2.046(2)	2.201(2)	2.053(2)	2.219(2)	2.080(2)	2.223(2)	2.114(3)	2.225(3)	2.149(3)	2.233(3)	2.157(3)	2.235(3)
Σ, °	43.60	63.75	45.02	68.08	50.52	68.79	60.90	70.29	66.31	70.74	68.56	70.54
Θ, °	124.47	198.15	129.35	216.06	143.62	221.00	170.69	228.08	186.97	229.90	193.20	230.65
ζ, Å	0.373	0.747	0.392	0.780	0.464	0.791	0.562	0.796	0.649	0.812	0.668	0.818
Compound	**2**											
Temp.	**100 K**		**240 K**		**300 K**		**353 K**		**403 K**		**423 K**	
Parameter	**Mn1**	**Mn2**	**Mn1**	**Mn2**	**Mn1**	**Mn2**	**Mn1**	**Mn2**	**Mn1**	**Mn2**	**Mn1**	**Mn2**
Mn-O, Å	1.871(2)	1.864(2)	1.872(2)	1.863(2)	1.873(2)	1.860(2)	1.864(2)	1.862(2)	1.862(3)	1.858(2)	1.860(3)	1.863(3)
Mn-N_{im}, Å	1.985(3)	2.095(3)	1.989(2)	2.100(2)	2.014(3)	2.099(3)	2.046(3)	2.099(3)	2.069(3)	2.096(3)	2.072(4)	2.100(4)
Mn-N_{am}, Å	2.045(3)	2.193(3)	2.053(2)	2.213(3)	2.087(3)	2.218(3)	2.122(3)	2.222(3)	2.149(4)	2.220(4)	2.154(4)	2.222(4)
Σ, °	43.19	68.82	46.12	68.82	53.14	69.75	61.31	71.38	67.56	71.81	68.55	72.38
Θ, °	125.73	215.36	132.81	215.36	149.32	220.22	172.39	226.58	191.11	230.89	193.61	230.63
ζ, Å	0.375	0.784	0.398	0.784	0.473	0.795	0.587	0.797	0.657	0.801	0.677	0.794

[1] Σ is the sum of the deviation from 90° of the 12 cis-angles in the coordination polyhedron. Θ is the sum of the deviation from 60° of the 24 trigonal angles of the projection of the octahedron onto the trigonal faces. ζ is the sum of deviation from individual M-N(or O) bond distances with respect to the mean M-X bond distance (the distance distortion parameter).

Besides the bond lengths changing, strong changes in the bond angles X-Mn-X are observed, which is well illustrated by the significant difference in the parameters of octahedral distortion (Σ and Θ) shown in Table 1. For complexes **1** and **2**, the octahedral distortion parameters for Mn(1) complexes at 100 and 240 K are close and corresponding to the Mn(III) ion in LS state [1]. Upon increasing the temperature from 240 to 423 K, the values of the octahedral distortion parameters for Mn(1) increase from ~45–46 to ~69 and from ~129–133 to ~193–194° for Σ and Θ, respectively. At 300K and higher, the octahedral distortion parameters for Mn(1) ions have intermediate values between ones for LS and HS Mn(III) ions, that may indicate on an incomplete gradual spin transition between LS (S = 1) and HS (S = 2) states of Mn(III) ion in complexes under consideration. At the same time, Mn(2) coordination polyhedra have the octahedral distortion parameters (Table 1) corresponding to the high-spin Mn(III) ion, which are not changed significantly with increasing of temperature from 100 to 423 K.

The crystal structures of **1** and **2**, besides electrostatic cation–anion interactions, are stabilized by intermolecular H-bonds and weak C-H ... π_{Ph} and C-H ... Hal intermolecular interactions. The key intermolecular interactions are non-covalent N–H· · · Hal interactions between $[Mn(5\text{-}Hal\text{-}sal_2323)]^+$ cations. Only Mn1 complexes that undergo a high-temperature spin transition are linked into chains with each other due to intermolecular N-H ... Cl(Br) hydrogen bonds along *c*-axes (Figure 2). The geometric parameters of these H-bonds are shown in Table 2.

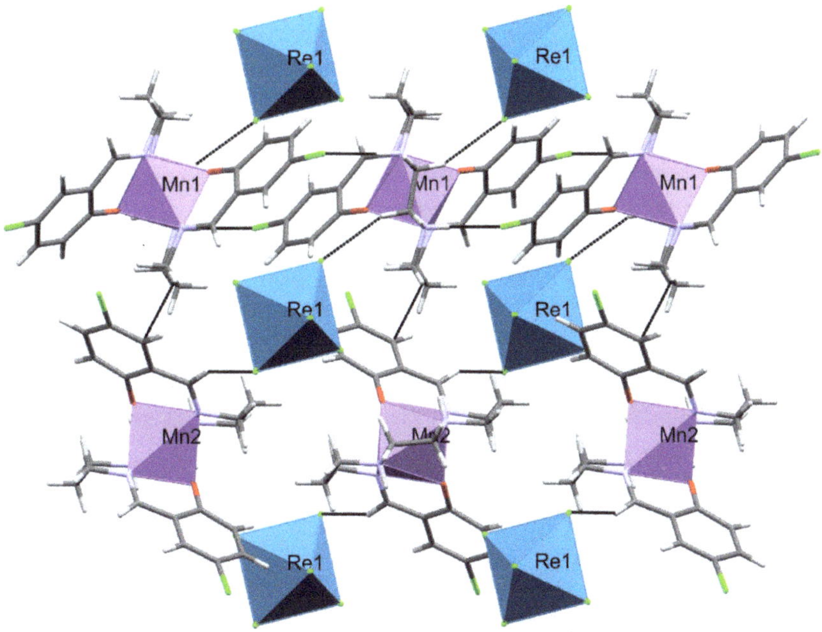

Figure 2. Projection of the fragment of the crystal structure of **1** (or **2**) on *ac* plane. Dashed lines show shortest intermolecular contacts.

Table 2. The geometric parameters of intermolecular H-bonds in crystal packing of **1** and **2**.

D-H ... A	d(H ... A), Å	d(D ... A), Å	<(D-H ... A), °
N-H ... Cl	2.70	3.41	141.5
N-H ... Br	2.77	3.48	140.5

The Mn2 complexes, which remain high-spin over the entire temperature range studied, weakly bound with neighboring the Mn2 complexes and $ReCl_6^{2-}$ anions due to C-H ... Cl(Br) contacts 2.77 (2.80 for **2**) Å and have also discrete intermolecular contacts C-H ... π_{Ph} 2.67 (2.67 for **2**) Å with the Mn1 complexes. Mn1 complexes have the same C-H ... Cl ($ReCl_6$) contacts 2.78 (2.80 for **2**) Å, which apparently do not influence the spin transition process in these salts. Thus, strong non-covalent intermolecular interactions between cations and anions that can play a significant role in the spin-state switching behavior are absent in crystal packing of **1** and **2** [1,6–12,29,30]. As a result, we can observe only an incomplete high-temperature spin transition in Mn1 complexes that are linked into chain by N-H ... Hal intermolecular H-bonds. The Mn and Re centers are well isolated from each other in the crystal structure, with the shortest intermolecular Mn...Mn, Mn...Re, and Re...Re distances being equal to 7.85(7.87 for **2**), 7.33(7.34), and 11.97(12.03) Å, respectively.

2.3. DC Magnetic Properties

Static magnetic susceptibility was measured on samples of complexes **1** and **2** in a field of 0.5 T at temperatures of 2–400 K. The phase purities of these bulk samples were confirmed by powder X-ray diffraction (Figure S1). Both compounds exhibit a very similar behavior (Figure 3). The $\chi_m T$ vs. T dependencies are close, and the curves remain the same upon heating and cooling. At temperature below 240 K, the $\chi_m T$ value slowly decreases, reaching 5.12 cm^3K/mol and 5.07 cm^3K/mol for **1** and **2**, respectively, at 100

K. These values are close to the calculated value (5.6 cm^3K/mol) for the sum of three magnetically non-interacting centers: Re(IV) ion with $\chi_m T$ = 1.6 cm^3K/mol (S = 3/2 with g = 1.80–1.90) [31,32] and two independent Mn (III) ions: Mn2 in high-spin (HS) state with $\chi_m T$ = 3.0 cm^3K/mol (S = 2, g = 2), and Mn1 in the low-spin (LS) state with $\chi_m T$ = 1.0 cm^3K/mol (S = 1, g = 2). There are good correlations between these magnetic data and the Mn-N bonds lengths at 100 K, which clearly indicate that the cations in complexes **1** and **2** are in different spin states: Mn1 is in the LS state, while Mn2 is in the HS state. With an increase in temperature from 240 K, a gradual increase in the $\chi_m T$ value is observed, which reaches values of 6.67 cm^3K/mol and 6.96 cm^3K/mol at 400 K for complexes **1** and **2**, respectively, indicating a gradual transition of the Mn1 cations in both complexes from the LS→HS states. However, by 400 K, complete spin transformation does not occur. An analysis of the crystal structures of **1** and **2** at temperatures above 400 K (423 K) showed that the Mn-N bond lengths continue to increase with a further increase in temperature. Thus, in both complexes at high temperatures, the Mn(III) cationic subsystem is in the high-spin state, while at low temperatures (below 240), an intermediate state is realized, in which half of the cations are in the high-spin state (Mn2) and the other (Mn1) in a low spin state.

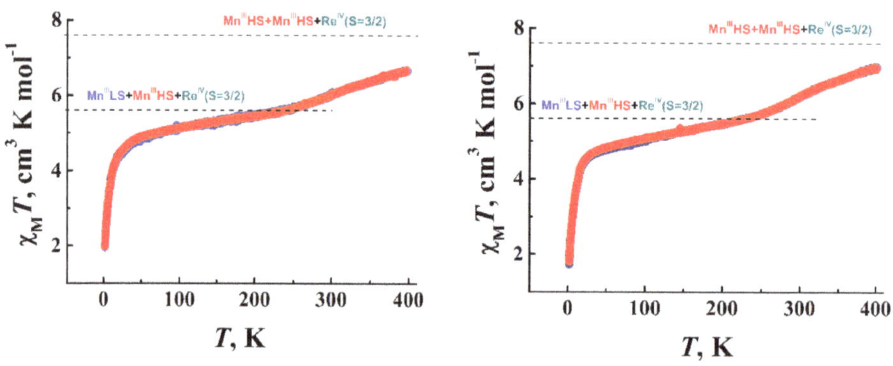

Figure 3. Temperature dependency of $\chi_m T$ product for complexes **1** (**a**) and **2** (**b**) upon cooling (blue) and heating (red).

Upon cooling below 100 K, the $\chi_m T$ values for these compounds continue to slowly decrease at first, and then, around 40-30 K, they decrease abruptly to reach 1.99 (**1**) and 1.80 cm^3K/mol (**2**) at 2 K. The abrupt decrease in the $\chi_m T$ values is mainly due to the strong magnetic anisotropy, which are observed in Re^{IV}- and [Mn^{III}($sal_2$323)]-based systems [31–35]. These $\chi_m T$ values at 2.0 K are comparable to the sum of the $\chi_m T$ values for the isolated Re(IV) complexes with halide ligands (approx. 1.0 cm^3K/mol) [15,16,32] and the Mn(III) complexes in mixed HS:LS = 1:1 state with the $sal_2$323 ligand family (approx. 0.7–0.9 cm^3K/mol) [9,11,36] at low temperature. Such a correspondence indicates that the antiferromagnetic coupling between the complexes is very weak in compounds **1** and **2**, which correlates with the strong isolation of Mn and Re centers in the crystal structure. It should be noted that the observed slow linear decrease in the $\chi_m T$ values upon cooling from 240 to 30 K is possibly associated with the contribution of the temperature-independent paramagnetism (TIP).

The field dependences of the magnetization (M vs. $\mu_0 H/T$, Figure 4) for complexes **1** and **2** were recorded at temperatures 2, 3, and 5 K in the field range of 0–5 T. For both complexes, the magnetization reaches a value of about 4.0 at 5.0 T without saturation. The low value of magnetization at 5.0 T, absence of saturation and the non-coincidence of M/μ_B vs. $\mu_0 H/T$ curves at different temperatures indicate the presence of considerable magnetic anisotropy in the complexes as a results to significant zero-field splitting effects inherent in the Re(IV) and Mn(III) ions [15–17,31–35,37–39].

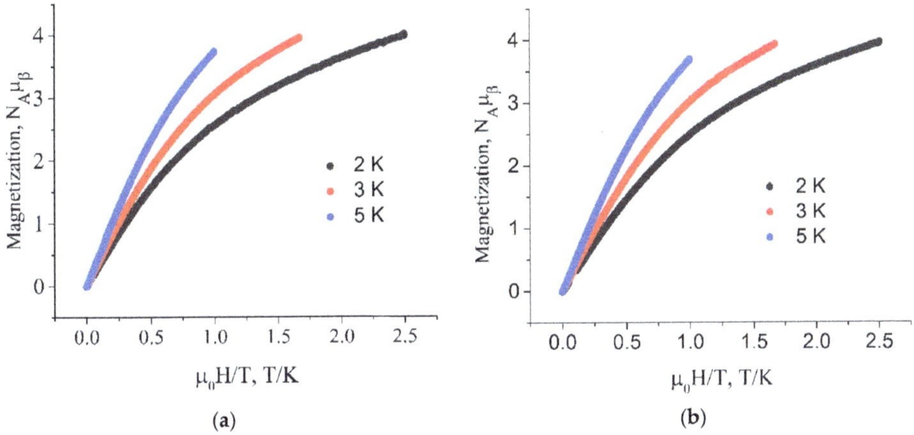

Figure 4. *Magnetization* vs. μ_0H/T plots for complexes **1** (**a**) and **2** (**b**) in the fields of 0–5 T and temperatures of 2, 3, and 5 K.

2.4. AC Magnetic Properties

In order to test whether the complexes **1** and **2** exhibit the properties of single-ion magnet (SIM), the ac susceptibility of both compounds was studied in the absence and at applied static magnetic field. The frequency dependences of the in-phase $\chi'(\nu)$ and non-zero out-of-phase $\chi''(\nu)$ signals of the ac susceptibility make it possible to test the relaxation mechanisms in magnetic systems [40,41]. Both compounds do not display the slow relaxation of the magnetization in the absence of the applied magnetic field; the χ'' signals are not observed. It is known that magnetic relaxation of SMMs is highly susceptible to quantum tunneling of magnetization (QTM), which can be suppressed by an external applied dc field. The field-dependencies of ac susceptibility (χ' and χ'') were recorded at 2 K in fields of 0–3000 Oe and two fixed frequencies (200 and 1000 Hz) in search for an optimal field to suppress the QTM effect. The value of the optimal magnetic field corresponds to the position of the maximum in the dependence of χ''(H). With an increase in the field, the χ'' increases and reaches some saturation at 3000 Oe (Figures S2 and S3). To test the relaxation behavior of compounds **1** and **2**, the ac susceptibility was studied in a field H_{dc} = 3000 Oe at frequencies of 100–10,000 Hz and temperatures of 2–3 K (Figure 5 and Figure S7). Frequency-dependent signals from both components of the ac susceptibility were observed at temperatures from 2.0 to 2.75 K, indicating field-induced slow relaxation of magnetization and, therefore, SIM phenomenon. However, the χ'' maxima lie above 10,000 Hz that is not feasible in our experimental set-up. Assuming that the relaxation of magnetization in **1** and **2** is Debye process driven by the thermal activation over the energy barrier U_{eff}, we estimated the values of energy barrier U_{eff} and τ_0, using the formula (1) [37,42].

$$\ln(\chi''/\chi') = \ln(\omega\tau_0) + U_{eff}/kT, \tag{1}$$

where ω = $2\pi\nu$–angular frequency, τ_0—preexponential factor in the Arrhenius law $\tau = \tau_0\exp(U_{eff}/kT)$, T—absolute temperature, τ—time relaxation, U_{eff}—the effective spin-reversal barrier, k—Boltzmann constant.

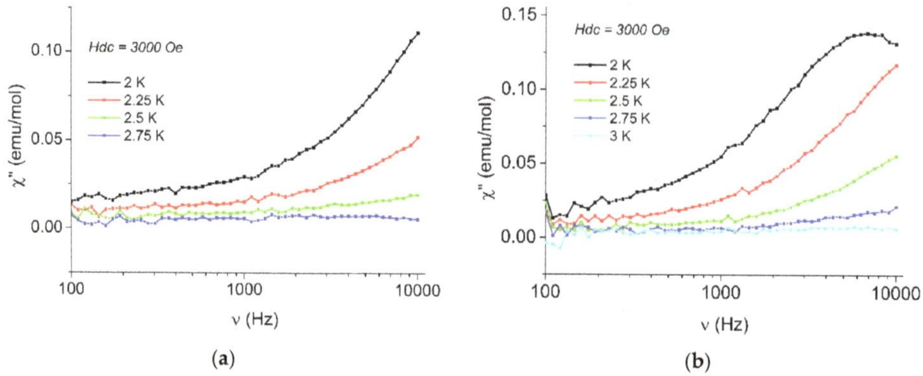

Figure 5. The frequency dependency of the χ'' for complexes **1** (**a**) and **2** (**b**) at temperatures of 2–3 K under H_{dc} = 3000 Oe.

The experimental plots $\ln(\chi''/\chi')$ vs. T^{-1} at different frequencies, their fitting by the formula (1) and values of parameters are presented in Figure 6. The values of the τ_0 parameter for **1** and **2** and other similar Re^{IV} complexes are close [16,17,38]. The average values of U_{eff} are 13(1) and 16(1) K for **1** and **2**, respectively. The U_{eff} values for **1** and **2** are similar between them, and close to that of compound $(Th_2imH^+)_2[ReCl_6]$ with diamagnetic photochromic cation Th_2imH^+ (U_{eff} = 12.21 before irradiation and 11.35 K after irradiation) [15]. At the same time, these values are higher than in the case of U_{eff} for $(Ph_4P)_2[ReI_6]$ and approximately two and three times lower than those for field-induced single-ion magnets $(Ph_4P)_2[ReX_6]$, X = F, Br, respectively [16,17]. Research on the hexahalorhenates $[Re^{IV}X_6]^{2-}$ (X = F, Cl, Br and I) has shown that the magnetic behavior of these anions depend greatly on the nature of the counterion [43]. For example, a study of the $[ReI_6]^{2-}$ ion with alkali metal cations, from Li^+ to Cs^+, has illustrated that none of these compounds is a SIM [39], same as in the case of hexahalorhenates with paramagnetic cations of ferrocenium [28] and metal oxazolidine nitroxides [43], whereas substitution of the perchlorate counterions in the Mn_6 SMM by $[Re^{IV}Cl_6]^{2-}$ led to the energy barrier for magnetization relaxation increasing by 30% [44].

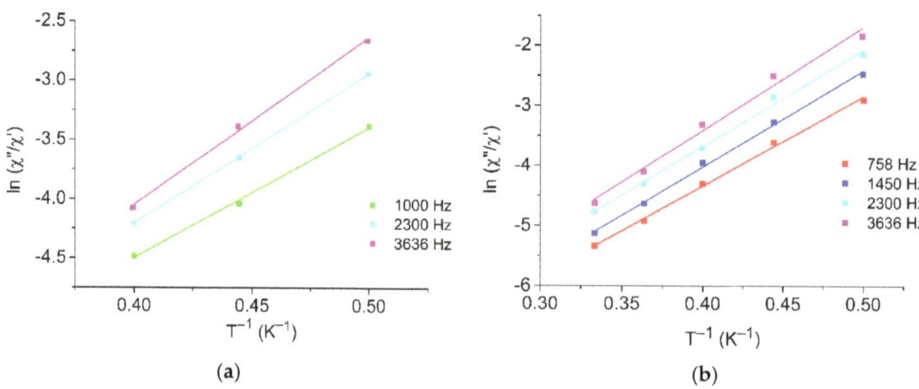

Figure 6. The plot of ratio $\ln(\chi''/\chi')$ vs. T^{-1} at different frequencies for **1** (**a**) and **2** (**b**). The color lines represent the fit by Equation (1) with values of parameters τ_0 = 8.1 10^{-10} s, 2.3 10^{-10} s, 1.0 10^{-10} s and U_{eff} = 11.0 K, 12.7 K, 14.2 K at ν = 1000, 2300, and 3636 Hz, respectively, for **1** and τ_0 =2.6 10^{-10} s, 1.1 10^{-10} s, 1.0 10^{-10} s, 5.5 10^{-11} s and U_{eff} = 15.0 K, 16.1 K, 16.2 K, 17.2 K at ν = 758, 1450, 2300, and 3636 Hz, respectively, for **2**.

3. Materials and Methods

The starting compounds and solvents were obtained from commercial sources and were used without further purification. The reactions related to the synthesis of the complexes were carried out in air. IR spectra were recorded on solid samples using a Bruker ALPHA spectrometer with the attenuated total reflectance (ATR) module. Elemental analyses were performed on a Vario MICRO cube (Elementar Analysensysteme GmbH, Langenselbold, Germany) equipment. Thermogravimetric analysis was carried out in a helium flow with a purity of 99.9999% using a STA 449 F3 Jupiter thermal analyzer. The heating rate was 2.0 °C/min.

3.1. Synthesis of [Mn(5-Cl-sal$_2$323)]$_2$[ReCl$_6$] (1)

5-Cl-salicylicaldehyde (0.5 mmol, 0.0783 g) was dissolved in acetonitrile-ethanol mixture (1:1, 15 mL) and added with vigorous stirring to N,N′-bis(aminopropyl)ethylenediamine (0.25 mmol, 0.0457 mL), previously dissolved in the same mixture of solvents (15 mL). After mixing the solutions, stirring was continued for 15 min. Manganese(II) nitrate tetrahydrate (0.25 mmol, 0.0627 g) and potassium hexachlororenate(IV) (0.125 mmol, 0.0596 g) were dissolved in the same solvent mixture (40 mL) with 5 mL distilled water added. After adding this solution to the ligand solution, the color changed from yellow to dark red. The solution continued to be stirred intensively for another 30 min and filtered out. Black single crystals precipitated from the solution within a week during the slow evaporation of the solvents. The crystals were filtered off and dried in air. Yield: 47%. Elemental analysis, calc. (%) for $C_{44}H_{52}Cl_{10}Mn_2N_8O_4Re$: C, 37.54, H, 3.72, N, 7.96; found: C, 36.96, H, 3.71, N, 7.85. IR (only intense bands, cm^{-1}): 3263, 2928, 2876, 1613, 1378, 1279, 1069, 985, 805, 704, 469 (Figure S4).

3.2. Synthesis of [Mn(5-Br-sal$_2$323)]$_2$[ReCl$_6$] (2)

The procedure for the synthesis of complex **2** is similar to the synthesis of complex **1**, except that 5-bromosalicylaldehyde was used instead of 5-chlorosalicylaldehyde. Yield: 52%. Elemental analysis, calc. (%) for $C_{44}H_{52}Br_2Cl_6Mn_2N_8O_4Re$: C, 33.33, H, 3.30, N, 7.07; found: C, 32.91, H, 3.32, N, 6.98. IR (only intense bands, cm^{-1}): 3249, 2928, 2871, 1613, 1374, 1279, 1071, 985, 828, 682, 461 (Figure S4).

The powder XRD analysis showed the phase purity of bulk samples of complexes **1** and **2** (Figure S1). Thermogravimetric analysis of both complexes showed that they are stable up to 220 °C (Figures S5 and S6).

3.3. Magnetic Measurements

Dc magnetic properties of powdered samples **1** and **2** were measured by a vibrating-sample magnetometer of a Cryogen Free Measurement System (Cryogenic Ltd., UK). The dependencies $M(T)$ were measured in cooling–heating conditions at T = 2.0–400 K in static magnetic field H = 0.5 T. The temperature change rate was 2 K/min. The magnetic moment of the samples took into account the contribution of the sample holder. Molar magnetic susceptibility χ was corrected taking into account a diamagnetic component of susceptibility according to the Pascal rule. Ac measurements carried out in a 3 Oe oscillating field on a magnetometer PPMS-9 (Quantum Design) in the absence and application of dc magnetic field.

3.4. X-ray Diffraction

X-ray diffraction data of **1** and **2** were collected at different temperatures (from 100 to 423 K) using an *Agilent XCalibur* diffractometer employing graphite monochromatic *MoK$_\alpha$* radiation (λ= 0.71073 Å). Experimental absorption corrections (SCALE3ABSPACK) were applied using the *CrysAlisPro* program [45] to all data. The structures were solved by direct methods and refined on F^2 by anisotropic full-matrix least-squares methods using *SHELXTL* program [46,47]. All non-hydrogen atoms were refined in the anisotropic approximation against F^2 for all reflections. Hydrogen-atom positions were obtained from difference Fourier syntheses and were refined with riding model constraints on their respective atoms. A summary of the crystallographic details of the X-ray analyses is given in Tables S1 and S2. Selected bond lengths and calculated using *OctaDist* software [48] octahedral distortion parameters are given in Table 1. The X-ray crystal structures data were deposited with the Cambridge Crystallographic Data Center (CCDC), with CCDC 2182503-2182508 and 2182509-2182514 reference codes for **1** and **2**, respectively. Powder X-ray diffraction patterns for **1** and **2** were registered on an Aeris diffractometer (Malvern PANalytical B.V., Netherlands) at room temperature. The comparison of powder XRD and single crystal data showed that polycrystalline samples **1** and **2** are monophase crystalline materials (Figure S1).

4. Conclusions

Multifunctional compounds based on spin-crossover complexes of the $[Mn^{III}(sal_2323)]^+$ family with the magnetic anion $[Re^{IV}Cl_6]^{2-}$: $[Mn(5\text{-}Cl\text{-}sal_2323)]_2[ReCl_6]$ (**1**) and $[Mn(5\text{-}Br\text{-}sal_2323)]_2[ReCl_6]$ (**2**) were synthesized. Complexes **1** and **2** are isostructural and contain in elemental unit two independent cations Mn1 and Mn2. The study of the crystal structures of the complexes at six temperatures (100, 240, 300, 350, 403, and 423 K) and their dc magnetic properties showed that cation Mn1 at 100 K is in the low-spin state with S = 1, and as the temperature rises to 423 K gradually passes into a high-spin state (S = 2), which is accompanied by an increase in the Mn-N bond lengths and a significant distortion of the $Mn1N_4O_2$ octahedral parameters (Table 1). In contrast to Mn1, cation Mn2 is in the HS state in the studied temperature range. Both components of ac susceptibility (χ' and χ'') exhibit frequency dependent signals when a constant magnetic field is applied. This behavior of the ac susceptibility indicates slow magnetic relaxation, meaning that complexes **1** and **2** are field-induced single-ion magnets. Until now, only one case of the manifestation of single-ion magnetic behavior of the $[Re^{IV}Cl_6]^{2-}$ anion has been known in the literature [15]. Compounds **1** and **2** are the first examples of the coexistence of a spin crossover at high temperatures and slow magnetic relaxation at low helium temperatures into the family of $[Mn^{III}(sal_2323)]$ cationic complexes. This result opens up prospects for creation of new multifunctional compounds combining spin transition and single molecular magnetism based on cationic complexes of the $[Mn^{III}(sal_2323)]^+$ with counterions–anionic single molecule magnets, proceeding from mononuclear complexes of Re(IV), Co(II), and Dy(III) with increased temperatures of magnetization barriers.

Supplementary Materials: The following supporting information can be downloaded at: https://www.mdpi.com/article/10.3390/ijms231911449/s1.

Author Contributions: Conceptualization, E.B.Y.; synthesis and characterization, A.V.T., A.V.K.; X-ray crystallography, formal analysis, D.V.K., G.V.S.; DC magnetometry experiment, formal analysis, A.I.D., M.V.Z.; AC magnetometry experiment, formal analysis, K.V.Z.; writing—original draft preparation, E.B.Y., D.V.K.; writing—review and editing, D.V.K., E.B.Y. and S.M.A.; supervision, project administration, E.B.Y. and S.M.A. All authors have read and agreed to the published version of the manuscript.

Funding: This research was funded by the Ministry of Science and Higher Education of the Russian Federation, grant number 075-15-2020-779.

Institutional Review Board Statement: Not applicable.

Informed Consent Statement: Not applicable.

Data Availability Statement: Not applicable.

Acknowledgments: This work was performed using the equipment of the Research Centre, IPCP RAS.

Conflicts of Interest: The authors declare no conflict of interest.

Sample Availability: Samples of compounds are available from the authors.

References

1. Olguin, J. Unusual metal centres/coordination spheres in spin crossover compounds. *Coord. Chem. Rev.* **2020**, *407*, 213148–213177. [CrossRef]
2. Panja, A.; Shaikh, N.; Gupta, S.; Butcher, R.; Banerjee, P. New Mononuclear Manganese(III) Complexes with Hexadentate (N_4O_2) Schiff Base Ligands: Synthesis, Crystal Structures, Electrochemistry, and Electron-Transfer Reactivity towards Hydroxylamine. *Eur. J. Inorg. Chem.* **2003**, 1540–1547. [CrossRef]
3. Morgan, G.G.; Murnaghan, K.D.; Muller-Bunz, H.; McKee, V.; Harding, C.J. A manganese(III) complex that exhibits spin crossover triggered by geometric tuning. *Angew. Chem. Int. Ed.* **2006**, *45*, 7192–7195. [CrossRef]
4. Martinho, P.N.; Gildea, B.; Harris, M.M.; Lemma, T.; Naik, A.D.; Meller-Bunz, H.; Keyes, T.E.; Garcia, Y.; Morgan, G.G. Cooperative Spin Transition in a Mononuclear Manganese(III) Complex. *Angew. Chem. Int. Ed.* **2012**, *51*, 12597–12601. [CrossRef] [PubMed]
5. Fitzpatrick, A.J.; Trzop, E.; Meller-Bunz, H.; Dirtu, M.M.; Garcia, Y.; Collet, E.; Morgan, G.G. Electronic *vs.* structural ordering in a manganese(III) spin crossover complex. *Chem. Commun.* **2015**, *51*, 17540–17543. [CrossRef] [PubMed]
6. Wang, S.; Li, Y.-J.; Ju, F.-F.; Xu, W.-T.; Kagesawa, K.; Li, Y.-H.; Yamashita, M.; Huang, W. The molecular and supramolecular aspects in mononuclear manganese(III) Schiff-base spin crossover complexes. *Dalton Trans.* **2017**, *46*, 11063–11077. [CrossRef]
7. Kazakova, A.V.; Tiunova, A.V.; Korchagin, D.V.; Shilov, G.V.; Yagubskii, E.B.; Zverev, V.N.; Yang, S.C.; Lin, J.-Y.; Lee, J.-F.; Maximova, O.V.; et al. The First Conducting Spin-Crossover Compound Combining a Mn^{III} Cation Complex with Electroactive TCNQ Demonstrating an Abrupt Spin Transition with a Hysteresis of 50K. *Chem. Eur. J.* **2019**, *25*, 10204–10213. [CrossRef]
8. Ghosh, S.; Bagchi, S.; Kamilya, S.; Mondal, A. Stepwise spin-state switching in a manganese(III) complex. *Dalton Trans.* **2020**, *49*, 14776–14780. [CrossRef]
9. Jakobsen, V.B.; Trzop, E.; Gavin, C.L.; Dobbelaar, E.; Chikara, S.; Ding, X.; Esien, K.; Muller-Bunz, H.; Felton, S.; Zapf, V.S.; et al. Stress-Induced Domain Wall Motion in a Ferroelastic Mn^{3+} Spin Crossover Complex. *Angew. Chem. Int. Ed.* **2020**, *59*, 13305–13312. [CrossRef]
10. Ghosh, S.; Bagchi, S.; Kamilya, S.; Mondal, A. Effect of ligand substituents and tuning the spin-state switching in manganese(III) complexes. *Dalton Trans.* **2021**, *50*, 4634–4642. [CrossRef]
11. Tiunova, A.V.; Kazakova, A.V.; Korchagin, D.V.; Shilov, G.V.; Zorina, L.V.; Simonov, S.V.; Zakharov, K.V.; Vasiliev, A.N.; Yagubskii, E.B. Abrupt Spin-State Switching in Mn(III) Complexes with BPh_4 Anion: Effect of Halide Substituents on Crystal Structure and Magnetic Properties. *Chem. Eur. J.* **2021**, *27*, 17609–17619. [CrossRef] [PubMed]
12. Jakobsen, V.B.; Trzop, E.; Dobbelaar, E.; Gavin, C.L.; Chikara, S.; Ding, X.; Lee, M.; Esien, K.; Muller-Bunz, H.; Felton, S.; et al. Domain Wall Dynamics in a Ferroelastic Spin Crossover Complex with Giant Magnetoelectric Coupling. *J. Am. Chem. Soc.* **2022**, *144*, 195–211. [CrossRef] [PubMed]
13. Dobbelaar, E.; Jakobsen, V.B.; Trzop, E.; Lee, M.; Chikara, S.; Ding, X.; Muller-Bunz, H.; Esien, K.; Felton, S.; Carpenter, M.A.; et al. Thermal and Magnetic Field Switching in a Two-Step Hysteretic Mn^{III} Spin Crossover Compound Coupled to Symmetry Breakings. *Angew. Chem. Int. Ed.* **2022**, *61*, e202114021. [CrossRef]
14. Jakobsen, V.B.; Chikara, S.; Yu, J.-X.; Dobbelaar, E.; Kelly, C.T.; Ding, X.; Weickert, F.; Trzop, E.; Collet, E.; Cheng, H.-P.; et al. Giant magnetoelectric coupling and magnetic-field-induced permanent switching in a spin crossover Mn(III) complex. *Inorg. Chem.* **2021**, *60*, 6167–6175. [CrossRef] [PubMed]
15. Gong, D.-P.; Chen, J.-F.; Zhao, Y.; Cao, D.-K. Bisthienylethene Th_2im and its complex $(Th_2imH)_2$ $[ReCl_6]$: Crystalline-phase photochromism, and photochemical regulation of luminescence and magnetic properties. *Dalton Trans.* **2016**, *45*, 3443–3449. [CrossRef] [PubMed]
16. Rojas-Dotti, C.; Sanchis-Perucho, A.; Orts-Arroyo, M.; Moliner, N.; González, R.; Lloret, F.; Martínez-Lillo, J. Field-Induced Single-Ion Magnet Phenomenon in Hexabromo- and Hexaiodorhenate(IV) Complexes. *Magnetochemistry* **2020**, *6*, 20. [CrossRef]
17. Pedersen, K.S.; Sigrist, M.; Sorensen, M.A.; Barra, A.-L.; Weyhermüller, T.; Piligkos, S.; Thuesen, C.A.; Vinum, M.G.; Mutka, H.; Weihe, H.; et al. $[ReF_6]^{2-}$: A Robust Module for the Design of Molecule-Based Magnetic Materials. *Angew. Chem. Int. Ed.* **2014**, *53*, 1351–1354. [CrossRef]
18. Kushch, N.D.; Buravov, L.I.; Kushch, P.P.; Shilov, G.V.; Yamochi, H.; Ishikawa, M.; Otsuka, A.; Shakin, A.A.; Maximova, O.V.; Volkova, O.S.; et al. Multifunctional Compound Combining Conductivity and Single-Molecule Magnetism in the Same Temperature Range. *Inorg. Chem.* **2018**, *57*, 2386–2389. [CrossRef]
19. Urtizberea, A.; Roubeau, O. Switchable slow relaxation of magnetization in the native low temperature phase of a cooperative spin-crossover compound. *Chem. Sci.* **2017**, *8*, 2290–2295. [CrossRef]
20. Feng, X.; Mathoniere, C.; Jeon, I.-R.; Rouzieres, M.; Ozarowski, A.; Aubrey, M.L.; Gonzalez, M.I.; Clerac, R.; Long, J.R. Tristability in a Light-Actuated Single-Molecule Magnet. *J. Am. Chem. Soc.* **2013**, *135*, 15880–15884. [CrossRef]
21. Mathoniere, C.; Lin, H.-J.; Siretanu, D.; Clerac, R.; Smith, J.M. Photoinduced Single-Molecule Magnet Properties in a Four-Coordinate Iron(II) Spin Crossover Complex. *J. Am. Chem. Soc.* **2013**, *135*, 9083–19086. [CrossRef] [PubMed]
22. Mossin, S.; Tran, B.L.; Adhikari, D.; Pink, M.; Heinemann, F.W.; Sutter, J.; Szilagyi, R.K.; Meyer, K.; Mindiola, D.J. A Mononuclear Fe(III) Single Molecule Magnet with a $3/2\leftrightarrow5/2$ Spin Crossover. *J. Am. Chem. Soc.* **2012**, *134*, 13651–13661. [CrossRef] [PubMed]
23. Gass, I.A.; Tewary, S.; Nafady, A.; Chilton, N.F.; Gartshore, C.J.; Asadi, M.; Lupton, D.W.; Moubaraki, B.; Bond, A.M.; Boas, J.F.; et al. Observation of Ferromagnetic Exchange, Spin Crossover, Reductively Induced Oxidation, and Field-Induced Slow Magnetic Relaxation in Monomeric Cobalt Nitroxides. *Inorg. Chem.* **2013**, *52*, 7557–7572. [CrossRef]

24. Cui, H.-H.; Wang, J.; Chen, X.-T.; Xue, Z.-L. Slow magnetic relaxation in five-coordinate spin-crossover cobalt(ii) complexes. *Chem. Commun.* **2017**, *53*, 9304–9307. [CrossRef]
25. Shao, D.; Deng, L.-D.; Shi, L.; Wu, D.-Q.; Wei, X.-Q.; Yang, S.-R.; Wang, X.-Y. Slow Magnetic Relaxation and Spin-Crossover Behavior in a Bicomponent Ion-Pair Cobalt (II) Complex. *Eur. J. Inorg. Chem.* **2017**, *2017*, 3862–3867. [CrossRef]
26. Palion-Gazda, J.; Machura, B.; Kruszynski, R.; Grancha, T.; Moliner, N.; Lloret, F.; Julve, M. Spin Crossover in Double Salts Containing Six- and Four-Coordinate Cobalt(II). *Inorg. Chem.* **2017**, *56*, 6281–6296. [CrossRef]
27. Yoon, J.H.; Lim, K.S.; Ryu, D.W.; Lee, W.R.; Yoon, S.W.; Suh, B.J.; Hong, C.S. Synthesis, Crystal Structures, and Magnetic Properties of Cyanide-Bridged W^VMn^{III} Anionic Coordination Polymers Containing Divalent Cationic Moieties: Slow Magnetic Relaxations and Spin Crossover Phenomenon. *Inorg. Chem.* **2014**, *53*, 10437–10442. [CrossRef]
28. González, R.; Chiozzone, R.; Kremer, C.; Guerra, F.; De Munno, G.; Lloret, F.; Julve, M.; Faus, J. Magnetic Studies on Hexahalorhenate(IV) Salts of Ferrocenium Cations $[Fe(C_5R_5)_2]_2[ReX_6]$ (R = H, CH_3; X = Cl, Br, I). *Inorg. Chem.* **2004**, *43*, 3013–3019. [CrossRef] [PubMed]
29. Gildea, B.; Gavin, L.C.; Murray, C.A.; Meller-Bunz, H.; Harding, C.J.; Morgan, G.G. Supramolecular Modulation of Spin Crossover Profile in Manganese(III). *Supramol. Chem.* **2012**, *24*, 641–653. [CrossRef]
30. Wang, S.; Xu, W.-T.; He, W.-R.; Takaishi, S.; Li, Y.-H.; Yamashita, M.; Huang, W. Structural insights into the counterion effects on the manganese(III) spin crossover system with hexadentate Schiff-base ligands. *Dalton Trans.* **2016**, *45*, 5676–5688. [CrossRef]
31. Chiozzone, R.; González, R.; Kremer, C.; De Munno, G.; Cano, J.; Lloret, F.; Julve, M.; Faus, J. Synthesis, Crystal Structure, and Magnetic Properties of Tetraphenylarsonium Tetrachloro(oxalato)rhenate(IV) and Bis(2,2′-bipyridine)tetrachloro(μ-oxalato)copper(II)rhenium(IV). *Inorg. Chem.* **1999**, *38*, 4745–4752. [CrossRef] [PubMed]
32. Martínez-Lillo, J.; Faus, J.; Lloret, F.; Julve, M. Towards multifunctional magnetic systems through molecular-programmed self assembly of Re(IV) metalloligands. *Coord. Chem. Rev.* **2015**, *289*, 215–237. [CrossRef]
33. Kühne, I.A.; Ozarowski, A.; Sultan, A.; Esien, K.; Carter, A.B.; Wix, P.; Casey, A.; Heerah-Booluck, M.; Keene, T.D.; Müller-Bunz, H.; et al. Homochiral Mn^{3+} Spin-Crossover Complexes: A Structural and Spectroscopic Study. *Inorg. Chem.* **2022**, *61*, 3458–3471. [CrossRef]
34. Ghosh, S.; Bagchi, S.; Kamilya, S.; Mehta, S.; Sarkar, D.; Herchel, R.; Mondal, A. Impact of counter anions on spin-state switching of manganese(III) complexes containing an azobenzene ligand. *Dalton Trans.* **2022**, *51*, 7681–7694. [CrossRef] [PubMed]
35. Barker, A.; Kelly, C.T.; Kühne, I.A.; Hill, S.; Krzystek, J.; Wix, P.; Esien, K.; Felton, S.; Müller-Bunz, H.; Morgan, G.G. Spin state solvomorphism in a series of rare $S = 1$ manganese(III) complexes. *Dalton Trans.* **2019**, *48*, 15560–15566. [CrossRef] [PubMed]
36. Qin, C.-Y.; Zhao, S.-Z.; Wu, H.-Y.; Li, Y.-H.; Wang, Z.-K.; Wang, Z.; Wang, S. The dynamic interplay between intramolecular and intermolecular interactions in mononuclear manganese(III) SCO complexes. *Dalton Trans.* **2021**, *50*, 5899–5910. [CrossRef]
37. Martínez-Lillo, J.; Mastropietro, T.F.; Lhotel, E.; Paulsen, C.; Cano, J.; De Munno, G.; Faus, J.; Lloret, F.; Julve, M.; Nellutla, S.; et al. Highly Anisotropic Rhenium(IV) Complexes: New Examples of Mononuclear Single-Molecule Magnets. *J. Am. Chem. Soc.* **2013**, *135*, 13737–13748. [CrossRef]
38. Feng, X.; Liu, J.-L.; Pedersen, K.S.; Nehrkorn, J.; Schnegg, A.; Holldack, K.; Bendix, J.; Sigrist, M.; Mutka, H.; Samohvalov, D.; et al. Multifaceted magnetization dynamics in the mononuclear complex $[Re^{IV}Cl_4(CN)_2]^{2-}$. *Chem. Commun.* **2016**, *52*, 12905–12908. [CrossRef]
39. González, R.; Chiozzone, R.; Kremer, C.; De Munno, G.; Nicolò, F.; Lloret, F.; Julve, M.; Faus, J. Magnetic Studies on Hexaiodorhenate(IV) Salts of Univalent Cations. Spin Canting and Magnetic Ordering in $K_2[ReI_6]$ with T_c=24 K. *Inorg. Chem.* **2003**, *42*, 2512–2518. [CrossRef]
40. Gatteschi, D.; Sessoli, R.; Villain, J. *Molecular Nanomagnets*; Oxford University Press: Oxford, UK, 2006.
41. Benelli, C.; Gatteschi, D. *Introduction to Molecular Magnetism*; Wiley-VCH: Weinheim, Germany, 2015.
42. Bartolomé, J.; Filoti, G.; Kuncser, V.; Schinteie, G.; Mereacre, V.; Anson, C.E.; Powell, A.K.; Prodius, D.; Turta, C. Magnetostructural correlations in the tetranuclear series of $\{Fe_3LnO_2\}$ butterfly core clusters: Magnetic and Mössbauer spectroscopic study. *Phys. Rev. B* **2009**, *80*, 014430. [CrossRef]
43. Pedersen, A.H.; Geoghegan, B.L.; Nichol, G.S.; Lupton, D.W.; Murray, K.S.; Martínez-Lillo, J.; Gass, I.A.; Brechin, E.K. Hexahalorhenate(IV) salts of metal oxazolidine nitroxides. *Dalton Trans.* **2017**, *46*, 5250–5259. [CrossRef] [PubMed]
44. Martinez-Lillo, J.; Cano, J.; Wernsdorfer, W.; Brechin, E.K. The Effect of Crystal Packing and ReIV Ions on the Magnetization Relaxation of $[Mn_6]$-Based Molecular Magnets. *Chem. Eur. J.* **2015**, *21*, 8790–8798. [CrossRef] [PubMed]
45. *CrysAlisPro, Version 1.171.33*; Oxford Diffraction Ltd.: Oxford, UK, 2010.
46. Sheldrick, G.M. *SHELXTL v. 6.14, Structure Determination Software Suite*; Bruker AXS: Madison, WI, USA, 2000.
47. Sheldrick, G.M. A short history of SHELX. *Acta Crystallogr. Sect. A* **2008**, *64*, 112–122. [CrossRef] [PubMed]
48. Ketkaew, R.; Tantirungrotechai, Y.; Harding, P.; Chastanet, G.; Guionneau, P.; Marchivie, M.; Harding, D.J. *OctaDist*: A tool for calculating distortion parameters in spin crossover and coordination complexes. *Dalton Trans.* **2021**, *50*, 1086–1096. [CrossRef] [PubMed]

International Journal of
Molecular Sciences

MDPI

Communication

How Do Intermolecular Interactions Evolve at the Nematic to Twist–Bent Phase Transition?

Katarzyna Merkel [1], Barbara Loska [1], Yuki Arakawa [2], Georg H. Mehl [3], Jakub Karcz [4] and Antoni Kocot [1,*]

1 Institute of Materials Engineering, Faculty of Science and Technology, University of Silesia, ul. 75 Pułku Piechoty, 41-500 Chorzów, Poland
2 Department of Applied Chemistry and Life Science, Graduate School of Engineering, Toyohashi University of Technology, Toyohashi 441-8580, Japan
3 Department of Chemistry, University of Hull, Hull HU6 7RX, UK
4 Faculty of Advanced Technologies and Chemistry, Military University of Technology, 00-908 Warszawa, Poland
* Correspondence: antoni.kocot@us.edu.pl; Tel.: +48-32-3497630

Abstract: Polarized beam infrared (IR) spectroscopy provides valuable information on changes in the orientation of samples in nematic phases, especially on the role of intermolecular interactions in forming the periodically modulated twist–bent phase. Infrared absorbance measurements and quantum chemistry calculations based on the density functional theory (DFT) were performed to investigate the structure and how the molecules interact in the nematic (N) and twist–bend (N_{TB}) phases of thioether dimers. The nematic twist–bend phase observed significant changes in the mean IR absorbance. On cooling, the transition from the N phase to the N_{TB} phase was found to be accompanied by a marked decrease in absorbance for longitudinal dipoles. Then, with further cooling, the absorbance of the transverse dipoles increased, indicating that transverse dipoles became correlated in parallel. To investigate the influence of the closest neighbors, DFT calculations were performed. As a result of the optimization of the molecular cores system, we observed changes in the square of the transition dipoles, which well corresponds to absorbance changes observed in the IR spectra. Interactions of molecules dominated by pairing were observed, as well as the axial shift of the core to each other.

Keywords: FTIR spectroscopy; DFT simulations; intermolecular interactions; liquid crystal dimers; twist–bend phase; self-assembling

1. Introduction

One of the main goals of soft matter theory is to understand the structure–property relationship of material systems so that the specific properties of the materials can be obtained through molecular design. Self-organization is the process by which molecules spontaneously arrange themselves into stable and ordered structures due to noncovalent interactions. This process can occur on all scales, ranging from the nanoscopic, involving atoms or molecules, to cosmic-sized objects. These spontaneous organizations can deliver information on mounting components, i.e., their shape, charge, polarizability, magnetic dipole, mass, etc., as these properties determine their interactions [1]. Molecular self-assembly studies provide essential information about the influence of intermolecular interactions on the structure of the investigated molecular system. Therefore, such studies are crucial when anisotropic molecular systems can lead to direction-dependent physical properties. One of the fascinating examples of molecular self-organization is the emergence of chirality. The formation of chiral superstructures and phases is an area of great interest from theoretical and practical perspectives [2]. In the biological context, helical structures are found in DNA and proteins; twisted beta sheets form helical columns [3], similar to silk and its synthetic analogs reported in [4,5]. Materials science examples include cholesteric liquid crystalline structures [6] and chiral smectic phases [7], which are formed by bent-core molecules. The most famous example of the last decade of creating chiral structures using nonchiral molecules is the (spatially modulated) twist–bend nematic phase (N_{TB}), which has a pitch length of several nanometers [8–12].

The experimental study of the structure of the N_{TB} phase has mainly been carried out with nonresonant (SAXS, WAXS) [13–15] and resonant X-ray scattering [10,11,16–19], freeze–fracture transmission electron and atomic force microscopy (FFTEM, AFM) [8,9,20–22], polarized Raman [23], infrared [24–26], and nuclear magnetic resonance spectroscopy (NMR) [27–31], as well as with optical methods under applied electric fields [32–38] and magnetic fields [39]. The formation of the N_{TB} structure has been modeled using rigid elements [15,40,41]. Some models [15,41] assume the nanophase segregation of the assembling elements: flexible central alkyl linker, terminal chains, and the mesogenic cores; molecular end groups with the flexible central spacers provide orientational freedom. Finally, the X-ray evidence of the half-molecular length periodicity along the N_{TB} helix led to a model based on the self-assembly of half-molecule-long blocks as the basic element of the N_{TB} phase [10,18,42]. The N_{TB} materials that have been most investigated are the cyanobiphenyl (CB)-based liquid crystal dimers (CBCnCB; n = 7, 9, 11) [8–11,13,16,21,23,27–33,36–39], where cyanobiphenyl units prefer an antiparallel arrangement due to dipole–dipole interactions, [43–46]. Up-to-date research has confirmed that molecular curvature is important for phase formation, but its stability increases as the bending angle decreases [47–52]. There have also been important reports suggesting that such factors also influence the N–N_{TB} phase transition as intramolecular torsion [17,31,53], conformational changes [54–58], bend angle fluctuations [59], the effect of free volume [54,60,61], as well as intermolecular interactions [25,62–66].

It has long been known that hydrogen bonds and the π–π intermolecular interactions play a crucial role in supramolecular biological phenomena such as base-pairing in double-stranded DNA, protein binding, cell–cell recognition, and viral infection [67,68]. A similar phenomenon is observed in the field of soft matter, where unconventional intermolecular interactions lead to the emergence of new supramolecular liquid crystal structures [69,70], such as the twist–bend phase (N_{TB}) [71] or the polar–twisted phase (N_{PT}) [72], as well as the newly discovered ferroelectric nematic phase (N_F) [73,74].

This work is devoted to observations of the intermolecular interactions and their impact on the structural changes that occur during the transition from the nematic phase to the twist–bend phase. For this purpose, infrared spectroscopy was an excellent tool for analyzing orientational order in a molecular system. In the case of conventional nematics, it is usually assumed that the transition dipole moments that correspond to a specific vibration are temperature-independent. As absorbance is related to the square of the transition dipole moment, the average absorbance is expected to be density-dependent. Density, in turn, is typically temperature-dependent. In the N_{TB} phase, however, the average absorbance of several molecular vibrations exhibits a specific behavior due to the molecular system's self-assembly. Experimental data indicate that short-range interactions grow significantly, and thus the intercorrelations between the transition dipoles of the interacting functional groups become more important. Moreover, the longitudinal dipoles' behavior differs from those of the transversal dipoles.

In the paper, we report combined FTIR and DFT studies of intermolecular interaction for two groups of dimers: a symmetric system with the thioether-linking groups (C-S-C) termed CBSCnSCB (n = 5, 7) and an asymmetric system with the ether- and thioether-linking groups, termed CBSCnOCB (n = 5, 7). A symmetric dimer with ether-linking groups (C-O-C) with the acronym CBOCnOCB (n = 7) is chosen for reference.

2. Results

2.1. Temperature Dependencies of the Absorbance

By measuring the absorbance in polarized light, the temperature dependencies of the absorbance components in the temperature range of the N and the N_{TB} phases could be directly analyzed [24–26]. The experiment will distinguish the different behavior of the absorbance components for the bands with longitudinal and transverse dipoles in the molecular frame. In an orientationally ordered material, the absorbance components are dependent on the angle between the alignment axis and the polarization direction of the incident beam. At a microscopic level, the infrared absorption depends on the angle between the molecular transition dipole moment μ_i of the particular absorption band and the polarization of the IR beam. The average IR absorbance $A_0 = (A_X + A_Y + A_Z)/3$ of the specific vibrational modes is determined by how the electric dipole moment of the system changes with the atomic oscillations. To the lowest order, the required quantities are proportional to the derivatives of the dipole moment with respect to the system's vibrational normal modes, i, evaluated at the equilibrium geometry. The IR absorbance of the i_{th} vibrational mode is given by [75]:

$$A_i = \int_{\nu 1}^{\nu 2} A(\nu)d\nu = \frac{n\pi}{3c}\left[\frac{d\mu_i}{dQ_i}\right]^2 \quad (1)$$

where n is the number of molecules per unit volume (molecular density) and μ_i is the molecule transition dipole moment for the normal coordinate, Q_i, corresponding to the ith mode.

The temperature dependence of the average absorbance, A_0, and the corresponding transition dipoles was analyzed for all dimers in the N and the N_{TB} phases. In the range of the N phase, the average absorbance increased with the molecular density, which confirmed that the transition dipole seemed to be constant in this temperature range. At temperatures below the N_{TB} transition, the behavior of symmetric and asymmetric dimers was different and, in some cases, was determined by cell windows [26].

A detailed analysis of the infrared spectra compared to those simulated for different conformers, along with the exact assignments of vibrations, is presented in a separate paper [76]. Several vibrational bands that belonged to the longitudinal and transverse transition dipole that showed significant dichroism of the band were selected to be analyzed. For the longitudinal transition dipole moment, the phenyl stretching band (νCC) at 1600 cm^{-1} and the C-H deformation vibration in the benzene plane at 1100 cm^{-1} (βCH) were selected. The band at 1100 cm^{-1} was a vibration that involved a sulfide group, so it may be a good indicator of the dimer's conformational change [76]. Figure 1 shows the temperature dependencies of the average absorbances, A_0, for the bands corresponding to the longitudinal transition dipole. We introduced the nematic phase trend line as a reference: the solid black line in Figure 1. There was a clear transition from the conventional N phase to the N_{TB} for all of the molecules, except for CBOC7OCB, which behaves like a classical calamitic molecule. This is attributed to its large opening angle and which is responsible for not forming the N_{TB} phase. The absorbances involving the biphenyl group (1600 cm^{-1} and 1100 cm^{-1} in Figure 1), and also the band at 2220 cm^{-1} (that involved a cyan group), show a significant decrease just after entering the N_{TB} phase (see Table 1). This indicates the structure rearrangement in the N_{TB} phase, resulting in the antiparallel orientation of the neighboring cyanobiphenyl groups. Both dipole–dipole interaction and London dispersion forces are likely involved in this process that stabilizes the N_{TB} phase. It seems the nanosegregation process proceeds along the Z-axis of the system. A similar conclusion has already been presented for cyanobiphenyl-based LCs, showing an overlap of the termini of the molecules [44–46,61]. Following the transition to the N_{TB}, a significant decrease in the average absorbance was measured, suggesting the presence of dipole correlations in this phase and the presence of other molecular interactions related to the change in the system's geometry.

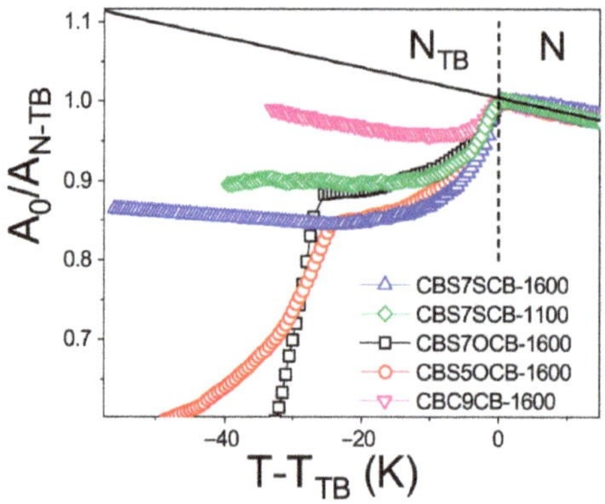

Figure 1. Normalized IR absorbance vs. temperature behavior of the dimers for the longitudinal dipole (μ): benzene ring vibration at 1600 cm^{-1} (νCC) and a deformation vibration of the C-H group in the benzene ring plane at 1100 cm^{-1} (βCH ip CB + $\nu_{as}C_{Ar}S$). □–CBSC7OCB, ○–CBSC5OCB, △–CBSC7SCB (1600 cm^{-1}), and ▽–CBC9CB; ◇–CBSC7SCB (1100 cm^{-1}); solid black line—nematic phase trend line. A_0—integral average absorbance (cm^{-1}), $A_{N\text{-}TB}$—integral absorbance in the transition from nematic to twist–bend phase (cm^{-1}).

Table 1. The absorbance changes on transition from the N to the N_{TB} with calculated ratio $\mu^2_{IMF}/\mu^2_{noIMF}$ in the presence of weak intermolecular interactions and local orientational order.

Vibration Frequency (cm^{-1})	$A_0/A_{N\text{-}TB}$	$\mu^2_{IMF}/\mu^2_{noIMF}$ *	$A_0/A_{N\text{-}TB}$	$A_0/A_{N\text{-}TB}$	$\mu^2_{IMF}/\mu^2_{noIMF}$ *	$A_0/A_{N\text{-}TB}$
	CBSC7SCB	**CBSC7SCB**	**CBC5OCB**	**CBC7OCB**	**CBC7OCB**	**CBC9CB**
520	1.25	1.60	1.42	1.48	1.54	–
810	1.22	1.64	1.32	1.33	1.41	1.05
1100	0.94	0.65	0.94	0.84	0.84	–
1250	–	–	0.89	0.71	0.80	–
1485	0.89	0.73	0.90	0.66	0.70	0.94
1600	0.86	0.63	0.78	0.60	0.61	0.93
2220	0.80	0.93	0.85	0.68	0.95	0.93

* DFT simulation—B3LYP/6-311 G (p, d).

We noticed that, for the CBC9CB dimer, there was a significantly smaller decrease in the mean absorbance for the band at 1600 cm^{-1} than for the other dimers. This indicates the importance of the sulfide group in this phenomenon.

Three bands were identified as a probe for transverse interaction: these are out of the benzene plane of the C-C vibration, which also involved a thioether bridge (γCC op CB + δCS; op—out of benzene plane) at 520 cm^{-1}; the second one was assigned to the out of plane deformation vibrations of the C-H groups at 811 cm^{-1} (γCH op CB), the complex band at 821 cm^{-1} for CBSCnOCB dimers (γCH op CB + ν_s $C_{Ar}O$; s–symmetric), and the in-plane deformation of the C-H groups at 1395 cm^{-1} (βCH ip CB with a sulfur linkage C-S-C). Figure 2 shows the temperature dependencies of the average absorbance, A_0, for the transversal transition dipole moment. We introduced the nematic phase trend line as a reference: the solid black line in Figure 2.

For a symmetric dimer (CBSC7SCB), the average absorbance of the transversal dipole is affected by the transition to the N_{TB} phase. However, in contrast to the longitudinal ones, they show a significant increase in their trend for all transversal dipoles. For the asymmetric dimers (CBSC7OCB), however, two steps of the temperature changes in the N_{TB} phase could clearly be distinguished (for transversal dipoles of sulfur linkages 811 and 521 cm^{-1}): the first extending almost 30 K below the N–N_{TB} transition temperature and the second below this range. In the first region, the absorbance continued to increase as the molecular density increased to nearly 30 K below the N–N_{TB} transition temperature. Then, the absorbances of the sulfur linkages (811 and 521 cm^{-1}) started to grow significantly. This was conveyed by the further absorbance decrease of the longitudinal dipole (Figure 1). Such behavior indicates the emergence of another interaction which causes so-called bond orientation and an escape from the uniaxial arrangement [26]. Please note the absorbance for CBC9CB is not affected by the N–N_{TB} transition. All the above can prove the leading role of the sulfur group in the emergence of bond orientations. They are involved in the π–π stacking interactions and in hydrogen bonds between neighbors. It should be pointed out that the vibration that involved the sulfur group (~811 cm^{-1}) showed an absorbance increase, while the other with the oxygen group (~820 cm^{-1}) did not.

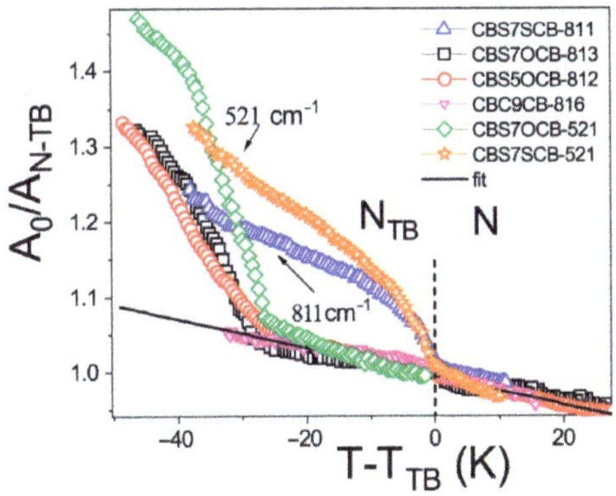

Figure 2. Normalized IR absorbance vs. temperature behavior of the dimers for the transversal dipole ($\mu_{\perp}$): 811 (813/816) cm^{-1} (γCH op CB for the rigid core with a sulfur linkage C−S−C): □-CBSC7OCB, ○-CBSC5OCB, △-CBSC7SCB, and ▽-CBC9CB; 520 cm^{-1} (γCC op CB + δCS): ◇-CBSC7OCB, ★-CBSC7SCB; solid black line—nematic phase trend line. A_0—integral average absorbance (cm^{-1}), $A_{N\text{-}TB}$—integral absorbance in the transition from nematic to twist–bend phase (cm^{-1}).

The latter result can be analyzed considering the rotation barrier between the cyanobiphenyl group's bond and the alkyl linker. This barrier is much higher for the oxygen group than for the sulfur one [12,76], so it can be assumed that the oxygen bridge remains in planar conformation. However, the sulfur bridge has conformational freedom and can escape from the phenyl plane [76]. It is reasonable to conclude that the origin of the absorbance increase is due to transversal order rearrangement and emergence phase biaxiality.

In light of the experimental results, we attempted to investigate possible intermolecular interactions resulting in bond orientation. To determine the molecular rearrangement on entering the N_{TB} phase, we performed density functional calculations (DFT) for the system, including the nearest lateral neighborhood. The calculated polarized IR spectra for such a system allow determining the transition dipole moments of particular vibration to control a specific type of interaction that might influence them.

2.2. Intermolecular Nonspecific Binding Energy in the System

Regarding the vibrations of individual molecular groups, we find the contributions of the couplings between vibrations for the rigid mesogenic cores within one molecule are negligible, in other words, hair-pin conformations are very rare [24–26]. On the other hand, the interactions of the mesogenic groups belonging to adjacent dimers turn out to be significant [25,26]. Therefore, to confirm the experimental observations and determine the molecular arrangement in the N_{TB} phase, DFT simulations were performed for the system, including the nearest lateral neighborhood. With the focus on following neighbor interactions and considering that DFT simulations for such large molecules are very time-consuming, calculations were directed at the CBSC7 and the structurally related CBOC7 monomers, judged to show the relevant features for adjacent interactions. First, the optimization of the single molecules was performed. The potentially stable conformers were determined based on calculating the energy barriers for the internal rotation of the cyanobiphenyl group (CB) and the rotation around the dihedral angle between the CB and the linker. All the technical details were described and discussed in a previous paper [76].

In the next step, the six monomer molecules consisting of the optimized molecule and the most energetically stable conformer [76] were arranged in a so-called sublayer (Figure 3a). The arrangement of the molecules in such a sublayer was created based on the detailed experimental data from resonant X-ray scattering measurements (TReXS) collected for these molecules [12].

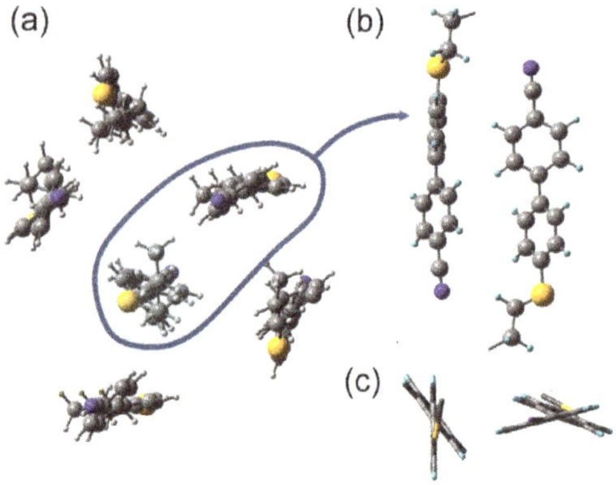

Figure 3. View of the CBSC7 monomer components. (**a**) Top view (X-Y plane) of the system with the initial arrangement of six molecules into a sublayer prepared for DFT calculations. IR spectra were calculated for a molecule (highlighted in blue) that is surrounded by other molecules. (**b**) The close-up view of a CBSC7 molecule pair after optimization using the B3LYP/6-311 G (d, p) method. (**c**) Top view (X-Y plane) of a pair of molecules (stick representation of the bonds).

We prepared two systems: one that contains only CBSC7 molecules and a hybrid system consisting of three CBSC7 and three CBOC7 molecules. Figure 3a shows, for example, a top view of the initial arrangement of the CBSC7 molecule selected for calculation. Molecules were arranged parallel to each other, the distance between them was set at approximately 5Å, and their cyan groups were arranged alternately. The system was optimized using the B3LYP/6-311 G (d, p) method. In the next stage for the optimized system, the atoms and bonds of the molecules located at the system's periphery were frozen. The frequencies and intensities of the vibrations were calculated for a central molecule that was surrounded by other molecules (Figure 3a, the selected molecule highlighted in blue). This approach allowed the calculation of the FTIR spectrum and, more precisely, the square of the dipole transition moment for a single molecule, taking into account the influence of the nearest surroundings. The binding energy was determined using the so-called supramolecular approach using the superposition error basis set (BSSE) correction utilizing the counterpoise method (CP). It was estimated at 30 and 37 kJ/mol (for the thioether and hybrid systems, respectively).

$$\Delta E^{CP} = E^{AB} - E^{A(AB)} - E^{B(AB)} \tag{2}$$

where E^{AB} is the energy of the complex, $E^{A(AB)}$—the energy of fragment A, which was calculated in the dimer base and, $E^{B(AB)}$—the energy of fragment B, which was calculated in the base of the dimer. Figure 3b,c show a group of the molecules into pairs.

2.3. Hypothetical Arrangement of the CBSC7SCB Molecules—DFT Modeling

In order to determine the molecular arrangement in the N_{TB} phase for the system, including the nearest lateral neighborhood, the geometry of the system was optimized based on the arrangement of the six interacting molecules. These are the monomers with a sulfur atom CBSC7, Figure 3a, and the hybrid system containing the CBSC7 and the CBOC7 molecules. After optimization, it was observed that the distance between the rigid mesogens decreased (ranging from 4.1 Å to 4.7 Å), which corresponds very well with the average lateral intermolecular distance determined from X-ray scattering measurements (~4.5 Å) for such molecules [12]. The differences are mainly due to ignoring/neglecting the thermal aspect in this DFT simulation. A group of the molecules were arranged into pairs and a shift of the molecular axes relative to each other was detected (see Figure 3b). As observed in numerous physical experiments, a nanosegregation into pseudolayers–biphenyls "stacking" next to each other, creating an aromatic pseudolayer, and the spacer tending to be parallel to form an aliphatic layer was observed as well. The molecules with a sulfide bridge are coaligned so that the phenyl planes of cyanobiphenyl remain in the so-called "T-shape" conformation (Figure 3c) relative to their neighbors. In the case of oxygen bridged molecules, a flatter "parallel-displaced" arrangement was observed. The segregation of the aromatic parts (partial overlapping) and an attraction of the CN group to the alkyl chains were observed. Molecular interdigitating also occurred [44,46].

In the next stage, the outer molecules were frozen, while the vibrational frequencies and transition dipole moments were calculated for the molecule that remained in the center (see Figure 3a). The theoretical IR spectrum for a surrounded molecule should reflect the situation in which the intermolecular forces (IMF) are considered. The IR spectrum of a molecule where the intermolecular forces were considered was compared with the spectrum of the isolated molecule and the experimental one in the N and N_{TB} phases (Figures 4 and 5).

For the longitudinal dipoles (bands at wavenumbers: 1000, 1100, 1485, 1600, and 2220 cm^{-1}), a decrease in the intensity of the bands for the system with the IMF when compared to a single monomer was detected. This result is consistent with the correlation along the Z-direction, visible in the IR experiments. An increase in the intensity for the transverse bands was also observed, i.e., for the bands at 520 and 811 cm^{-1}. This result implies a correlation of the transverse dipoles in the direction perpendicular to the nematic order (so call bond ordering).

For bands with transversal dipoles (520, 811 cm^{-1}), an increase in the intensity of the bands by approximately 50% relative to an isolated molecule was detected. For the vibrations in the longitudinal dipoles (1100, 1485, 1600 cm^{-1}), the decrease was approximately 40%. The slightest change in intensity was obtained for the CN vibrations at 2220 cm^{-1} (about 10%). These values align with the trend of absorbances observed in experimental spectra at the transition from the N to the N_{TB} phase. Table 1 summarizes the absorbance changes in the transition from the N to the N_{TB} phase (related to the square of the corresponding transition dipole moments) with a calculated ratio $\mu^2_{IMF}/\mu^2_{noIMF}$ in the presence of weak intermolecular interactions and local orientational order.

We note that the interactions between several molecules do not yet fully explain the behavior of a group of molecules. Thermodynamic and probabilistic considerations related to entropy and enthalpy changes must be considered to determine this behavior fully.

Hence, our current calculations are to be viewed as the first step toward more advanced simulations using periodic density functional theory modeling, which we plan to perform in the future.

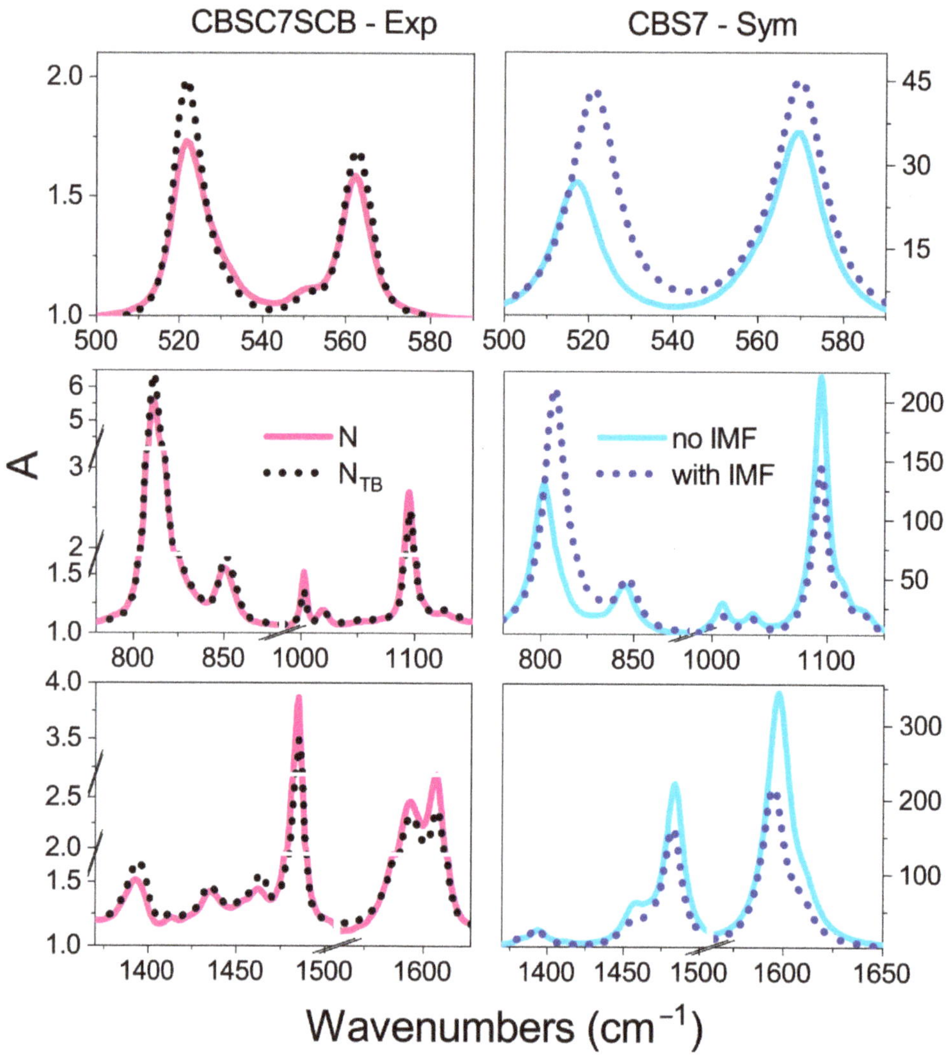

Figure 4. Comparison of the theoretical and experimental spectra for the symmetric dimer. Left Panel: Experimental spectra of the CBS7SCB dimer, solid magenta line—the nematic phase, short black dotted line—the twist–bend nematic phase (spectrum is represented for 5 µm cell). Right Panel: Simulations for the CBS7 monomer: solid blue line—spectra for an isolated molecule (no IMF), short navy dotted line—a system of six interacting molecules (with IMF).

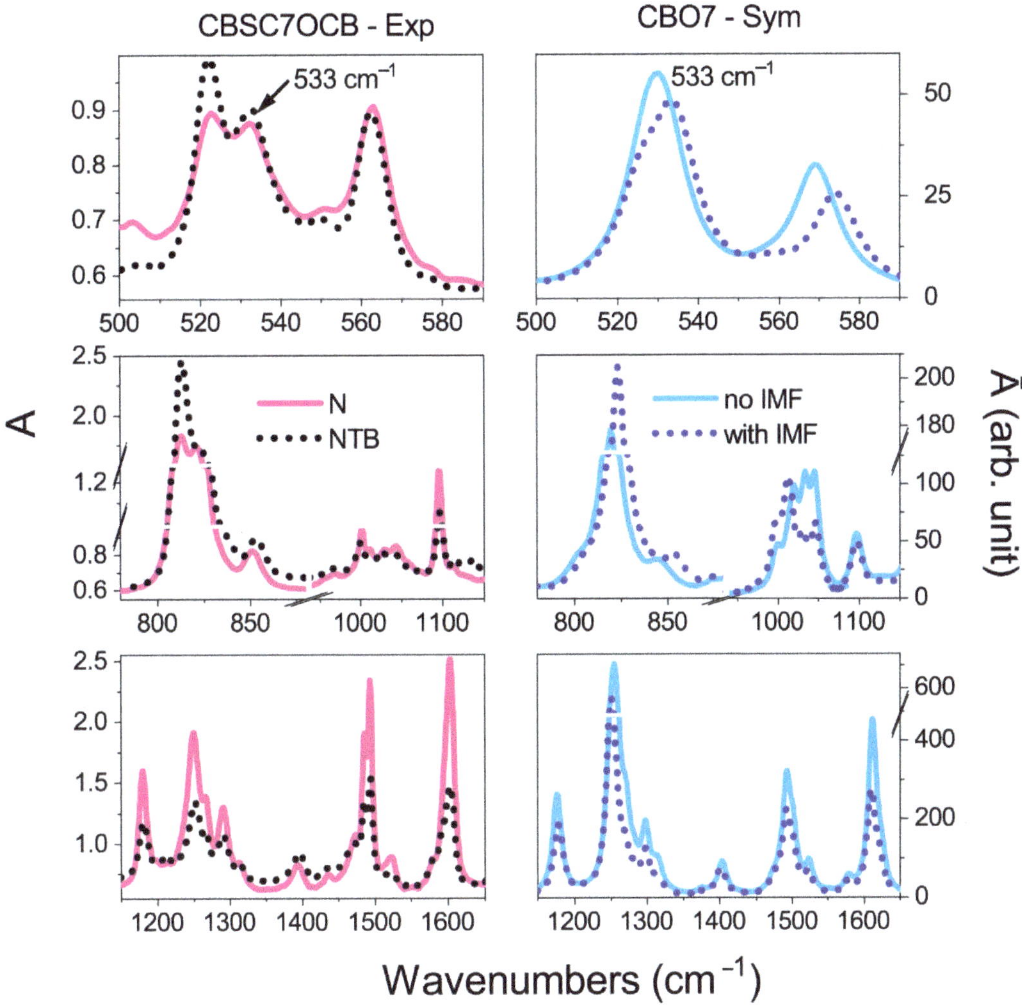

Figure 5. Comparison of the theoretical and experimental spectra for the asymmetric dimer. Left Panel: Experimental spectra of the CBS7OCB dimer, solid magenta line—the nematic phase, short black dotted line—the twist–bend nematic phase (spectrum is represented for 5 μm cell). Right Panel: Simulations for the CBO7 monomer: solid blue line—spectra for an isolated molecule (no IMF), short navy dotted line—a system of six interacting molecules (with IMF).

3. Discussion

Based on the analysis of the molecular structure, a model of the geometrical arrangement of the molecules in the N_{TB} phase was developed (Figure 6). In this model, the molecules follow the shape of the helix formed by the director, with overlapping the CB groups formed. This is due to increased bond and dipole correlations, as shown through the optimization of the geometry of the monomer system. We found that for the investigated dimers, the hydrogen bonds between the CN group and the hydrogen atoms of the spacer, the hydrogen bonds between the sulfur atom and the hydrogen atom of the benzene ring, and the interactions of the π–π orbitals of the benzene rings all play an essential role. The presence of intermolecular interactions in the N_{TB} phase is clearly visible. The model and the experimental results for intermolecular interactions are consistent with the reports of other measurements obtained using the TReXS (Tender Resonant X-ray Scattering) method [11,12]. Table 2 shows the summary of the determined molecular parameters.

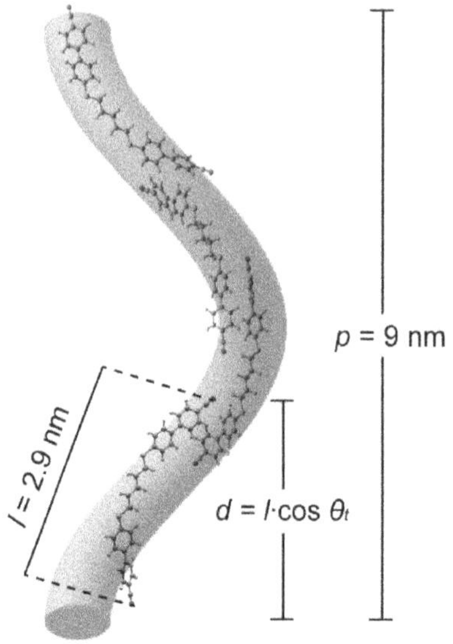

Figure 6. Proposed hypothetical arrangement of the thioether dimers in the twist–bend nematic phase. The helix pitch was computed as $p = 2\pi/q$, where q is the wave vector.

Table 2. Summary of the determined molecular parameters: the length of the molecule (l), the tilt angle (θ_t), the helix pitch (p), the effective length of the molecule (d), and the number of molecules per helix pitch (p/d).

CBACnBCB	θ_t (°)	p (nm)	l (nm) *	$d = l\cos\theta_t$	p/d
CBC9CB	25.6	8.15 [10]	2.88	2.6	3.1
CBSC7SCB	33	9.1 [11,12]	2.90	2.43	3.7
CBSC7OCB	15.6	11.5 [12]	2.90	2.79	4.1

*—lengths of the molecules were determined for the optimized molecules using the B3LYP/6-31 G (d, p) method (CBC9CB–U conf., CBSC7SCB–F conf., CBSC7OCB–F conf., see paper [76]).

4. Materials and Methods

4.1. Materials

Liquid crystal dimers were investigated based on the cyanobiphenyl (CB) mesogenic groups. We represent symmetrical dimers by the general abbreviation CBAC7BCB with A = B, where A and B = C, S or O (see Figure 7a). In the asymmetric dimers with the acronym CBSCnOCB (n = 5,7), the mesogens were linked to an alkyl spacer on one side by a thioether bridge and the other by an ether one. The primary CBC9CB compound was synthesized as described in Refs. [77–80]. All details on the synthesis of thioether dimers and the preliminary result of differential scanning calorimetry (DSC) and X-ray scattering studies are summarized in papers [81,82]. Figure 7a shows the molecular structure of the CB dimers.

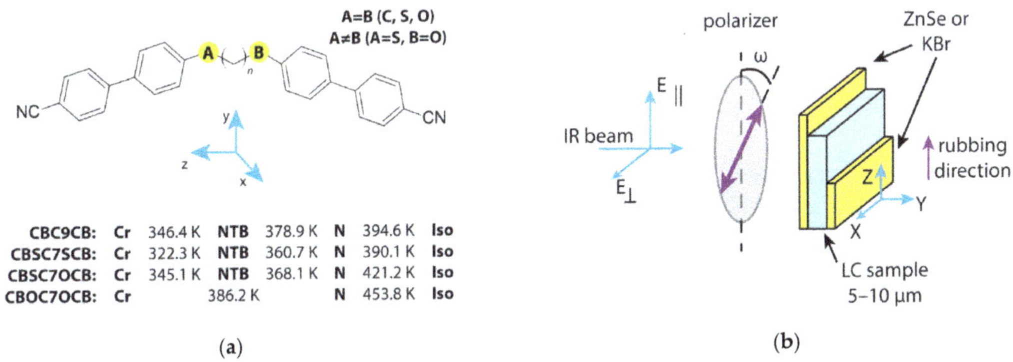

Figure 7. (**a**) Transition temperature and molecular structure of the cyanobiphenyl dimers with a molecular frame of reference: z—long axis (bowstring), x—axis normal to the bent plane, y—bow arrow axis. (**b**) Schematic of the polarized infrared transmission technique at a normal incidence of light. The planar cell's laboratory frame (X, Y, Z). In the nematic phase, Z was the axis along and X was perpendicular to the optical axis (the optical axis coincided with the rubbing direction). In the N_{TB} phase, Z coincided with the helix axis.

4.2. Infrared Spectroscopy

The planar-aligned cells were prepared between two optically polished zinc selenide (ZnSe) windows. The windows were spin-coated with a SE-130 commercial polymer aligning agent (Nissan Chemical Industries, Ltd., Pasadena, TX, USA) in order to obtain a homogeneous alignment. The cells were assembled with a parallel positioning of the rubbing direction, and Mylar foil was used as a spacer to provide 2 μm and 5 μm separation. The thickness of the cells was determined by the measurements on the interference fringes using a spectrometer interfaced with a PC (Avaspec-2048). The samples were capillary filled by heating an empty cell in the N phase, five degrees below the transition to the isotropic phase. The quality of the alignment was tested using polarizing microscopy. The textures of the samples were monitored using a polarizing microscope (Olympus BX56). The spectra were acquired using a Fourier infrared spectrometer (Agilent Cary 670) with a resolution of 1 cm^{-1}, and these spectra are averaged over 32 scans. The experiment was conducted in the transmission mode with a polarized IR beam. An IR-KRS5 grid polarizer was used to polarize the IR beam. The IR spectra were measured as a function of the polarizer rotation angle in the wavenumber range 500–4000 cm^{-1}. To keep the absorbance in the linear regime, we combined the spectrum of a 5 μm sample with a spectrum of 2 μm near strong bands of the spectra (810 cm^{-1}).

The measurements were performed using slow cooling and heating at a rate of 0.5 K/min. For the samples with a thioether bridge (CBSC7SCB, CBSC7OCB), an additional measurement was taken with a faster cooling rate (4 K/min) in a temperature range from a few degrees above the N–N_{TB} transition until the solid phase (glass or crystal) was obtained. A faster cooling rate near the N–N_{TB} transition temperature significantly increased the width of the N_{TB} phase range for sulfide-bridged samples, so the later results were selected in the data analysis. The temperature of the samples was stabilized using a PID temperature controller with an accuracy of 2 mK.

Figure 7b shows the configuration of the infrared measurements using the polarized transmission technique. These measurements enabled the orientation of the transition dipole moment of the bands to be determined with respect to the long molecular axis and the temperature dependencies of the absorbance of the samples. To determine all components of absorbance (A_X, A_Y, and A_Z), it was necessary to measure two cells with different orientations: planar (homogeneous) and homeotropic. Unfortunately, in the case of the tested materials, i.e., for the cyanobiphenyl dimers, it was extremely difficult to obtain a good homeotropic alignment. Therefore, to calculate the mean absorbance, assuming that the material was uniaxial and $A_X = A_Y$, the mean absorbance was determined as $A_0 = (2A_X + A_Z)/3$. The absorbance components were determined as being the area bound by the contour of the given band using Bio-Rad Win-IR Pro version 2.96e. In the case of complex bands that contained more vibrations, they were separated using the Origin Pro 2021 software using the Pearson VII fit.

4.3. Density Functional Theory Calculations

In this work, the molecules' electronic structure calculations were performed using the Gaussian09 software package (version E.01) [83]. The molecular structures, binding energy, harmonic vibrational force constants, absolute IR intensities, and components of the transition dipole moments were calculated using the density functional theory (DFT). The Becke's three-parameter exchange functional combined with the Lee, Yang, and Parr correlation functional was applied (B3LYP) with the polarization basis set (6-311 (d, p)). [84,85].

Information about the components of the transition dipole moment for a specific vibration enables the parallel and perpendicular components of the spectral density to be calculated. The parallel component of the absorption coefficient was calculated as the square of the component of the transition dipole moment along the axis that coincided with the long axis of the dimer $|\mu_z|^2 = \mu_{\|}^2$. To determine the perpendicular component of the spectral density, the sum of the squares of the transition dipole moments along the vertical directions was used $|\mu_x|^2 + |\mu_y|^2 = \mu_{\perp}^2$. The direction of the transition dipole moment was determined according to the molecular reference system (Figure 7a).

The theoretical vibrational frequencies were scaled by one coefficient equal to 0.98 to simplify the comparison with the experiment, and the Gaussian profile was used with a 7 cm^{-1} full width at half maximum (FWHM). The results were visualized using GaussView 6.

5. Conclusions

The study's most striking finding was the evidence that the intermolecular forces evolved on the transition from the nematic (N) phase to the twist–bend (N_{TB}) phase. Changes in the behavior of intermolecular interactions were observed by significant differences in the values of transition dipole moments for selected vibrational bands. The longitudinally induced dipoles of the CB group showed negative correlations due to the antiparallel mesogen arrangement, while the perpendicular dipoles were positively correlated; in other words, they increased. To explain this phenomenon of such self-organization, the DFT modeling was performed for the system, taking into account the nearest side neighborhood. Results that is most consistent with the experimental data is a system of several molecules that exhibits a clustering of the rigid CB cores and nonspecific weak intermolecular interactions. These interactions are mainly associated with the π–π orbitals interactions of the aromatic rings and the sulfur atoms. However, the formation of weak hydrogen bonds H···S/O and H···N may also be present. The nonspecific intermolecular interactions resulted in a significant bond ordering in the N_{TB} phase.

Based on the DFT simulation for the groups of interacting molecules, along with the experimental data from absorbance measurements, as well as X-ray resonance scattering, a model for the packing of the dimer molecules in the twist–bend phase was developed with overlapping of the rigid cyanobiphenyl cores stabilizing the N_{TB} phase and thus affecting the helical pitch.

Author Contributions: Conceptualization: K.M. and A.K.; Synthesis: Y.A., G.H.M. and J.K.; Methodology: K.M. and A.K.; Investigation: B.L. and K.M.; Formal Analysis: K.M.; Writing—original draft: K.M. and A.K.; Writing—review and editing: K.M., G.H.M. and A.K.; Visualization: B.L. and K.M.; Supervision: K.M. and A.K. All authors have read and agreed to the published version of the manuscript.

Funding: This research was funded by National Science Centre, Poland for grants no. 2018/31/B/ST3/03609 and no. 2020/39/O/ST5/03460.

Institutional Review Board Statement: Not applicable.

Informed Consent Statement: Not applicable.

Data Availability Statement: Raw data linked to this publication has been deposited on RepOD. And are available at the link: https://repod.icm.edu.pl/dataset.xhtml?persistentId=doi%3A10.18150%2FWJUUXU, doi:10.18150/WJUUXU.

Acknowledgments: Authors K.M. and A.K. thank the National Science Centre for funding through grant no. 2018/31/B/ST3/03609. B.L. thanks the National Science Centre for funding through project no. 2020/39/O/ST5/03460. G.H.M. acknowledges Diamond Light Source, UK for support through projects SM28688 and SM30755 at station B23. All DFT calculations were carried out with the Gaussian09 program using the PL-Grid Infrastructure on the ZEUS and Prometheus clusters in Academic Computer Center CYFRONET AGH (AGH University of Science and Technology) in Cracow, Poland.

Conflicts of Interest: The authors declare no conflict of interest.

References

1. Whitesides, G.M.; Grzybowski, B. Self-Assembly at All Scales. *Science* **2002**, *295*, 2418–2421. [CrossRef]
2. Yan, F.; Hixson, C.A.; Earl, D.J. Self-Assembled Chiral Superstructures Composed of Rigid Achiral Molecules and Molecular Scale Chiral Induction by Dopants. *Phys. Rev. Lett.* **2008**, *101*, 157801–157804. [CrossRef]
3. Chothia, C. Conformation of twisted β-pleated sheets in proteins. *J. Mol. Biol.* **1973**, *75*, 295–302. [CrossRef]
4. Muraoka, T.; Cui, H.; Stupp, S.I. Quadruple Helix Formation of a Photoresponsive Peptide Amphiphile and Its Light-Triggered Dissociation into Single Fibers. *J. Am. Chem. Soc.* **2008**, *130*, 2946–2947. [CrossRef]
5. Li, L.; Jiang, H.; Messmore, B.W.; Bull, S.R.; Stupp, S.I. A Torsional Strain Mechanism To Tune Pitch in Supramolecular Helices. *Angew. Chem. Int. Ed.* **2007**, *119*, 5977–5980. [CrossRef]
6. De Gennes, P.G.; Prost, J. *The Physics of Liquid Crystals*, 2nd ed.; Oxford University Press: New York, NY, USA, 1993.
7. Reddy, R.A.; Tschierske, C. Bent-core liquid crystals: Polar order, superstructural chirality and spontaneous desymmetrization in soft matter systems. *J. Mater. Chem.* **2006**, *16*, 907–961. [CrossRef]
8. Chen, D.; Porada, J.H.; Hooper, J.B.; Klittnick, A.; Shen, Y.; Tuchband, M.R.; Korblova, E.; Bedrov, D.; Walba, D.M.; Glaser, M.A.; et al. Chiral heliconical ground state of nanoscale pitch in a nematic liquid crystal of achiral molecular dimers. *Proc. Natl. Acad. Sci. USA* **2013**, *110*, 15931–15936. [CrossRef]
9. Borshch, V.; Kim, Y.K.; Xiang, J.; Gao, M.; Jákli, A.; Panov, V.P.; Vij, J.K.; Imrie, C.T.; Tamba, M.G.; Mehl, G.H.; et al. Nematic twist-bend phase with nanoscale modulation of molecular orientation. *Nat. Commun.* **2013**, *4*, 2635–2638. [CrossRef]
10. Zhu, C.; Tuchband, M.R.; Young, A.; Shuai, M.; Scarbrough, A.; Walba, D.M.; Maclennan, J.E.; Wang, C.; Hexemer, A.; Clark, N.A. Resonant carbon k-edge soft x-ray scattering from lattice-free heliconical molecular ordering: Soft dilative elasticity of the twist-bend liquid crystal phase. *Phys. Rev. Lett.* **2016**, *116*, 147803–147809. [CrossRef]
11. Cruickshank, E.; Salamończyk, M.; Pociecha, D.; Strachan, G.J.; Storey, J.M.D.; Wang, C.; Feng, J.; Zhu, C.; Gorecka, E.; Imrie, C.T. Sulfur-linked cyanobiphenyl-based liquid crystal dimers and the twist-bend nematic phase. *Liq. Cryst.* **2019**, *46*, 1595–1609. [CrossRef]
12. Cao, Y.; Feng, J.; Nallapaneni, A.; Arakawa, Y.; Zhao, K.; Zhang, H.; Mehl, G.H.; Zhu, C.; Liu, F. Deciphering helix assembly in the heliconical nematic phase via tender resonant X-ray scattering. *J. Mater. Chem. C* **2021**, *9*, 10020–10028. [CrossRef]
13. Cestari, M.; Diez-Berart, S.; Dunmur, D.A.; Ferrarini, A.; de la Fuente, M.R.; Jackson, D.J.B.; Lopez, D.O.; Luckhurst, G.R.; Perez-Jubindo, M.A.; Richardson, R.M.; et al. Phase behavior and properties of the liquid-crystal dimer 1″,7″-bis(4-cyanobiphenyl-4′-yl) heptane: A twist-bend nematic liquid crystal. *Phys. Rev. E* **2011**, *84*, 031704-20. [CrossRef]
14. Adlem, K.; Čopič, M.; Luckhurst, G.R.; Mertelj, A.; Parri, O.; Richardson, R.M.; Snow, B.D.; Timimi, B.A.; Tuffin, R.P.; Wilkes, D. Chemically induced twist-bend nematic liquid crystals, liquid crystal dimers, and negative elastic constants. *Phys. Rev. E* **2013**, *88*, 022503-8. [CrossRef]
15. Stevenson, W.D.; Zou, H.-X.; Zeng, X.-B.; Welch, C.; Ungar, G.; Mehl, G.H. Dynamic calorimetry and XRD studies of the nematic and twist-bend nematic phase transitions in a series of dimers with increasing spacer length. *Phys. Chem. Chem. Phys.* **2018**, *20*, 25268–25274. [CrossRef]
16. Salamończyk, M.; Vaupotič, N.; Pociecha, D.; Wang, C.; Zhu, C.; Gorecka, E. Structure of nanoscale-pitch helical phases: Blue phase and twist-bend nematic phase resolved by resonant soft X-ray scattering. *Soft Matter* **2017**, *13*, 6694–6699. [CrossRef]
17. Stevenson, W.; Ahmed, Z.; Zeng, X.; Welch, C.; Ungar, G.; Mehl, G. Molecular organization in the twist–bend nematic phase by resonant X-ray scattering at the Se K-edge and by SAXS, WAXS and GIXRD. *Phys. Chem. Chem. Phys.* **2017**, *19*, 13449–13454. [CrossRef]

18. Tuchband, M.R.; Paterson, D.A.; Salamończyk, M.; Norman, V.A.; Scarbrough, A.N.; Forsyth, E.; Garcia, E.; Wang, C.; Storey, J.M.; Walba, D.M.; et al. Distinct differences in the nanoscale behaviors of the twist–bend liquid crystal phase of a flexible linear trimer and homologous dimer. *Proc. Natl Acad. Sci. USA* **2019**, *116*, 10698–10704. [CrossRef]
19. Saha, R.; Feng, C.; Welch, C.; Mehl, G.H.; Feng, J.; Zhu, C.; Gleeson, J.; Sprunt, S.; Jákli, A. The interplay between spatial and heliconical orientational order in twist-bend nematic materials. *Phys. Chem. Chem. Phys.* **2021**, *23*, 4055–4063. [CrossRef]
20. Chen, D.; Nakata, M.; Shao, R.; Tuchband, M.R.; Shuai, M.; Baumeister, U.; Weissflog, W.; Walba, D.M.; Glaser, M.A.; Maclennan, J.E.; et al. Twist-bend heliconical chiral nematic liquid crystal phase of an achiral rigid bent-core mesogen. *Phys. Rev. E.* **2014**, *89*, 022506. [CrossRef]
21. Gorecka, E.; Salamonczyk, M.; Zep, A.; Pociecha, D.; Welch, C.; Ahmed, Z.; Mehl, G.H. Do the short helices exist in the nematic TB phase? *Liq. Cryst.* **2015**, *42*, 1–7. [CrossRef]
22. Knežević, A.; Dokli, I.; Sapunar, M.; Sĕgota, S.; Baumeister, U.; Lesac, A. Induced smectic phase in binary mixtures of twist-bend nematogens. *Beilstein J. Nanotechnol.* **2018**, *9*, 1297–1307. [CrossRef]
23. Zhang, Z.; Panov, V.P.; Nagaraj, M.; Mandle, R.J.; Goodby, J.W.; Luckhurst, G.R.; Jones, J.C.; Gleeson, H.F. Raman scattering studies of order parameters in liquid crystalline dimers exhibiting the nematic and twist-bend nematic phases. *J. Mater. Chem. C* **2015**, *3*, 10007–10016. [CrossRef]
24. Merkel, K.; Loska, B.; Welch, C.; Mehl, G.H.; Kocot, A. Molecular biaxiality determines the helical structure–infrared measurements of the molecular order in the nematic twist-bend phase of difluoro terphenyl dimer. *Phys. Chem. Chem. Phys.* **2021**, *23*, 4151–4160. [CrossRef]
25. Merkel, K.; Loska, B.; Welch, C.; Mehl, G.H.; Kocot, A. The role of intermolecular interactions in stabilizing the structure of the nematic twist-bend phase. *RSC Adv.* **2021**, *11*, 2917–2925. [CrossRef]
26. Kocot, A.; Loska, B.; Arakawa, Y.; Merkel, K. Structure of the twist-bend nematic phase with respect to the orientational molecular order of the thioether-linked dimers. *Phys. Rev. E* **2022**, *105*, 044701-10. [CrossRef]
27. Beguin, L.; Emsley, J.W.; Lelli, M.; Lesage, A.; Luckhurst, G.R.; Timimi, B.A.; Zimmermann, H. The chirality of a twist–bend nematic phase identified by NMR spectroscopy. *J. Phys. Chem. B* **2012**, *116*, 7940–7951. [CrossRef]
28. Emsley, J.W.; Lesot, P.; Luckhurst, G.R.; Meddour, A.; Merlet, D. Chiral solutes can seed the formation of enantiomorphic domains in a twist-bend nematic liquid crystal. *Phys. Rev. E* **2013**, *87*, 040501. [CrossRef]
29. Emsley, J.W.; Lelli, M.; Lesage, A.; Luckhurst, G.R. A comparison of the conformational distributions of the achiral symmetric liquid crystal dimer cb7cb in the achiral nematic and chiral twist-bend nematic phases. *J. Phys. Chem. B* **2013**, *117*, 6547–6557. [CrossRef]
30. Jokisaari, J.P.; Luckhurst, G.R.; Timimi, B.A.; Zhu, J.F.; Zimmermann, H. Twist-bend nematic phase of the liquid crystal dimer cb7cb: Orientational order and conical angle determined by xe-129 and h-2 nmr spectroscopy. *Liq. Cryst.* **2015**, *42*, 708–721.
31. Ivšić, T.; Vinković, M.; Baumeister, U.; Mikleušević, A.; Lesac, A. Towards understanding the NTB phase: A combined experimental, computational and spectroscopic study. *RSC Adv.* **2016**, *6*, 5000–5007. [CrossRef]
32. Panov, V.; Balachandran, R.; Nagaraj, M.; Vij, J.K.; Tamba, M.G.; Kohlmeier, A.; Mehl, G.H. Microsecond linear optical response in the unusual nematic phase of achiral bimesogens. *Appl. Phys. Lett.* **2011**, *99*, 261903. [CrossRef]
33. Meyer, C.; Luckhurst, G.R.; Dozov, I. Flexoelectrically driven electroclinic effect in the twist-bend nematic phase of achiral molecules with bent shapes. *Phys. Rev. Lett.* **2013**, *111*, 067801. [CrossRef]
34. Sebastián, N.; Tamba, M.G.; Stannarius, R.; de la Fuente, M.R.; Salamonczyk, M.; Cukrov, G.; Gleeson, J.T.; Sprunt, S.; Jakli, A.; Welch, C.; et al. Mesophase structure and behaviour in bulk and restricted geometry of a dimeric compound exhibiting a nematic–nematic transition. *Phys. Chem. Chem. Phys.* **2016**, *18*, 19299. [CrossRef]
35. Merkel, K.; Kocot, A.; Vij, J.K.; Shanker, G. Distortions in structures of the twist bend nematic phase of a bent-core liquid crystal by the electric field. *Phys. Rev. E* **2018**, *98*, 022704-8. [CrossRef]
36. Merkel, K.; Kocot, A.; Welch, C.; Mehl, G.H. Soft modes of the dielectric response in the twist–bend nematic phase and identification of the transition to a nematic splay bend phase in the CBC7CB dimer. *Phys. Chem. Chem. Phys.* **2019**, *21*, 22839–22848. [CrossRef]
37. Merkel, K.; Welch, C.; Ahmed, Z.; Piecek, W.; Mehl, G.H. Dielectric response of electric-field distortions of the twist-bend nematic phase for LC dimers. *J. Chem. Phys.* **2019**, *151*, 114908. [CrossRef]
38. Meyer, C.; Blanc, C.; Luckhurst, G.R.; Davidson, P.; Dozov, I. Biaxiality-driven twist-bend to splay-bend nematic phase transition induced by an electric field. *Sci. Adv.* **2020**, *6*, eabb8212. [CrossRef]
39. Challa, P.K.; Borshch, V.; Parri, O.; Imrie, C.T.; Sprunt, S.N.; Gleeson, J.T.; Lavrentovich, O.D.; Jakli, A. Twist-bend nematic liquid crystals in high magnetic fields. *Phys. Rev. E* **2014**, *89*, 060501. [CrossRef]
40. Mandle, R.J.; Goodby, J.W. Progression from nano to macro science in soft matter systems: Dimers to trimers and oligomers in twist-bend liquid crystals. *RSC Adv.* **2016**, *6*, 34885–34893. [CrossRef]
41. Tuchband, M.R.; Shuai, M.; Graber, K.A.; Chen, D.; Zhu, C.; Radzihovsky, L.; Klittnick, A.; Foley, L.M.; Scarbrough, A.; Porada, J.H.; et al. Double-Helical Tiled Chain Structure of the Twist-Bend Liquid Crystal phase in CB7CB. *arXiv* **2017**, arXiv:1703.10787v1.
42. Salamończyk, M.; Vaupotič, N.; Pociecha, D.; Walker, R.; Storey, J.M.D.; Imrie, C.T.; Wang, C.; Zhu, C.; Gorecka, E. Multi-level chirality in liquid crystals formed by achiral molecules. *Nat. Commun.* **2019**, *10*, 1922. [CrossRef]
43. Leadbetter, A.J.; Frost, J.C.; Gaughan, J.P.; Gray, G.W.; Mosley, A. The structure of smectic a phases of compounds with cyano end groups. *J. Phys. France* **1979**, *40*, 375–380. [CrossRef]

44. Goodby, J.W.; Davis, E.J.; Mandle, R.J.; Cowling, S.J. Nano-Segregation and Directed Self-Assembly in the Formation of Functional Liquid Crystals. *Isr. J. Chem.* **2012**, *52*, 863–880. [CrossRef]
45. Dwivedi, K.K.; Dwivedi, M.K.; Tiwa, S.N. Study of Intermolecular Interactions in Liquid Crystals: Para-butyl-p′-cyano-biphenyl. *J. Cryst. Process Tech.* **2014**, *4*, 31–38. [CrossRef]
46. Sims, M.T.; Abbott, L.C.; Goodby, J.W.; Moore, J.N. Shape segregation in molecular organization: A combined X-ray scattering and molecular dynamics study of smectic liquid crystals. *Soft Matter* **2019**, *15*, 7722–7732. [CrossRef]
47. Mandle, R.J.; Goodby, J.W. Does Topology Dictate the Incidence of the Twist-Bend Phase? Insights Gained from Novel Unsymmetrical Bimesogens. *Chem. Eur. J.* **2016**, *22*, 18456–18464. [CrossRef]
48. Mandle, R.J. The dependency of twist-bend nematic liquid crystals on molecular structure: A progression from dimers to trimers, oligomers and polymers. *Soft Matter* **2016**, *12*, 7883–7901. [CrossRef]
49. Paterson, D.A.; Gao, M.; Kim, Y.-K.; Jamali, A.; Finley, K.L.; Robles-Hernández, B.; Diez-Berart, S.; Salud, J.; de la Fuente, M.R.; Timimi, B.A.; et al. Understanding the twist-bend nematic phase: The characterization of 1-(4-cyanobiphenyl-40-yloxy)-6-(4-cyanobiphenyl-40-yl)hexane (CB6OCB) and comparison with CB7CB. *Soft Matter* **2016**, *12*, 6827–6840. [CrossRef]
50. Greco, C.; Marini, A.; Frezza, E.; Ferrarini, A. From the Molecular Structure to Spectroscopic and Material Properties: Computational Investigation of a Bent-Core Nematic Liquid Crystal. *ChemPhysChem* **2014**, *15*, 1336–1344. [CrossRef]
51. Mandle, R.J.; Archbold, C.T.; Sarju, J.P.; Andrews, J.L.; Goodby, J.W. The Dependency of Nematic and Twist-bend Mesophase Formation on Bend Angle. *Sci. Rep.* **2016**, *6*, 36682. [CrossRef]
52. Tomczyk, W.; Longa, L. Role of molecular bend angle and biaxiality in the stabilization of the twist-bend nematic phase. *Soft Matter* **2020**, *16*, 4350–4357. [CrossRef]
53. Lesac, A.; Baumeister, U.; Dokli, I.; Hameršak, Z.; Ivšić, T.; Kontrec, D.; Viskić, M.; Knežević, A.; Mandle, R.J. Geometric aspects influencing N-N TB transition-implication of intramolecular torsion. *Liq. Cryst.* **2018**, *45*, 1101–1110. [CrossRef]
54. Abberley, J.P.; Storey, J.M.D.; Imrie, C.T. Structure-property relationships in azobenzene-based twist-bend nematogens. *Liq. Cryst.* **2019**, *46*, 2102–2114. [CrossRef]
55. Sebastián, N.; Robles-Hernández, B.; Diez-Berart, S.; Salud, J.; Luckhurst, G.R.; Dunmur, D.A.; López, D.O.; de la Fuente, M.R. Distinctive dielectric properties of nematic liquid crystal dimers. *Liq. Cryst.* **2017**, *44*, 177–190. [CrossRef]
56. Archbold, C.T.; Mandle, R.J.; Andrews, J.L.; Cowling, S.J.; Goodby, J.W. Conformational landscapes of bimesogenic compounds and their implications for the formation of modulated nematic phases. *Liq. Cryst.* **2017**, *44*, 2079–2088. [CrossRef]
57. Pocock, E.E.; Mandle, R.J.; Goodby, J.W. Molecular shape as a means to control the incidence of the nanostructured twist bend phase. *Soft Matter* **2018**, *14*, 2508–2514. [CrossRef]
58. Pocock, E.E.; Mandle, R.J.; Goodby, J.W. Experimental and Computational Study of a Liquid Crystalline Dimesogen Exhibiting Nematic, Twist-Bend Nematic, Intercalated Smectic, and Soft Crystalline Mesophases. *Molecules* **2021**, *26*, 532–544. [CrossRef]
59. Ferrarini, A. The twist-bend nematic phase: Molecular insights from a generalized Maier–Saupe theory. *Liq. Cryst.* **2017**, *44*, 45–57. [CrossRef]
60. Mandle, R.J.; Goodby, J.W. A twist-bend nematic to an intercalated, anticlinic, biaxial phase transition in liquid crystal bimesogens. *Soft Matter* **2016**, *12*, 1436–1443. [CrossRef]
61. Goodby, J.W. Free volume, molecular grains, self-organization, and anisotropic entropy: Machining materials. *Liq. Cryst.* **2017**, *44*, 1755–1763.
62. Jansze, S.M.; Martínez-Felipe, A.; Storey, J.M.D.; Marcelis, A.T.M.; Imrie, C.T. A Twist-Bend Nematic Phase Driven by Hydrogen Bonding. *Angew. Chem. Int. Ed.* **2015**, *54*, 643–646. [CrossRef]
63. Martinez-Felipe, A.; Cook, G.; Abberley, J.P.; Walker, R.; Storey, J.M.D.; Imrie, C.T. An FT-IR spectroscopic study of the role of hydrogen bonding in the formation of liquid crystallinity for mixtures containing bipyridines and 4-pentoxybenzoic acid. *RSC Adv.* **2016**, *6*, 108164–108179. [CrossRef]
64. Walker, R.; Pociecha, D.; Abberley, J.P.; Martinez-Felipe, A.; Paterson, D.A.; Forsyth, E.; Lawrence, G.B.; Henderson, P.A.; Storey, J.M.D.; Gorecka, E.; et al. Spontaneous chirality through mixing achiral components: A twist-bend nematic phase driven by hydrogen-bonding between unlike components. *Chem. Commun.* **2018**, *54*, 3383–3386. [CrossRef]
65. Knežević, A.; Sapunar, M.; Buljan, A.; Dokli, I.; Hameršak, Z.; Kontrec, D.; Lesac, A. Fine-tuning the effect of p–p interactions on the stability of the NTB phase. *Soft Matter* **2018**, *14*, 8466–8474. [CrossRef]
66. Alaasar, M.; Tschierske, C. Nematic phases driven by hydrogen-bonding in liquid crystalline nonsymmetric dimers. *Liq. Cryst.* **2019**, *49*, 124–130. [CrossRef]
67. Cragg, P.J. *Supramolecular Chemistry: From Biological Inspiration to Biomedical Applications*, 1st ed.; Springer: Dordrecht, The Netherlands, 2010.
68. Asensio, J.L.; Arda, A.; Canada, F.J.; Jimenez-Barbero, J. Carbohydrate-aromatic interactions. *Acc. Chem. Res.* **2013**, *46*, 946–954. [CrossRef]
69. Mingos, D.M.P. Structure and Bonding. In *Liquid Crystalline Functional Assemblies and Their Supramolecular Structures*; Kato, K., Ed.; Springer: Berlin/Heidelberg, Germany, 2008; Volume 128. [CrossRef]
70. Kato, T.; Gupta, M.; Yamaguchi, D.; Kian, P.G.; Nakayama, M. Supramolecular Association and Nanostructure Formation of Liquid Crystals and Polymers for New Functional Materials. *Bull. Chem. Soc. Jpn.* **2021**, *94*, 357–376. [CrossRef]
71. Walker, R.; Pociecha, D.; Salamończyk, M.; Storey, J.M.D.; Górecka, E.; Imrie, C.T. Supramolecular liquid crystals exhibiting a chiral twist-bend nematic phase. *Mater. Adv.* **2020**, *1*, 1622–1630. [CrossRef]

72. Vanakaras, A.G.; Photinos, D.J. A molecular theory of nematic–nematic phase transitions in mesogenic dimers. *Soft Matter* **2016**, *12*, 2208–2220. [CrossRef]
73. Lavrentovich, O.D. Ferroelectric nematic liquid crystal, a century in waiting. *Proc. Natl. Acad. Sci. USA* **2020**, *117*, 14629–14631. [CrossRef]
74. Chen, X.; Korblova, E.; Dong, D.; Wei, X.; Shao, R.; Radzihovsky, L.; Glaser, M.A.; Maclennan, J.E.; Bedrov, D.; Walba, D.M.; et al. First-principles experimental demonstration of ferroelectricity in a thermotropic nematic liquid crystal: Polar domains and striking electro-optics. *Proc. Natl. Acad. Sci. USA* **2020**, *117*, 14021–14031. [CrossRef]
75. Wilson, E.B.; Decius, J.C.; Cross, P.C. *Molecular Vibrations*; McGraw-Hill: New York, NY, USA, 1955.
76. Kocot, A.; Loska, B.; Arakawa, Y.; Mehl, G.H.; Merkel, K. Study of the experimental and simulated vibrational spectra together with conformational analysis for thioether cyanobiphenyl-based liquid crystal dimers. *Int. J. Mol. Sci.* **2022**, *23*, 8005–8016. [CrossRef]
77. Barnes, P.J.; Douglass, A.G.; Heeks, S.K.; Luckhurst, G.R. An enhanced odd-even effect of liquid crystal dimers Orientational order in the α,ω-bis(4′-cyanobiphenyl-4-yl)alkanes. *Liq. Cryst.* **1993**, *13*, 603–613. [CrossRef]
78. Carvalho, J.; Cruz, C.; Figueirinhas, J.L.; Tamba, M.G.; Kohlmeier, A.; Mehl, G.H. Proton and deuterium NMR study of the CBC9CB dimer system. *J. Phys. Chem. B* **2019**, *123*, 1442–1451. [CrossRef]
79. Aluculesei, A.; Cachitas, H.; Carvalho, J.; Chavez, F.V.; Figueirinhas, J.L.; Sebastião, P.J.; Cruz, C.; Tamba, M.G.; Kohlemeir, A.; Mehl, G.H. ^{1}H NMR study of molecular order and dynamics in the liquid crystal CB-C9-CB. *Phys. Chem. Chem. Phys.* **2019**, *21*, 4523–4537. [CrossRef]
80. Krzyżewska, K.; Jaroch, T.; Maranda-Niedbała, A.; Pociecha, D.; Górecka, E.; Ahmed, Z.; Welch, C.; Mehl, G.H.; Proń, A.; Nowakowski, R. Supramolecular organization of liquid-crystal dimers-bis-cyanobiphenyl alkanes on HOPG by scanning tunneling microscopy. *Nanoscale* **2018**, *10*, 16201–16210. [CrossRef]
81. Arakawa, Y.; Komatsu, K.; Tsuji, H. Twist-bend nematic liquid crystals based on thioether linkage. *New J. Chem.* **2019**, *43*, 6786–6793. [CrossRef]
82. Arakawa, Y.; Ishida, Y.; Tsuji, H. Ether- and Thioether-Linked Naphthalene-Based Liquid-Crystal Dimers: Influence of Chalcogen Linkage and Mesogenic-Arm Symmetry on the Incidence and Stability of the Twist–Bend Nematic Phase. *Chem. Eur. J.* **2020**, *26*, 3767–3775. [CrossRef]
83. Frisch, M.J.; Trucks, G.W.; Schlegel, H.B.; Scuseria, G.E.; Robb, M.A.; Cheeseman, J.R.; Scalmani, G.; Barone, V.; Mennucci, B.; Petersson, G.A.; et al. *Gaussian 09, Revision E.01*; Gaussian, Inc.: Wallingford, UK, 2009.
84. Becke, A.D. Density-functional exchange-energy approximation with correct asymptotic behavior. *Phys. Rev. A* **1988**, *38*, 3098. [CrossRef]
85. Hertwig, R.H.; Koch, W. On the parameterization of the local correlation functional. What is Becke-3-LYP? *Chem. Phys. Lett.* **1997**, *268*, 345. [CrossRef]

International Journal of
Molecular Sciences

MDPI

Article

New Organic–Inorganic Salt Based on Fluconazole Drug: TD-DFT Benchmark and Computational Insights into Halogen Substitution

Hela Ferjani [1,*], Rim Bechaieb [2,3], Marzook Alshammari [4], O. M. Lemine [5] and Necmi Dege [6]

1 Chemistry Department, College of Science, Imam Mohammad Ibn Saud Islamic University (IMISU), Riyadh 11623, Saudi Arabia
2 Selective Organic & Heterocyclic Synthesis-Biological Activity Evaluation, Département de Chimie, Faculté des Sciences de Tunis, Université de Tunis El Manar, Tunis 2092, Tunisia
3 Laboratoire de Chimie Théorique, Sorbonne Universités, UPMC Univ Paris 06, UMR 7616, F-75005 Paris, France
4 The National Center for Nanotechnology and Semiconductors, KACST, Riyadh 11442, Saudi Arabia
5 Department of Physics, College of Sciences, Imam Mohammad Ibn Saud Islamic University (IMISU), Riyadh 11623, Saudi Arabia
6 Department of Physics, Faculty of Arts and Sciences, Ondokuz Mayıs University, 55139 Samsun, Turkey
* Correspondence: hhferjani@imamu.edu.sa

Abstract: In this study, we report the synthesis of a new organic–inorganic molecular salt of the clinically used antifungal drug fluconazole, (H2Fluconazole).$SnCl_6.2H_2O$. By detailed investigation and analysis of its structural properties, we show that the structure represents a 0D structure built of alternating organic and inorganic zig-zag layers along the crystallographic c-axis and the primary supramolecular synthons in this salt are hydrogen bonding, F···π and halogen bonding interactions. Magnetic measurements reveal the co-existence of weak ferromagnetic behavior at low magnetic field and large diamagnetic contributions, indicating that the synthesized material behaves mainly as a diamagnetic material, with very low magnetic susceptibility and with a band gap energy of 3.6 eV, and the salt is suitable for semiconducting applications. Extensive theoretical study is performed to explain the acceptor donor reactivity of this compound and to predict the Cl-substitution effect by F, Br and I. The energy gap, frontier molecular orbitals (FMOs) and the different chemical reactivity descriptors were evaluated at a high theoretical level. Calculations show that Cl substitution by Br and I generates compounds with more important antioxidant ability and the intramolecular charge transfer linked to the inorganic anion.

Keywords: crystal structure; fluconazole drug; salt; spectroscopy; optical and magnetic studies; DFT; TD-DFT; chemical reactivity descriptors; ICT

1. Introduction

The great technological potential of hybrid organic–inorganic materials as low-cost solar-cell absorbers, phosphors, and light-emitting diodes [1–4] has renewed interest in this large and well-known family of materials. The synthetic versatility and structural definition of these crystalline solids, which are typically formed through solution-state self-assembly, drive research in this field. Many possible combinations of organic amine (A), metal (B = Sn or Pb) and halides (X) within the $A_aB_bX_{3b+a}$ hybrid material framework provide access to a diverse set of compositions. These material structures are formed from MX_6 octahedra sharing corners, edges, or faces to form three-dimensional structures, two-dimensional systems, one-dimensional chains, or zero-dimensional clusters that are separated by organic cations [5–8]. Among these materials, the halogenostannate hybrid compounds containing protonated amine cations draw considerable attention, because of their interesting physical and chemical properties such as magnetic, electroluminescence, photoluminescence, and conductivity, which potentially lead to technological advances [9–12]. Apart from that, these compounds have gained attention due to their interesting antibacterial and antifungal properties against a wide range of bacterial and fungal species [13–19]. Several studies with fascinating results on antitumor activities against a diverse panel of human tumor cell lines have been reported [20–22]. In general, the structure of these compounds and the type of organic cations have a significant impact on their biochemical activity.

To date, many halogenostannate hybrid materials have been successfully assembled. However, there is a dearth of research on the interactions between organic and inorganic units. In comparison with other organic–inorganic coordination compounds, the rational design and assembly of organic halogenostannate, especially non-covalent connection, remains a difficult task for halogenostannate chemists. Several factors must be considered to achieve this goal, including the coordination geometry of the metal ions, the nature of the organic ligands, the solvent, pH, and reaction temperature. Ligand selection is the most crucial of these factors.

Fluconazole (1-(2,4-difluorophenyl)-1, 1-bis [(1H-1,2,4-triazol-1-yl) methyl] ethanol) (Scheme 1) is a 1,2,4-triazole derivative that is a commonly used as antifungal medicine but also a good flexible ligand to build metal–organic materials with optical properties and medical applications [23,24]. This ligand has several donor/acceptor groups (difluoro phenyl, triazole, and hydroxyl groups) that can form non-covalent interactions to stabilize the supramolecular architecture. A search in the Cambridge Structural Database (CSD version 5.42, November 2020) revealed 128 entries relating to metal complexes including fluconazole, compared to 112 entries relating to fluconazole molecules directly linked to the metal, with one structure reported by our group as the first organic–inorganic (linked by non-covalent interactions) salt base on bismuth-fluconazole [25].

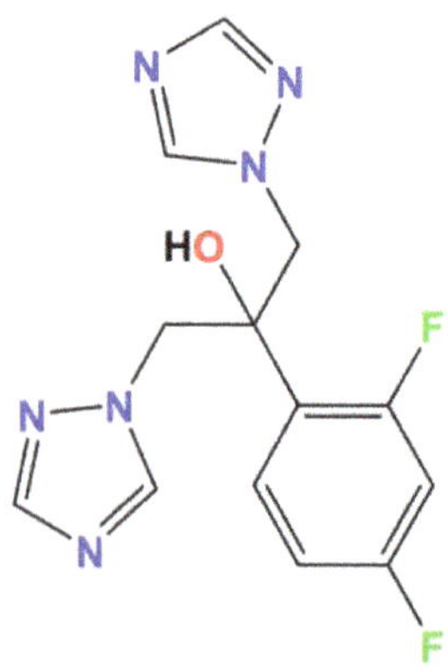

Scheme 1. Molecular structure of fluconazole.

In the context of this current study, we report the synthesis, detailed crystallographic analysis, vibrational study, and the optical and magnetic properties of a new hybrid organic–inorganic salt, $(C_{13}H_{14}F_2N_6O).SnCl_6.2H_2O$. In addition, this study is supported by an extensive computational investigation using the density functional theory (DFT) and time dependent density functional theory (TD-DFT). In the first step, we performed a theoretical benchmark that is more complete than any previously published data of this kind of systems, from the point of view of the number of functionals evaluated. Then, we retained the most suitable one to investigate the optical and electronic properties, as well as the frontier molecular orbitals (FMOs) and the energy gap between the highest occupied molecular orbital (HOMO) and the lowest unoccupied molecular orbital (LUMO). Then, we tried, to study the Cl-substitution effect to predict the gap energy and chemical descriptor.

2. Results and Discussion

2.1. Crystal Structure of (H2Fluconazole).SnCl6.2H_2O

The single-crystal data and the refinement results are presented in Table 1. Atomic coordinates and equivalent isotropic displacement parameters are provided in Table S1 in the Supporting Information (SI). Selected interatomic distances and bond angles are presented in Table S2 in the Supporting Information (SI).

Table 1. Crystallographic data and structure refinement parameters of (H2Fluconazole).$SnCl_6$.2H_2O.

Empirical Formula	($C_{13}H1_4F_2N_6O$).$SnCl_6$.2H_2O
Formula weight (g/mol)	675.72
Crystal system, space group	monoclinic, $P2_1/n$
a (Å)	8.0328 (5)
b (Å)	29.4011 (13)
c (Å)	10.3323 (7)
β (°)	100.901 (5)
V (Å^3)	2396.2 (2)
Z	4
μ (mm^{-1})	1.78
Dx (Mg m^{-3})	1.873
F(000)	1328
Crystal size (mm)	0.72 × 0.64 × 0.58
Crystal habit	Colorless, Prism
θmin/θmax (deg)	1.4/26.8
Measured reflections	13,858
Independent reflections	5043
Observed refl. with I > 2σ(I)	4495
Rint	0.046
Data/restraints/parameters	5043/0/289
R[F^2 > 2σ(F^2)]	0.037
wR (F^2)	0.088
GooF = S	1.12
Δρmax/Δρmin (e.Å^{-3})	0.53/−0.95

The (H2Fluconazole).$SnCl_6$.2H_2O salt crystallizes in the monoclinic system $P2_1/n$ space group with one anion $[SnCl_6]^{2-}$, one (H2Fluconazole)$^{2+}$ cation and two water molecules in the asymmetric unit (Figure 1). Table S1 lists some selected bond lengths and angles of this structure. These bond lengths and angles are in good agreement with those observed in similar compounds based on fluconazole cation and hexachlorostannate (IV) [23,26–29]. The (H2Fluconazole)$^{2+}$ cations interact with $[SnCl_6]^{2-}$ anions and the water molecules through strong intermolecular O-H···O (cyan dashed lines), N-H···O (orange dashed lines) and N-H···Cl (purple dashed lines) hydrogen bonds as shown in Figure 2a (Table 2). Additionally, the water molecule is bonded to the adjacent $[SnCl_6]^{2-}$ anions through O-H···Cl (Green dashed lines) hydrogen bonds (Table 2, Figure 2a). In addition to these strong hydrogen bonds, the cations are connected to the adjacent anions and water molecules via weak C-H···X (X = Cl and O) contacts (green dashed lines in Figure 2b). In this structure, H2Fluconazole randomly adopts anti-gauche (A-G) conformations [30]. The existence of intramolecular C-H···X (X = F, O and N) hydrogen bonds causes this conformation (Table 2, Figure 2b). Figure 3a shows how $[SnCl_6]^{2-}$ anions interact with H2Fluconazole cations via Cl1/5···Cg1/2/3 interactions. The d(Cl···Cg) distances are dCl1···Cg2 = 3.468 (2) Å, dCl5···Cg1 = 3.613 (2) Å and dCl5···Cg3 = 3.712 (2) Å, where Cg1 and Cg2 are the centroids of the triazole rings and Cg3 is the centroid of difluophenyl ring (purple dashed lines). Moreover, due to the inversion center, the H2Fluconazole cations are related to each other by halogen bonding, F···π, and π-stacking interactions. However, weak π-stacking interactions (Cg3···Cg3 = 4.325 (2) Å) were observed between the centroids of the difluorophenyl rings of the (H2Fluconazole) cations (red dashed lines in Figure 3b). In addition, the fluorine atom forms weak halogen bonds. From a geometric point of view, these F···F contacts can be classified as type I (d(F2···F2 = 2.961 (3) Å, Ang(C-F···F) = 100.80(16)°) purple dashed lines in Figure 3b) [31–36]. Additionally, the F atoms engage in a non-covalent interaction with the nitrogen atoms of the nearby. H2Fluconazole cations (d(F2···N2 = 3.134 (4) Å; blue dashed lines in Figure 3b). Furthermore, through C-F···π interactions, the fluorine atom interacts with adjacent triazole and difluorophenyl rings (dF1···Cg1 = 3.283 (3) Å, and dF1···Cg3 = 3.552 (3) Å, green dashed lines in Figure 3b). According to the geometry, the C-F bond is perpendicular or parallel to the aromatic ring, or, as in this instance, the C-F···π interaction take the form of T-shaped halogen bonding with Cg1 (C-F1···Cg1 = 171.31°) and π···π (lone-pair···π) with Cg3 (C-F1···Cg3 = 85.3°). Calculation on the electrostatic map was performed at the HSE theoretical level (Figure S1). This figure confirms the below Hirshfeld surface analysis and enrichment ratio calculation. indeed, we can see that the red region is the negative extreme which is mainly located in the inorganic part and the blue is the positive extreme and located in the organic part with slightly positive region located in the center of the difluorphenyl group interacting with the negative equatorial region of F. Finally, the interplay between hydrogen and halogen bonding as well as F···π and π-stacking interactions results in the formation of endless zig-zag chains in the [100] direction (Figure 4).

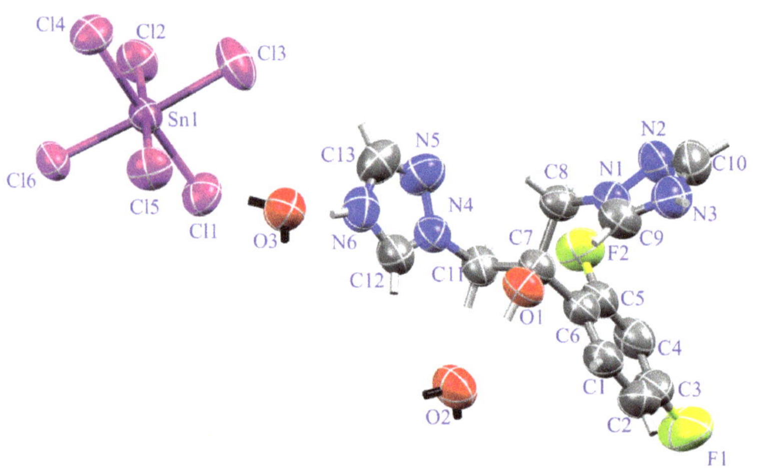

Figure 1. Asymmetric unit of (H2Fluconazole).SnCl6.2H_2O with 50% probability ellipsoids.

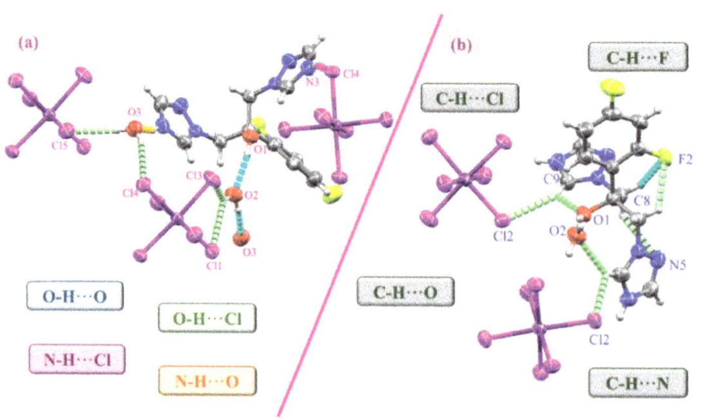

Figure 2. Hydrogen bond interactions in (H2Fluconazole).$SnCl_6$.2H_2O.

Table 2. Intermolecular interactions in (H2Fluconazole).$SnCl_6$.2H_2O.

D-H···A	D-H (Å)	H···A (Å)	D···A (Å)	D-H···A(°)
O1-H1···O2	0.82	1.94	2.750 (3)	169
O2-H2A···O3 ^{ii}	0.85	2.01	2.820(4)	160
O2-H2A···Cl1 ^{ii}	0.85	2.81	3.522(3)	142
O2-H2B···Cl3 ^{i}	0.85	2.77	3.395(3)	132
O3-H3A···Cl5	0.85	2.37	3.214(3)	177
C12-H12···Cl2 ^{i}	0.93	2.72	3.359 (4)	127
C12-H12···O2	0.93	2.52	3.284 (4)	140
C11-H11A···F2	0.97	2.29	2.921 (4)	122
C11-H11B···Cl3 ^{ii}	0.97	2.93	3.546 (3)	122
C8-H8A···F2	0.97	2.40	3.038 (4)	123
C8-H8B···Cl6 ^{iii}	0.97	2.86	3.471 (3)	122
C8-H8B···Cl1 ^{iii}	0.97	2.90	3.704 (3)	141
C8-H8B···N5	0.97	2.57	3.221 (5)	124
C9-H9···Cl2 ^{iv}	0.93	2.70	3.566 (3)	155
C9-H9···O1	0.93	2.34	2.813 (4)	111
C10-H10···Cl6 ^{v}	0.93	2.93	3.855 (4)	175
N6-H6···O3	0.85 (5)	1.91 (5)	2.736 (4)	162 (6)
N3-H3···Cl4 ^{iv}	0.85 (5)	2.44 (5)	3.240 (3)	157 (4)

Symmetry codes: (i) x − 1, y, z; (ii) x − 1/2, −y + 1/2, z + 1/2; (iii) x − 1/2, −y + 1/2, z − 1/2; (iv) x − 3/2, −y + 1/2, z − 1/2; (v) −x + 3/2, y + 1/2, −z + 1/2.

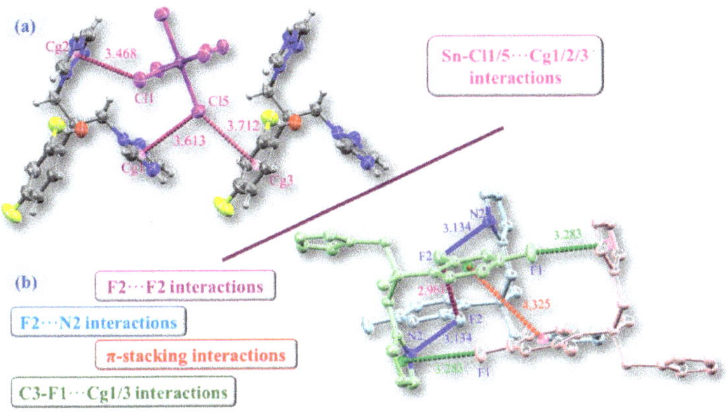

Figure 3. Details of the halogen bonding, X···π, π-stacking interactions in (H2Fluconazole).$SnCl_6$.2H_2O (with 50% and 20% probability ellipsoids in (**a**) and (**b**), respectively).

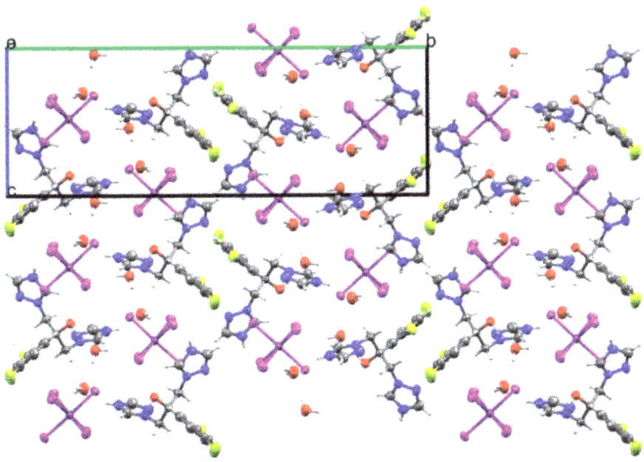

Figure 4. Overall packing of unit cell of (H2Fluconazole).$SnCl_6$.2H_2O viewed along the c-axis.

2.2. Hirshfeld Surface Analysis and Contact Enrichment Ratio

The Hirshfeld surfaces mapped over d_{norm} as well as the 2D fingerprint plots of the abundant contacts in the structure are presented in Figure 5. It should be noted that regions of intense red color are located over the oxygen, carbon, and nitrogen atoms of organic cations and water molecules (Figure 5a). Hirshfeld surface (HFs) maps (Figure 5b) show that the most abundant contacts are the O/N/C-H···Cl hydrogen interactions, which account for 44.5% of the total HFs maps. The non-covalent H···H interactions occupy 16.6% of the total HFs maps indicating high contribution of triazole and difluorophenyl rings in the stabilization of the crystal packing (Figure 5c). The relative contributions of all contacts governing the structure are summarized in Figure 6.

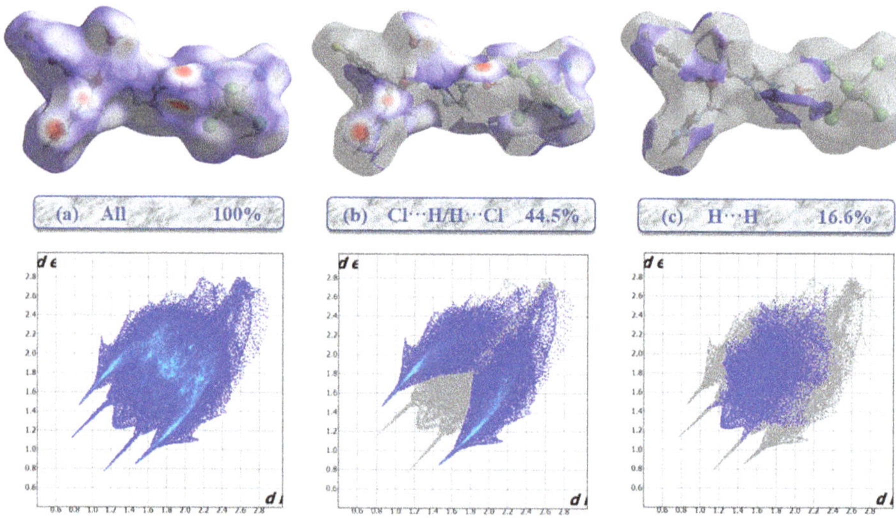

Figure 5. Two-dimensional fingerprint plots order with a d_{norm} view of (**a**) All (100%), (**b**) Cl···H/H···Cl (44.5%) and (**c**) H···H (16.6%) contacts in (H2Fluconazole).$SnCl_6$.$2H_2O$.

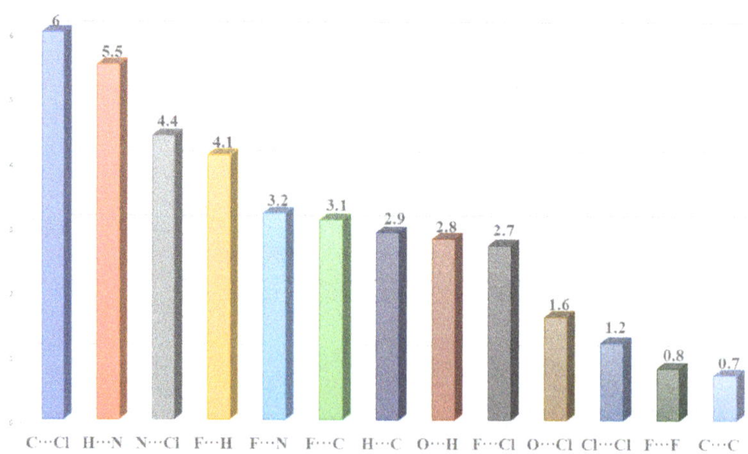

Figure 6. The relative contribution of intermolecular contacts in (H2Fluconazole).$SnCl_6$.$2H_2O$ to the Hirshfeld surface.

The enrichment ratios (E), which are the ratios of actual contacts in the crystal to random contacts (theoretical proportion), were calculated using surface contact data from the Hirshfeld surface analysis, with E greater than one indicating a favored contact between pairs of elements and E less than one indicating a disfavored contact. The enrichment ratios of (H_2Fluconazole).$SnCl_6.2H_2O$ (Table 3) shows that the F···N halogen bonding and F···π interactions (E_{FN} = 3.33 and E_{FC} = 3.16) are the most favoured contact in the crystal packing followed by the H···Cl hydrogen bond contacts (E_{HCl} = 1.55), C···C π-stacking interactions (E_{CC} = 1.55), C···Cl contacts (E_{CCl} = 1.45), O···Cl (E_{OCl} = 1.18) and N···Cl contacts (E_{NCl} = 1.09), respectively. Theoretical and experimental equilibrium geometry of the (H2Fluconazole).SnCl6.2H2O listed in Table S4.

Table 3. Analysis of contacts on the Hirshfeld surface. The % of contact types (SX) between chemical species is given followed by random contacts (%) and their enrichment ratio (E). The major contacts as well as the major enriched ones are highlighted in bold characters.

Atoms	Cl	H	C	N	O	F
			(%) Contacts from fingerprint plots			
Cl	1.2	-	-	-	-	-
H	44.5	16.6	-	-	-	-
C	6	2.9	0.7	-	-	-
N	4.4	5.5	0	0	-	-
O	1.6	2.8	0	0	0	-
F	2.7	4.1	3.1	3.2	0	0.8
Surface (%)	**30.8**	**46.5**	**6.7**	**6.54**	**2.2**	**7.35**
			Random contacts % (R_{XX} and R_{XY})			
Cl	9.84	-	-	-	-	-
H	28.64	21.6	-	-	-	–
C	4.13	6.32	0.45	-	-	-
N	4.03	6.09	0.88	0.043	-	-
O	1.35	2.05	0.29	0.28	0.048	-
F	4.52	6.84	0.98	0.96	0.32	0.54
			Enrichment ratio E (E_{XX} and E_{XY})			
Cl	0.13	-	-	-	-	-
H	**1.55**	0.77	-	-	-	-
C	**1.45**	0.46	**1.55**	-	-	-
N	**1.09**	0.9	0	0	-	-
O	**1.18**	**1.36**	0	0	0	-
F	0.59	0.6	**3.16**	**3.33**	0	

2.3. FT-IR and Raman Spectral Analysis

The solid-state IR and Raman spectra for (H2Fluconazole).$SnCl_6.2H_2O$ (Figure 7 and Figure S2, respectively) are consistent with their crystal structure. Table 4 shows the band assignments in the IR and Raman spectra, which are based on the spectra from earlier studies of fluconazole cation and $SnCl_6$ cluster [23–25,37–39] as well as quantum chemical computations.

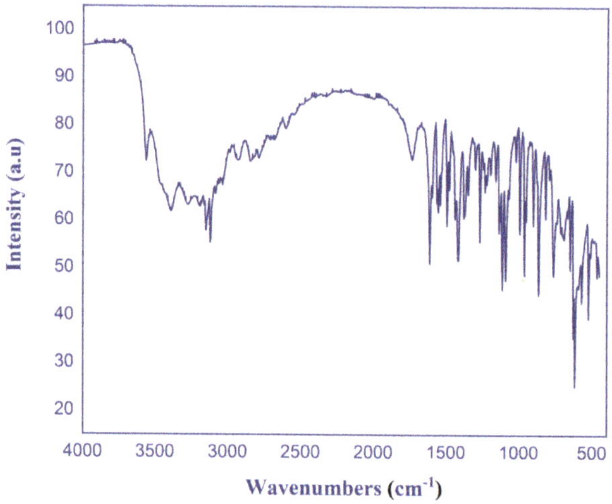

Figure 7. Experimental IR absorption spectrum of (H2Fluconazole).$SnCl_6$.$2H_2O$ recorded at room temperature.

Table 4. Attribution of the experimental and calculated IR frequencies (cm^{-1}) of (H2Fluconazole).$SnCl_6$.$2H_2O$.

Experimental	HSEH1PBE	Attribution
495-547	450-572	skeletal deformation
613	611	C-H out of plane
677	695	O-H out of plane
749	755	N-H out of plane of triazole group
837	818	C-H bending of triazole
904	897	C-H bending of benzene
948	965	C-C bending
1004	1001	C-N bending
1071	1085	C-O stretch mode
1103	1115	C-H and N-H bending of triazole
1257	1278	C-F bending mode
1408	1415	C-N stretch of triazole ring
1479	1490	C = C stretch 2,4-difluorobenzyl group
1530	1554	C = N stretch of triazole ring
1620	1607	C = N stretch of triazole ring
1631	1740	Benzene ring deformation
2411	2850	N-H stretch of triazole ring
2989	2928	C-H stretch of propane backbone
3156	3120	C-H stretch of propane backbone
3339	3150	C-H stretch (antisymmetric) of propane backbone
3412	3190	N-H stretch of triazole
3497	3274	O-H stretch mode
3603	3390	νs(OH) of crystallization water
3607	3560	νas(OH) of crystallization water

Internal Modes of H2Fluconazole Cation

The stretching vibrations of the different functional groups interconnected by a series of hydrogen bonds in the crystal are represented in the high frequency region, which ranges from 3560 to 2600 cm^{-1}. However, the presence of crystallization water in the structure is indicated by the strong signals centered at roughly 3560 and 3390 cm^{-1}, which are assigned to the O-H stretching vibrations. The O-H stretching modes of the tertiary alcohol and N-H groups appeared as a weak band centered at 3274 and 3190 cm^{-1}, respectively. In addition, the spectrum shows strong intensity bands in the 3150–3120 cm^{-1} spectral range which are attributed to C-H stretching modes of the triazole ring. These individual frequencies are slightly shifted because of the presence of intermolecular C-H/π interactions, as was discussed in the analysis of the crystal structure. The absorption bands of the ν_a(CH2) and ν_s(CH2) group in the complexes are observed in the frequency range 2928–2850 cm^{-1}.

Asymmetric and symmetric bending vibrations of δ(N-H) and δ(C-H), as well as valence vibrations of ν(C-O), ν(C = C), ν(C-N), ν(C = N) and ν(C-F) generally appear coupled and can be observed in the 1607–1085 cm^{-1} range. Finally, N-H, O-H and C-H out-of-plane bending modes are attributed to the bands in the 1000–500 cm^{-1} range.

The scaled IR spectrum using HSEH1PBE functional associated to LANL2DZ pseudopotential and 6-311++G** basis in the considered region is in good agreement with the experimental data (Table 4).

2.4. UV–Visible Spectroscopic Study

The optical absorption spectrum (300–700 nm) of (H2Fluconazole).$SnCl_6.2H_2O$ (Figure 8a) reveals a broad absorption band across the ultraviolet region up to the absorption edge onset at 408 nm (3.03 eV). Two less pronounced bands can be seen in the range of 200–300 nm at 216 nm (5.74 eV) and 266 nm (4.66 eV). These two bands are produces by the charge transfer between the conjugated rings in H2Fluconazole [38,40].

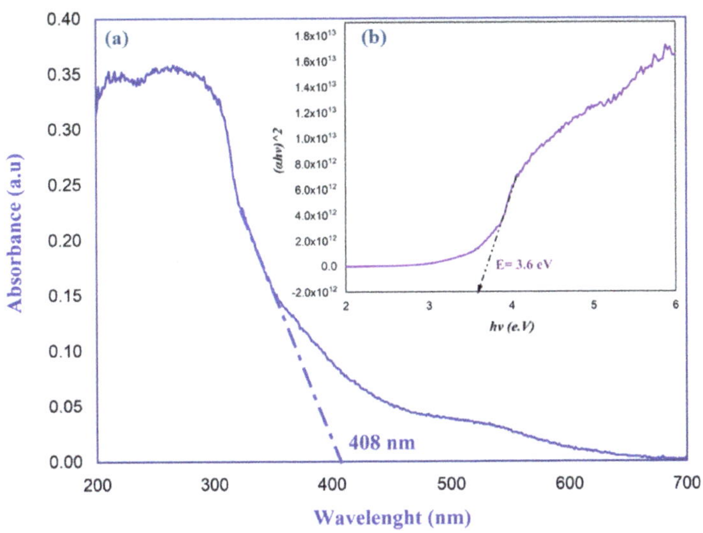

Figure 8. Solid-state UV-visible absorption spectra (**a**) and the Kubelka-Munk transformation of absorption data (**b**) of (H2Fluconazole).$SnCl_6.2H_2O$.

The material's diffuse reflectance was measured in order to estimate its optical band gap by changing the absorption spectrum into a Tauc plot Figure 8b, according to the Kubelka–Munk function: $F(R) = (1 - R)^2/2R$, in which R is the experimentally observed reflectance. In a $F(R)^2$ vs. photon energy plot, the energy band gap is determined as the point where the energy axis intersects with the extrapolated linear component of the absorption edge. The direct band gap value measured is 3.60 eV, which is typical with the compound's semiconductor characteristics.

2.5. Magnetic Study

The magnetic hysteresis loops measured at 10 and 300 K are shown in Figure 9. The magnetization curves indicate the co-existence of two components: weak ferromagnetic and large diamagnetic behavior. The M-H curve at room temperature shows mainly a diamagnetic contribution with small magnetic susceptibility. The diamagnetic contributions (i.e., the linear part) at high magnetic field and the ferromagnetic behavior as shown in the inset of Figure 9. The temperature dependence of the magnetic susceptibility measured at H = 50 kOe is shown in Figure 10. The diamagnetic behavior is confirmed again by the negative magnetic susceptibility at high field. It can be concluded from the magnetic measurement that the synthesized 0D organic–inorganic hybrid material behaves as a diamagnetic material, with very low magnetic susceptibility. Previous reports showed a ferromagnetic behavior for two-dimensional organic–inorganic hybrids [41,42].

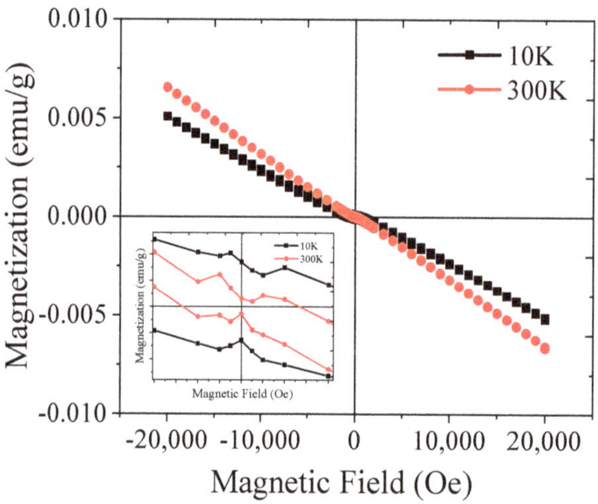

Figure 9. M-H curves at 10 and 300 K.

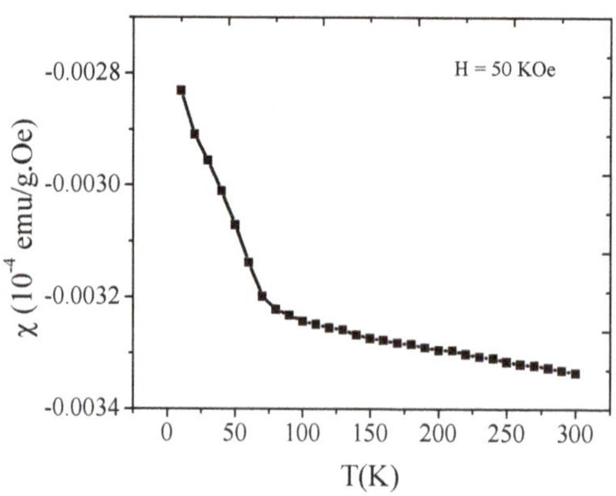

Figure 10. Magnetic susceptibility χ(T) measured at 50 KOe.

To reveal the origin of the ferromagnetism in this sample, the magnetization as a function of temperature measured in the field-cooled (FC) and zero-field-cooled (ZFC) configurations at an applied magnetic field of 100 Oe was measured as shown in Figure 11.

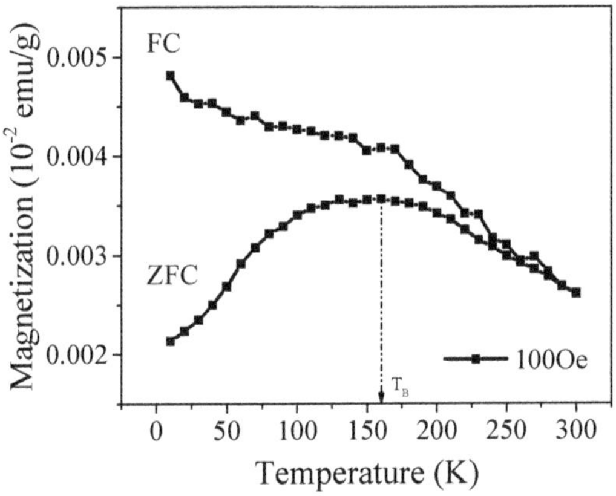

Figure 11. ZFC/FC curves.

The FC and ZFC curves show obvious bifurcation, which indicate the existing of magnetic nano clusters with blocking temperature of 160 K. Furthermore, the magnetization is not inversely proportional to T in the ZFC plot, indicating the non-existence of superparamagnetic nanoparticles. Therefore, the overall magnetization contributed to this system is originated from two sources. The first source of magnetization is the free carriers located in defect states where the density of magnetic contribution increases with the increase in localized carriers at low temperature. The second source of magnetization is the contribution of the magnetic nanoparticle exist in this sample as shown in the ZFC/FC measurements.

2.6. TD-DFT Study

Despite the difficult performance of benchmarks for this kind of systems, especially due to the requested resources, TD-DFT methods have become useful tools used by theorists and experiments to examine, understand, and predict the optical and chemical behavior of different types of organic and inorganic compounds [43–46] due to the remarkable ratio of accuracy and computational time.

As mentioned above, we have used nine density functionals to investigate the optical and chemical behavior of the studied compound.

In the first step, we focused on the evaluation of different functional performance by the comparison of calculated energy gap (Eg) with the experimental value (3.6 eV).

As we see in Figure 12, the B3LYP, CAM-B3LYP, LC-wHPBE and wB97XD functionals fail to recover the correct Eg. This is mainly due to the weak bonds of water molecules.

For the five functionals employing PBE correlation, PBE1PBE underestimates the Eg by 0.8 eV while PBEh1PBE overestimates it by 1.2 eV. However, the OHSE1PBE, OHSE2PBE and HSE1PBE functionals effectively reproduce the energy gap with error at approximately 8% (3.9 vs. 3.6 eV).

Since the HSE1PBE functional is the recommended one for the study of this type of systems, it will be retained in the next step for the description of the electron excited state transitions, the examination of the different chemical reactivity descriptors and the prediction of the halogen-substitution effect on the chemical behavior of studied compound.

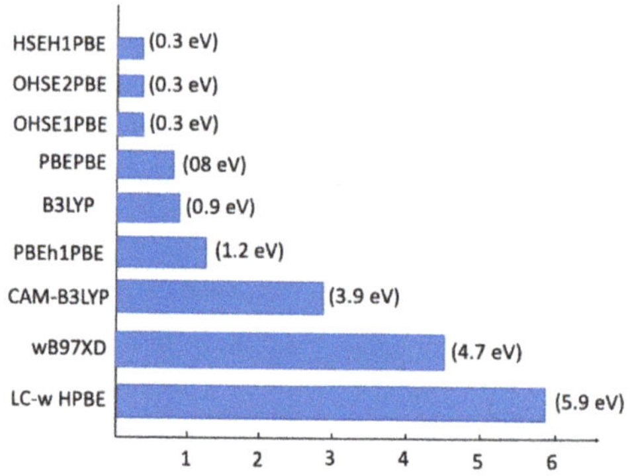

Figure 12. Absolute error between calculated energy gap (Eg) using different functionals and experimental value (3.6 eV).

To investigate the electronic and optical properties, as well as the UV-visible spectra and the way that any molecule interacts with other species, is related to the inspection of the highest occupied molecular orbital (HOMO) and the lowest unoccupied molecular orbital (LUMO). The higher energy of LUMO as an electron acceptor represents the higher resistance to accept an electron. While the low energy of the HOMO indicates the lower ability to donate an electron. The UV–visible spectra were calculated using the HSE1PBE functional and listed in Table S3. Close inspection of the properties of the electronic transitions shows that the first transition is mainly the promotion of a single electron from the HOMO to the LUMO with a predicted oscillator strength of 0.0012. The second transition occurs in UV, with a relatively high oscillator strength calculated to be 0.0265. Finally, a transition in the UV region calculated to occurs at 219 nm. This latter is a combination from a promotion of one electron from HOMO−10 to LUMO+2 and a second transition from HOMO−8 to LUMO+2. It should be noted that theoretical calculations are in good agreement with those measured experimentally in UV the region, whereas we note a low accuracy in the near UV-visible region. This may be due to the multireference character of the transition.

To better understand and predict the chemical behavior of molecules, several chemical reactivity descriptors, which been have detailed in our previous works [25,47], are widely used.

These descriptors are related to frontier molecular orbitals (FMOs) and evaluated using HOMO and LUMO energy values according to the following equations [48,49]:

$$\text{Energy gap (Eg), Eg} = \text{E(LUMO)} - \text{E(HOMO)} \quad (1)$$

$$\text{First Ionization energy (I), I} = -\text{E}_{(\text{HOMO})} \quad (2)$$

$$\text{Affinity (A), A} = -\text{E}_{(\text{LUMO})} \quad (3)$$

$$\text{Chemical hardness } (\eta),\ \eta = \frac{(\text{E}_{\text{LUMO}} - \text{E}_{\text{HOMO}})}{2} \quad (4)$$

$$\text{Chemical potential } (\mu),\ \mu = \frac{(\text{E}_{\text{HOMO}} + \text{E}_{\text{LUMO}})}{2} \quad (5)$$

$$\text{Electrophilicity } (\omega),\ \omega = \frac{\mu^2}{2\eta} \quad (6)$$

$$\text{Electronegativity } (\chi),\ \chi = \frac{(\text{I} + \text{A})}{2} \quad (7)$$

$$\text{Softness (S), S} = \frac{1}{2\eta} \quad (8)$$

The energy gap (Eg) between the HOMO and the LUMO is a critical parameter to demonstrate the bioactivity and the intermolecular charge transfer [50]. A molecule with a small Eg is more polarizable and is generally associated with high chemical reactivity and low kinetic stability due to ease of charge transfer. It is also a soft molecule, which is confirmed by the low softness (S) value.

The low Eg of (H2Fluconazole).$SnCl_6$.$2H_2O$ (Eg = 3.6 eV) is a proof of the intramolecular charge transfer (ICT), which supports the antioxidant ability of the studied compound.

We will now examine the Cl-substitution effect on the electronic properties, that is, the exchange of Cl by F, Br or I. The main electronic data for the different. (H2Fluconazole).$SnCl_6$.$2H_2O$ compounds are reported in Table 5. At first glance, it can be seen that Eg decreases with the atomic number of halogens.

Table 5. Global reactivity descriptor values (in eV) of (H2Fluconazole).SnX_6.$2H_2O$; (X = F, Cl, Br, and I).

	F (Z = 9)	Cl (Z = 17)	Br (Z = 35)	I (Z = 53)
ELUMO	−1.9	−3.2	−3.4	−4.3
EHOMO	−7.4	−7.2	−6.5	−6.0
Eg	5.5	3.9	3.1	1.6
Ionis	7.4	7.2	6.5	6.0
Elec. Aff	1.9	3.2	3.4	4.3
Hardness	2.7	2.0	1.6	0.8
Softness	0.2	0.3	0.3	0.6
Chem. Pot	−4.7	−5.2	−5.0	−5.2
Electrophilicity	4.0	6.9	7.8	16.5
Electronegativity	4.7	5.2	5.0	5.2

As a large Eg is preferred for higher stability, we can conclude that the compound resulting from the Cl substitution by F is more stable and thus less reactive. The Cl substitution by I leads to a less stable compound and eventually more important intramolecular charge transfer within the molecule and so a significant antioxidant ability.

This trend was is consistent with those of other chemical reactivity descriptors.

The contour surfaces of the FMOs are depicted in Figure 13. Differently to Br and I, in the case of the compound resulting from the Cl substitution by F, the HOMO and the LUMO are localized on the organic cation. More precisely, the electron concentration of HOMO is on the 2,4-difluorophenyl ring of the H2Fluconazolium molecule, while the LUMO is localized on one of the triazolium rings without any electron concentration on inorganic anion.

In the case of (H2Fluconazole).SnX_6.$2H_2O$; (X = Cl, Br, I) compounds, the HOMO and the LUMO are localized only in the inorganic anion. Hence, Eg and other chemical reactivity descriptors are primarily linked to the inorganic anion due its importance in the activity of the molecule.

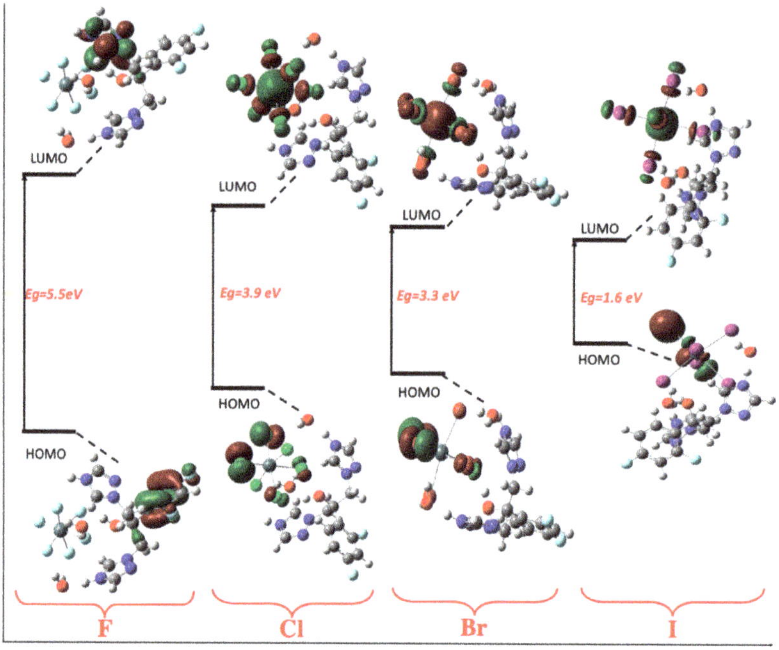

Figure 13. Frontier molecular orbitals and energy gap (Eg in eV) of (H2Fluconazole).SnX_6.$2H_2O$; (X = F, Cl, Br and I).

3. Materials and Methods

3.1. Salt Preparation

All the starting reagents were acquired commercially and were used without further purification. All the synthesis processes were carried out in the open air. A mixture of tin chloride (1 mmol, 0.074 g) and fluconazole (2 mmol, 0.1 g) was dissolved in concentrated hydrochloric acid (10 mL) and stirred for 2 h at 60 °C. The saturated colorless solution was slowly cooled and evaporated at ambient temperature. Colorless crystals suitable for X-ray diffraction analysis were obtained after 1 day.

3.2. X-ray Structure Determination

Single-crystal X-ray diffraction experiments were carried out on a STOE IPDS 2 diffractometer equipped with a graphite monochromator. Data collection and unit cell refinement for the datasets were performed using the X-AREA, data reduction and integration were performed by X-RED32 [51]. The crystal structures were solved by using Olex2 [52] software and the model obtained was refined by full-matrix least-squares on F2 (SHELXL2018/3) [53].

Anisotropic displacement parameters were used to refine all non-hydrogen atoms. A riding model was used to place hydrogen atoms at calculated positions. The new salt structure's crystallographic information was deposited in the CSD [54] with the code CCDC 2012995.

3.3. Physicochemical Characterization

The solid-state infrared spectrum was recorded using a Bruker Tensor 27 FT-IR spectrometer with scanning range set from 400 to 4000 cm^{-1}. The room temperature Raman spectrum was recorded in the region of 500–100 cm^{-1} with a RENISHAW RAMAN in Via Microscope with an excitation wavelength of 532 nm. The room-temperature UV–visible absorption spectrum of the polycrystalline powder of (H2Fluconazole).$SnCl_6$.H_2O was measured using a PerkinElmer Lambda 750 spectrophotometer in the wavelength range 300–700 nm, and the BaSO4 plate was used as reference. Magnetic characterizations were performed in a Quantum Design MPMS-5S SQUID magnetometer. Zero-field-cooled and field-cooled (ZFC-FC) curves were recorded at a magnetic field of 100 Oe.

3.4. Hirshfeld Surface Analysis (HS)

Hirshfeld surface analysis is a useful approach for evaluating weaker intermolecular interactions such as hydrogen and halogen bonds and other weak contacts. The Hirshfeld surface [55] and the decomposed two-dimensional (2D) fingerprint plots [55] were computed with CrystalExplorer17 [56]. Furthermore, the enrichment ratios (E) [57] of the intermolecular contacts were calculated to estimate the likelihood of two chemical species interacting.

3.5. Theoretical Calculations

Computations have been carried out with the Gaussian 16 program [58]. The asymmetric unit of (H2Fluconazole).$SnCl_6$.$2H_2O$ crystal (Figure 1) was used to start the geometry optimization and frequency calculations. To perform our benchmark, we have employed several hybrid functionals. We started by using the famous Becke Three-Parameter Hybrid functional (B3LYP) [59,60], then we tested two long-range corrected functionals-CAM-B3LYP [61] and LC-wHPBE [62,63]. We also employed the wB97XD [64] functional including the dispersion correction (Grimme's D2 dispersion model). In addition, we tested five functionals employing PBE correlation-PBE1PBE [65] and PBEH1PBE [66]; the HSE functionals OHSE1PBE and OHSE2PBE; and the recommended version of the full Heyd-Scuseria-Erzerhof functional HSE1PBE given by Scuseria and coworkers [67–71], which have been proved as suitable functionals for solid-state calculations. In these calculations, the LANL2DZ pseudopotential [72] was used to describe the Sn and I atoms and the 6-311++G** basis set was used to describe F, O, N, C, H, Cl and Br atoms.

To treat all the elements on an equal footing, additional calculations with the 6-31+G* basis set for the light atoms and LANL2DZ for the Sn atom were performed. The quality of the basis set does not affect the UV-visible spectrum and the different chemical reactivity descriptors studied. All geometries were fully optimized in the C1 symmetry group. All structures present a real minimum in their potential energy surface, and this was confirmed by the lack of imaginary frequencies.

TD-DFT calculations were performed at the optimized geometries using the same levels of theory. These extensive calculations were performed to examine the optical behavior, the chemical reactivity descriptors and the reliability of various functionals.

4. Conclusions

In this work, a new salt of the antifungal drug fluconazole, (H2Fluconazole).$SnCl_6$.$2H_2O$, was synthesized and characterized by single-crystal XRD, Hirshfeld surface analysis, Raman, FT-IR and UV-visible spectroscopies. This new compound crystallizes with a monoclinic crystal structure in the $P2_1/n$ space group and represents an 0D structure built of alternating organic and inorganic zig-zag layers along the crystallographic c-axis. The nature of intermolecular interactions in the title compound was also confirmed and quantified using HS analysis and enrichment ratio calculations. IR and Raman techniques confirm all vibrational modes of molecular groups. The optical properties of (H2Fluconazole).$SnCl_6$.$2H_2O$ at the absorption edge revealed less pronounced bands at 216 and 266 nm, respectively. The direct band gap energy measured was 3.6 eV. The magnetic studies show that at low magnetic fields, weak ferromagnetic contributions coexist with large diamagnetic contributions, implying that the synthesized material behaves primarily as a diamagnetic material, with very low magnetic susceptibility. Despite the extensive DFT calculations in conjunction with large basis sets, only the HSE functionals, i.e., OHSE1PBE, OHSE2PBE and HSE1PBE, reproduce the Eg effectively. From a comparative point of view, HSE1PBE agrees better with the measured experimental values, where the largest error is approximately 16% in the H-O bond. According to our study, we recommend the HSE1PBE functional to study similar systems. HSE1PBE provides accurate data about the antioxidant ability of the synthetized crystal and predicts the Cl-substitution effect on the chemical behavior. A significant intermolecular charge transfer (ICT) was found in the case of Cl, Br and I. Finally, FMOs show that the ICT process is mainly linked to the inorganic anion part.

Supplementary Materials: The following supporting information can be downloaded at: https://www.mdpi.com/article/10.3390/ijms23158765/s1.

Author Contributions: Conceptualization, H.F. and R.B.; methodology, H.F. and R.B.; software, H.F. and R.B.; validation, H.F., O.M.L. and R.B.; formal analysis, M.A. and N.D.; investigation, H.F. and R.B.; data curation, M.A. and N.D.; writing—original draft preparation, H.F., O.M.L. and R.B.; writing—review and editing, H.F. and R.B.; visualization, M.A. and N.D.; supervision, H.F. and R.B.; project administration, H.F.; funding acquisition, H.F. All authors have read and agreed to the published version of the manuscript.

Funding: This research received no external funding.

Institutional Review Board Statement: Not applicable.

Informed Consent Statement: Not applicable.

Data Availability Statement: The raw/processed data generated in this work are available upon request from the corresponding author.

Acknowledgments: The authors extend their appreciation to the Deanship of Scientific Research at Imam Mohammad Ibn Saud Islamic University for funding this work through Research Group no. RG-21-09-68. We thank Tarek Trabelsi for further discussion and the scientific support.

Conflicts of Interest: The authors declare no conflict of interest.

References

1. Allendorf, M.D.; Bauer, C.A.; Bhakta, R.; Houk, R. Luminescent metal-organic frameworks. *Chem. Soc. Rev.* **2009**, *38*, 1330–1352. [CrossRef]
2. Saparov, B.; Mitzi, D.B. Organic–inorganic perovskites: Structural versatility for functional materials design. *Chem. Rev.* **2016**, *116*, 4558–4596. [CrossRef] [PubMed]
3. Wang, M.-S.; Guo, G.-C. Inorganic-organic hybrid white light phosphors. *Chem. Commun.* **2016**, *52*, 13194–13204. [CrossRef] [PubMed]
4. Shan, Q.; Song, J.; Zou, Y.; Li, J.; Xu, L.; Xue, J.; Dong, Y.; Han, B.; Chen, J.; Zeng, H. High performance metal halide perovskite light-emitting diode: From material design to device optimization. *Small* **2017**, *13*, 1701770. [CrossRef] [PubMed]
5. Ferjani, H. Structural, Hirshfeld surface analysis, morphological approach, and spectroscopic study of new hybrid iodobismuthate containing tetranuclear 0D cluster $Bi_4I_{16}\cdot 4(C_6H_9N_2)\ 2(H_2O)$. *Crystals* **2020**, *10*, 397. [CrossRef]
6. Ferjani, H.; Boughzala, H. New hybrid material: $(C_3H_6N_3)_4Bi_2Cl_{10}$. Synthesis, structural study and spectroscopic behavior. *Russ. J. Inorg. Chem.* **2018**, *63*, 349–356. [CrossRef]
7. Ferjani, H.; Boughzala, H.; Driss, A. Poly[bis(1-carbamoylguanidinium) [tri-μ-chlorido-dichloridobismuthate (III)]]. *Acta Crystallogr. Sect. E Struct. Rep. Online* **2012**, *68*, m615. [CrossRef]
8. Ferjani, H.; Boughzala, H.; Driss, A. Synthesis, Crystal Structure, and Characterization of a New Organic-Inorganic Hybrid Material. *J. Crystallogr.* **2013**, *2013*, 658939. [CrossRef]
9. Fu, P.; Huang, M.; Shang, Y.; Yu, N.; Zhou, H.-L.; Zhang, Y.-B.; Chen, S.; Gong, J.; Ning, Z. Organic-Inorganic layered and hollow tin bromide perovskite with tunable broadband emission. *ACS Appl. Mater. Interfaces* **2018**, *10*, 34363–34369. [CrossRef]
10. Lassoued, M.S.; Osman, H.H.; Abdelbaky, M.S.M.; Lassoued, A.; Ammar, S.; Ben Salah, A.; Gadri, A.; García-Granda, S. Synthesis, crystal structure, DFT (B3LYP/LanL2DZ) and photoluminescence study of new stanate (IV) based inorgan-ic-organic hybrid. *J. Phys. Chem. Solids* **2018**, *121*, 177–185. [CrossRef]
11. Spanopoulos, I.; Hadar, I.; Ke, W.; Guo, P.; Sidhik, S.; Kepenekian, M.; Even, J.; Mohite, A.D.; Schaller, R.D.; Kanatzidis, M.G. Water-Stable 1D Hybrid Tin(II) Iodide Emits Broad Light with 36% Photoluminescence Quantum Efficiency. *J. Am. Chem. Soc.* **2020**, *142*, 9028–9038. [CrossRef] [PubMed]
12. Mathlouthi, M.; Valkonen, A.; Rzaigui, M.; Smirani, W. Structural characterization, spectroscopic, thermal, AC conductivity and dielectric properties and antimicrobial studies of $(C_8H_{12}N)_2[SnCl_6]$. *Phase Transit.* **2017**, *90*, 399–414. [CrossRef]
13. Matela, G.; Aman, R.; Sharma, C.; Chaudhary, S. Reactions of tin- and triorganotin(IV) isopropoxides with thymol derivative: Synthesis, characterization and in vitro antimicrobial screening. *J. Serb. Chem. Soc.* **2013**, *78*, 1323–1333. [CrossRef]

14. Amir, M.K.; Khan, S.; Zia-ur-Rehman; Shah, A.; Butler, I.S. Anticancer activity of organotin(IV) carboxylates. *Inorg. Chim. Acta* **2014**, *423*, 14–25. [CrossRef]
15. Kaiba, A.; Geesi, M.H.; Riadi, Y.; Ibnouf, E.O.; Aljohani, T.A.; Guionneau, P. A new Organic—Inorganic hybrid compound $(NH_3(CH_2)_2C_6H_5)_2[SnCl_6]$: Crystal structure, characterization, Hirshfeld surface analysis, DFT calculation, vibrational properties and biological evaluation. *J. Solid State Chem.* **2021**, *304*, 122587. [CrossRef]
16. Sedaghat, T.; Naseh, M.; Khavasi, H.R.; Motamedi, H. Synthesis, spectroscopic investigations, crystal structures and antibacterial activity of 3-(3-hydroxypyridin-2-ylamino)-1-phenylbut-2-en-1-one and its diorganotin (IV) complexes. *Polyhedron* **2012**, *33*, 435–440. [CrossRef]
17. Joshi, R.; Kumari, A.; Singh, K.; Mishra, H.; Pokharia, S. New diorganotin(IV) complexes of Schiff base derived from 4-amino-3-hydrazino-5-mercapto-4H1,2,4-triazole: Synthesis, structural characterization, density functional theory studies, atoms-in-molecules analysis and antifungal activity. *Appl. Organomet. Chem.* **2019**, *33*, 4894. [CrossRef]
18. Shah, F.A.; Sabir, S.; Fatima, K.; Ali, S.; Qadri, I.; Rizzoli, C. Organotin(iv) based anti-HCV drugs: Synthesis, characterization and biochemical activity. *Dalton Trans.* **2015**, *44*, 10467–10478. [CrossRef]
19. Nath, M.; Vats, M.; Roy, P. Design, spectral characterization, anti-tumor and anti-inflammatory activity of trorganotin (IV) hydroxycarboxulates, apoptosis inducers: In vitro assessment of inductiom of apoptosis by enzyme, DNA fragmentation, acridin orange and comet assays. *Inorg. Chim. Acta* **2014**, *423*, 70–82. [CrossRef]
20. Mjos, K.D.; Orvig, C. Metallodrugs in medicinal inorganic chemistry. *Chem. Rev.* **2014**, *114*, 4540–4563. [CrossRef]
21. Gasser, G.; Otto, I.; Metzler-Nolte, N.J. Organometallic Anticancer Compounds. *Med. Chem.* **2011**, *54*, 3–25. [CrossRef] [PubMed]
22. Zhang, C.X.; Lippard, S.J. New metal complexes as potential therapeutics. *Curr. Opin. Chem. Biol.* **2003**, *7*, 481–489. [CrossRef]
23. Dayo Owoyemi, B.C.; da Silva, C.C.; Souza, M.S.; Diniz, L.F.; Ellena, J.; Carneiro, R.L. Fluconazole: Synthesis and structural characterization of four new pharmaceutical cocrystal forms. *Cryst. Growth Des.* **2019**, *19*, 648–657. [CrossRef]
24. Gonnade, R.G.; Sangtani, E. Polymorphs and cocrystals: A comparative analysis. *J. Indian Inst. Sci.* **2017**, *97*, 193–226. [CrossRef]
25. Ferjani, H.; Bechaieb, R.; El-Fattah, W.A.; Fettouhi, M. Broad-band luminescence involving fluconazole antifungal drug in a lead-free bismuth iodide perovskite: Combined experimental and computational insights. *Spectrochim. Acta Part A Mol. Biomol. Spectrosc.* **2020**, *237*, 118354. [CrossRef]
26. Ali, M.M.; Ali, M.; Gaballa, A.S.; El-Korashy, S.A.; Teleb, S.M. Preparation, spectroscopic, characterizations and biological studies of new charge transfer complexes formed between fluconazole drug with various acceptors. *Bioorganic Chem.* **2021**, *115*, 105190. [CrossRef]
27. Zhou, B.; Liu, H. Tetrakis (4-chloroanilinium) hexachloridostannate (IV) dichloride. *Acta Crystallogr. Sect. E Struct. Rep. Online* **2012**, *68*, m782. [CrossRef]
28. Surov, A.O.; Voronin, A.P.; Vasilev, N.A.; Churakov, A.V.; Perlovich, G.L. Cocrystals of fluconazole with aromatic carboxylic acids: Competition between anhydrous and hydrated solid forms. *Cryst. Growth Des.* **2019**, *20*, 1218–1228. [CrossRef]
29. Ferjani, H. Crystal structure, optical property, and Hirshfeld surface analysis of bis [1-(prop-2-en-1-yl)-1H-imidazol-3-ium] hexachloridostannate (IV). *Acta Crystallogr. Sect. E Crystallogr. Commun.* **2020**, *76*, 1624–1628. [CrossRef]
30. Yanai, T.; Tew, D.P.; Handy, N.C. A new hybrid exchange-correlation functional using the Coulomb-attenuating method (CAM-B3LYP). *Chem. Phys. Lett.* **2004**, *393*, 51–57. [CrossRef]
31. Auffinger, P.; Hays, F.A.; Westhof, E.; Ho, P.S. Halogen bonds in biological molecules. *Proc. Natl. Acad. Sci. USA* **2004**, *101*, 16789–16794. [CrossRef] [PubMed]
32. Scholfield, M.R.; Vander Zanden, C.M.; Carter, M.; Ho, P.S. Halogen bonding (X-bonding): A biological perspective. *Protein Sci.* **2013**, *22*, 139–152. [CrossRef] [PubMed]
33. Chopra, D.; Row, T.N.G. Role of organic fluorine in crystal engineering. *Cryst. Eng. Comm.* **2011**, *13*, 2175. [CrossRef]
34. Scheiner, S. The F-Halogen Bond: Conditions for its Existence. *J. Phys. Chem. A* **2020**, *124*, 7290–7299. [CrossRef]
35. Eskandari, K.; Lesani, M. Does Fluorine Participate in Halogen Bonding? *Chem. A Eur. J.* **2015**, *21*, 4739–4746. [CrossRef]
36. Pavan, M.S.; Prasad, K.D.; Row, T.N.G. Halogen bonding in fluorine: Experimental charge density study on intermolecular F···F and F···S donor–acceptor contacts. *Chem. Commun.* **2013**, *49*, 7558. [CrossRef]
37. Gupta, A.K.; Foley, K.A.; Versteeg, S.G. New antifungal agents and new formulations against dermatophytes. *Mycopathologia* **2017**, *182*, 127–141. [CrossRef]
38. Naveed, S. UV spectrophotometric assay method for the determination of fluconazole capsules. *Open Access Libr. J.* **2015**, *2*, 68301. [CrossRef]
39. Nalintya, E.; Kiggundu, R.; Meya, D. Evolution of cryptococcal antigen testing: What is new? *Curr. Fungal Infect. Rep.* **2016**, *10*, 62–67. [CrossRef]
40. Zayed, M.; El-Shall, M.; Azime, M.A. Spectrophotometric Determination of Fluconazole, Voriconazole and Butoconazole nitrate by Ion-Pair Formation with Rose Bengal Reagent. *Egypt. J. Chem.* **2017**, *60*, 1177–1188. [CrossRef]
41. Derbel, M.A.; Turnbull, M.M.; Naïli, H.; Rekik, W. A new mixed halide 2D hybrid perovskite: Structural, thermal, optic and magnetic properties. *Polyhedron* **2020**, *175*, 114220. [CrossRef]
42. Bochalya, M.; Kumar, S. Magnetocaloric effect in 2D-alkylammonium copper halides layered inorganic-organic systems. *J. Appl. Phys.* **2020**, *127*, 055501. [CrossRef]

43. Zárate, X.; Schott, E.; Carey, D.M.-L.; Bustos, C.; Arratia-Pérez, R. DFT study on the electronic structure, energetics and spectral properties of several bis (organohydrazido(2-)) molybdenum complexes containing substituted phosphines and chloro atoms as ancillary ligands. *J. Mol. Struct. Theochem* **2010**, *957*, 126–132. [CrossRef]
44. Xu, Z.; Li, Y.; Zhang, W.; Yuan, S.; Hao, L.; Xu, T.; Lu, X. DFT/TD-DFT study of novel T shaped phenothiazine-based organic dyes for dye-sensitized solar cells applications. *Spectrochim. Acta Part A Mol. Biomol. Spectrosc.* **2019**, *212*, 272–280. [CrossRef] [PubMed]
45. Wei, J.; Song, P.; Ma, F.; Saputra, R.M.; Li, Y. Tunable linear and nonlinear optical properties of chromophores containing 3, 7-(di) vinylquinoxalinone core by modification of receptors moieties. *Opt. Mater.* **2020**, *99*, 109580. [CrossRef]
46. Wang, L.; Zhang, J.; Duan, Y.-C.; Pan, Q.-Q.; Wu, Y.; Geng, Y.; Su, Z.-M. Theoretical insights on the rigidified dithiophene effects on the performance of near-infrared cis-squaraine-based dye-sensitized solar cells with panchromatic absorption. *J. Photochem. Photobiol. A Chem.* **2019**, *369*, 150–158. [CrossRef]
47. Ferjani, H.; Bechaieb, R.; Dege, N.; El-Fattah, W.A.; Elamin, N.Y.; Frigui, W. Stabilization of supramolecular network of fluconazole drug polyiodide: Synthesis, computational and spectroscopic studies. *J. Mol. Struct.* **2022**, *1263*, 133192. [CrossRef]
48. Gümüş, H.P.; Tamer, Ö.; Avcı, D.; Atalay, Y. Quantum chemical calculations on the geometrical, conformational, spectroscopic and nonlinear optical parameters of 5-(2-Chloroethyl)-2, 4-dichloro-6-methylpyrimidine. *Spectrochim. Acta Part A Mol. Biomol. Spectrosc.* **2014**, *129*, 219–226. [CrossRef]
49. Tamer, Ö.; Bhatti, M.H.; Yunus, U.; Avcı, D.; Atalay, Y.; Nadeem, M.; Shah, S.R.; Helliwell, M. Structural, spectroscopic, nonlinear optical and electronic properties of calcium *N*-phthaloylglycinate: A combined experimental and theoretical study. *J. Mol. Struct.* **2016**, *1125*, 315–322. [CrossRef]
50. Hajji, M.; Kouraichi, C.; Guerfel, T. Modelling, structural, thermal, optical and vibrational studies of a new organic–inorganic hybrid material $(C_5H_{16}N_2)Cd_{1.5}Cl_5$. *Bull. Mater. Sci.* **2017**, *40*, 55–66. [CrossRef]
51. Stoe, C. *X-Area (Version 1.18) and X-Red32 (Version 1.04)*; Stoe & Cie: Darmstadt, Germany, 2002.
52. Dolomanov, O.V.; Bourhis, L.J.; Gildea, R.J.; Howard, J.A.; Puschmann, H. OLEX2: A complete structure solution, refinement and analysis program. *J. Appl. Crystallogr.* **2009**, *42*, 339–341. [CrossRef]
53. Sheldrick, G.M. Crystal structure refinement with SHELXL. *Acta Crystallogr. Sect. C Struct. Chem.* **2015**, *71*, 3–8. [CrossRef]
54. Allen, F.H. The Cambridge Structural Database: A quarter of a million crystal structures and rising. *Acta Crystallogr. Sect. B Struct. Sci.* **2002**, *58*, 380–388. [CrossRef] [PubMed]
55. Spackman, M.A.; McKinnon, J.J. Fingerprinting intermolecular interactions in molecular crystals. *Cryst. Eng. Comm.* **2002**, *4*, 378–392. [CrossRef]
56. Turner, M.; McKinnon, J.; Wolff, S.; Grimwood, D.; Spackman, P.; Jayatilaka, D.; Spackman, M. *CrystalExplorer17*; The University of Western Australia: Perth, Australia, 2017.
57. Jelsch, C.; Ejsmont, K.; Huder, L. The enrichment ratio of atomic contacts in crystals, an indicator derived from the Hirshfeld surface analysis. *IUCrJ* **2014**, *1*, 119–128. [CrossRef]
58. Frisch, M.J.; Trucks, G.W.; Schlegel, H.B.; Scuseria, G.E.; Robb, M.A.; Cheeseman, J.R.; Scalmani, G.; Barone, V.; Petersson, G.A.; Nakatsuji, H. *Gaussian 16*; Gaussian, Inc.: Wallingford, CT, USA, 2016.
59. Becke, A.D. Density-functional thermochemistry. IV. A new dynamical correlation functional and implications for exact-exchange mixing. *J. Chem. Phys.* **1996**, *104*, 1040–1046. [CrossRef]
60. Lee, C.; Yang, W.; Parr, R.G. Development of the Colle-Salvetti correlation-energy formula into a functional of the electron density. *Phys. Rev. B* **1988**, *37*, 785. [CrossRef]
61. Vydrov, O.A.; Scuseria, G.E. Assessment of a long-range corrected hybrid functional. *J. Chem. Phys.* **2006**, *125*, 234109. [CrossRef]
62. Vydrov, O.A.; Scuseria, G.E.; Perdew, J.P. Tests of functionals for systems with fractional electron number. *J. Chem. Phys.* **2007**, *126*, 154109. [CrossRef]
63. Chai, J.-D.; Head-Gordon, M. Long-range corrected hybrid density functionals with damped atom–atom dispersion corrections. *Phys. Chem. Chem. Phys.* **2008**, *10*, 6615–6620. [CrossRef]
64. Perdew, J.P.; Burke, K.; Ernzerhof, M. Generalized gradient approximation made simple. *Phys. Rev. Lett.* **1996**, *77*, 3865. [CrossRef] [PubMed]
65. Ernzerhof, M.; Perdew, J.P. Generalized gradient approximation to the angle-and system-averaged exchange hole. *J. Chem. Phys.* **1998**, *109*, 3313–3320. [CrossRef]
66. Heyd, J.; Scuseria, G.E.; Ernzerhof, M. Hybrid functionals based on a screened Coulomb potential. *J. Chem. Phys.* **2003**, *118*, 8207–8215. [CrossRef]
67. Heyd, J.; Scuseria, G.E. Efficient hybrid density functional calculations in solids: Assessment of the Heyd-Scuseria-Ernzerhof screened Coulomb hybrid functional. *J. Chem. Phys.* **2004**, *121*, 1187–1192. [CrossRef]
68. Heyd, J.; Peralta, J.E.; Scuseria, G.E.; Martin, R.L. Energy band gaps and lattice parameters evaluated with the Heyd-Scuseria-Ernzerhof screened hybrid functional. *J. Chem. Phys.* **2005**, *123*, 174101. [CrossRef]
69. Henderson, T.M.; Izmaylov, A.F.; Scalmani, G.; Scuseria, G.E. Can short-range hybrids describe long-range-dependent properties? *J. Chem. Phys.* **2009**, *131*, 044108. [CrossRef]
70. Izmaylov, A.F.; Scuseria, G.E.; Frisch, M.J. Efficient evaluation of short-range Hartree-Fock exchange in large molecules and periodic systems. *J. Chem. Phys.* **2006**, *125*, 104103. [CrossRef]

71. Lee, Y.S.; Ermler, W.C.; Pitzer, K.S. Ab initio effective core potentials including relativistic effects. I. Formalism and applications to the Xe and Au atoms. In *Molecular Structure and Statistical Thermodynamics: Selected Papers of Kenneth S Pitzer*; World Scientific: Singapore, 1993; pp. 112–127.
72. Chiodo, S.; Russo, N.; Sicilia, E. LANL2DZ basis sets recontracted in the framework of density functional theory. *J. Chem. Phys.* **2006**, *125*, 104107. [CrossRef]

International Journal of
Molecular Sciences

MDPI

Article

Study of the Experimental and Simulated Vibrational Spectra Together with Conformational Analysis of Thioether Cyanobiphenyl-Based Liquid Crystal Dimers

Antoni Kocot [1], Barbara Loska [1], Yuki Arakawa [2], Georg H. Mehl [3] and Katarzyna Merkel [1,*]

1 Institute of Materials Engineering, Faculty of Science and Technology, University of Silesia, ul. 75. Pułku Piechoty, 41-500 Chorzów, Poland; antoni.kocot@us.edu.pl (A.K.); barbara.loska@us.edu.pl (B.L.)
2 Department of Applied Chemistry and Life Science, Graduate School of Engineering, Toyohashi University of Technology, Toyohashi 441-8580, Japan; arakawa@tut.up
3 Department of Chemistry, University of Hull, Hull HU6 7RX, UK; g.h.mehl@hull.ac.uk
* Correspondence: katarzyna.merkel@us.edu.pl; Tel.: +48-32-349-7630

Abstract: Infrared spectroscopy (IR) and quantum chemistry calculations that are based on the density functional theory (DFT) have been used to study the structure and molecular interactions of the nematic and twist-bend phases of thioether-linked dimers. Infrared absorbance measurements were conducted in a polarized beam for a homogeneously aligned sample in order to obtain more details about the orientation of the vibrational transition dipole moments. The distributions to investigate the structure and conformation of the molecule dihedral angle were calculated. The calculated spectrum was compared with the experimental infrared spectra and as a result, detailed vibrational assignments are reported.

Keywords: FTIR spectroscopy; DFT simulations; intermolecular interactions; liquid crystal dimers; spatially modulated phases

1. Introduction

Soft matter physics encompasses materials that are easily deformed by thermal fluctuations and external forces, such as, for example, liquid crystals (LC) [1]. Undoubtedly the terms "crystal" and "liquid" are mutually exclusive, but it turns out that their merge perfectly reflects the uniqueness of the material to which it refers. Liquid crystals can be broadly defined as any state of matter with some degree of positional or orientational organization intermediate between the isotropic liquid state and a crystalline solid with long-range positional and orientational order in three dimensions. They can be classified, according to the source behind the formation of the mesophases, into lyotropic and thermotropic [2]. The first group, lyotropic liquid crystals, emerges when amphiphilic mesogens, composed of a flexible hydrophobic chain and a polar head group (ionic or nonionic), are dissolved in a suitable solvent at an appropriate temperature, pressure, and concentration [3]. The second group comprises thermotropic liquid crystals, where the existence of mesophase is exclusively controlled by temperature, and the composition of the forming material is not changing with temperature. In this article, we will narrow the discussion to only thermotropic liquid crystalline phases, primarily nematic and twist-bend phases (N_{TB}). Without any doubt, the discovered nematic twist-bend phase is different from 3D liquids known to date because it exhibits a macroscopic chirality while formed from chemically achiral, bent-core-like molecules.

Vibrational spectroscopy has become one of the most informative tools in the study of liquid crystal materials [4–13]. An analysis of the intensity and positional changes of the vibrational bands makes it possible to identify the liquid crystal phases and to calculate the orientational order parameters. Unfortunately, for large molecules, it is virtually impossible to reliably define the vibrational fundamentals without any theoretical input; therefore, determining the vibrational frequencies using computational methods is extremely important. These methods are helpful in interpreting the experimental vibrational spectra for large molecules. They also give information about the shape of the bands and the orientation of the transition dipole moment in the molecular coordinate system.

In this research, we studied the liquid crystal dimers that are based on cyanobiphenyl mesogens. These dimers typically contain two rigid terminal groups that are chemically linked to each other by a flexible spacer with an odd number of methylene units [14–16]. Interest in liquid crystal dimers is high because of their extraordinary flexoelectric [17–19] and electro-optical properties [20–23] and their ability to form modulated nematic phases (N_{TB}—twist-bend, N_{SB}—splay-bend phase) [24–31]. This unique twist-bend phase is, in a sense, an example of the spontaneous formation of a helical structure similar to chiral helices found in DNA and proteins; twisted beta sheets form helical columns that are similar to silk and some modern synthetic analogs. The twist-bend modulated nematic phase (NTB) has a helical structure with a pitch length of several nanometers [21,25,32–34]. The structure of the NTB phase has primarily been studied using non-resonant (SAXS, WAXS) [35–37] and resonant X-ray scattering [32,33,38–41] as well as using polarized Raman [42], infrared [43–46] and nuclear magnetic resonance spectroscopy (NMR) [47–51]. Current experimental results show that the formation of the N_{TB} phase is very sensitive to changes in the shape of the molecules [52–57]; therefore, the key is to carefully define the relationship between the molecular parameters of the molecules and the occurrence of the modulated nematic phase. For this purpose, many authors have used molecular modeling methods, mainly the density functional theory (DFT) [28,29,33,34,52,58–61] or molecular dynamics simulations (MD) [62–64] to analyze structural and conformational changes of the dimers. However, no analysis of the conformational changes of molecules in the transition from the nematic to the twist-bend nematic phase based on a comparison of simulated and experimental vibrational spectra has been performed so far. Polarized measurements of the vibrational spectra (Raman and infrared) are extremely useful in studies of the supramolecular systems that primarily arise through intermolecular interactions such as hydrogen bonding [45,65–70] and π-π stacking [71]. The analysis of the vibrational spectra for such supramolecular systems in the first stage requires the assignment of the bands in the spectrum with regard to the individual functional groups in the molecule, which the modulated nematic phases are clearly an example of.

Here, we report on a spectroscopic and theoretical study of two groups of dimers: a symmetric one that contains the thioether-linking groups (C-S-C) and ether-linking groups (C-O-C) and an asymmetric one that contains both the ether- and thioether-linking groups. By using the polarized infrared absorbance method for a homogeneously aligned sample, it was possible to obtain information about the orientation of the transition dipole moments of a molecule. This information was compared with the theoretically calculated cartesian components of the vibrational transition dipole moment for specific vibrations and has proven to be very useful in the precise assignment of the bands in experimental spectra.

2. Results

2.1. Molecular Structures

The rotational potential barriers of the molecules were obtained using the relaxed potential energy surface scan method with the molecular geometry optimized in order to find the most probable conformations that occurred in the tested materials. In order to find the most stable conformation of a dimer, the optimization of the geometry was performed in a few steps. All of the possible conformations of the dimers that were considered were defined by the values of the dihedral angles φ_1–φ_4 (Figure 1b). In the first stage, the energy barriers to the internal rotation of the cyanobiphenyl (torsional angles φ_3 and φ_4) were determined. In a further step, the energy barriers to the rotation around the dihedral angle (φ_1, φ_2) between the cyanobiphenyl and the linker were determined. The approximate potential energy functions were calculated at intervals of ten. For the calculations, the torsional angles (φ_1–φ_4, each in turn) were fixed at arbitrarily selected values while the other geometrical parameters were optimized, after which relaxed potential energy scans were performed. This procedure enabled the values of the torsion angles for which the minimum energy was obtained to be determined. As the energy barrier to the internal rotation in the alkyl chain is very small (approx. 1 kJ/mol), no other linker/alkyl chain conformations than the all-trans were considered. In the ordered phases, in principle, the alkyl chains can take up all of the possible conformations, and therefore, adopting all of the all-trans conformations is a better representation of the average molecular shape of the dimers [72]. In the next step, taking into account the values of torsion angles that were determined, a full optimization of the geometry was performed for all of the dimers.

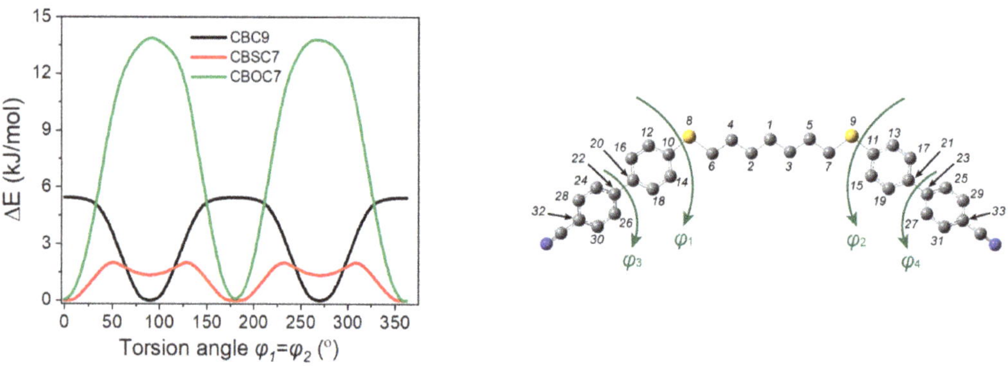

(**a**) (**b**)

Figure 1. (**a**) The calculated potential energy functions for the torsional motion of the cyanobiphenyl dimers relative to the minimum value (φ_1 or φ_2); black solid line—CBC9CB dimer, red solid line—CBS7SCB dimer, green solid line—CBO7OCB dimer. (**b**) Definition of the investigated dihedral angles of the dimers.

Dihedral Angle Distributions and Conformations

Figure 1 shows the calculated rotational potential barriers. First, the torsion angle of the cyanobiphenyl was calculated (φ_3, φ_4). The calculated value of the angle for the minimum energy was 40°, and the potential barriers to this twisted conformation were 8.71 kJ/mol for $\varphi = 0°$ and 8.98 kJ/mol (for $\varphi = 90°$). The value of the φ_1/φ_2 angle is the most crucial for determining the bend angle of the molecule. There are a number of conformations for the CBSCnSCB and CBSCnOCB dimers, which are important for calculating the orientational order, primarily due to possible dihedral angle (φ_1, φ_2): C_{Al}-S-C_{Ar}-C_{Ar} of the sulfur bridge for CBSCnSCB and also C_{Al}-O-C_{Ar}-C_{Ar} (Ar—aromatic ring, Al—alkyl) of the oxygen bridge for CBSCnOCB. The conformational energies of the individual molecules were calculated with respect to those dihedral angles. Both angles were found to have a minimum energy for the "zero" angle (planar conformation) in contrast to the corresponding angle for CBCnCB: C_{Al}-C_{Al}-C_{Ar}-C_{Ar}, which was close to 90° with a potential barrier of 5.5 kJ/mol (upright conformation). The energy barrier to the rotation of the C_{Al}-O-C_{Ar}-C_{Ar} bridge was high enough (ΔU~13.7 kJ/mol) to enable only the planar conformer to be possible. In the case of the sulfur bridge (C_{Al}-S-C_{Ar}-C_{Ar}), however, the barrier was much smaller (ΔU~2.1 kJ/mol), and therefore, the intermolecular interactions might have influenced the molecular structure, which is in agreement with Y. Cao et al. [34]. This means that for the dimers that contain sulfur, more than one energetically stable conformation is possible. As the energy barrier to the internal rotation in the alkyl chain is very small (approx. 1 kJ/mol), no spacer other than the all-trans was considered. In the ordered phases, in principle, the alkyl chains can occupy all of the possible conformations, and, therefore, adopting all of the all-trans conformations is a better representation of the average molecular shape of the dimers [72].

Taking into account the designated torsion angles, a full optimization was performed for each conformation of the dimers. Based on the coordinates of the atoms in space and the coordinates and lengths of the vectors corresponding to the arms of the molecules, the opening angles of the molecules were estimated for all of the probable conformations of the molecules (Figure 2). Table 1 shows the torsion angle values and opening angle values for the fully optimized geometry of all of the dimers.

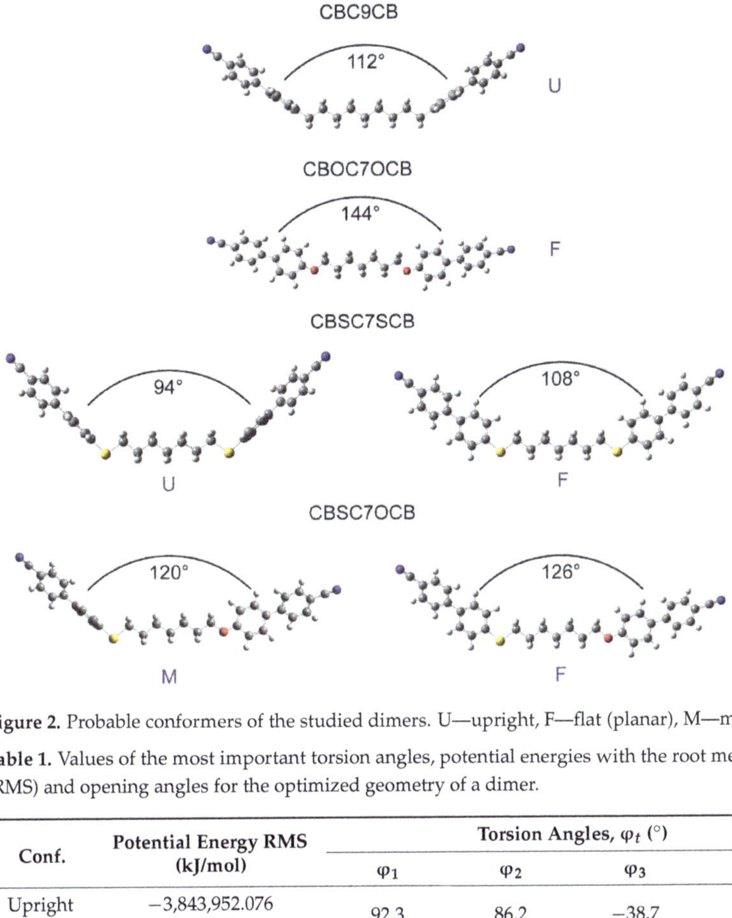

Figure 2. Probable conformers of the studied dimers. U—upright, F—flat (planar), M—mixed.

Table 1. Values of the most important torsion angles, potential energies with the root mean square (RMS) and opening angles for the optimized geometry of a dimer.

Sample	Conf.	Potential Energy RMS (kJ/mol)	Torsion Angles, φ_t (°)				Opening Angle (°)
			φ_1	φ_2	φ_3	φ_4	
CBC9CB	Upright (U)	−3,843,952.076 0.014	92.3	86.2	−38.7	−38.5	112
CBSC7SCB	Upright (U)	−5,521,962.5193 0.0035	−92.9	−92.8	−39.0	−39.0	94
	Flat (F)	−5,521,963.9252 0.0024	5.8	5.8	−37.2	−37.2	108
CBSC7OCB	Mixed (M)	−4,673,985.9556 0.0077	−96.1	−179.1	−38.8	−37.2	120
	Flat (F)	−4,673,986.6314 0.0037	−176.3	−180.0	−37.4	−37.6	126
CBOC7OCB	Flat (F)	−3,826,009.436 0.020	−179.3	−179.2	−37.3	−37.3	144

The bend of a molecule is necessary for the formation of the twist-bend phase. The largest opening angle (144°) was obtained for the CBOC7OCB molecule. This caused its shape to be closer to a calamitic molecule than to a bent (banana) molecule, which means that it did not meet the conditions for the formation of the twist-bend phase, which was confirmed experimentally via the POM, DSC and FTIR measurements [46,59,73].

2.2. Molecular Vibrations

To perform an in-depth analysis of the experimental spectra, the density functional theory (DFT) was used to calculate the theoretical IR spectra for an isolated molecule. Generally, however, there are no absolute assignments of the IR frequencies, and for low symmetry molecules, the correlations of the calculated and experimental frequencies are primarily made by comparing bands that have similar frequencies, assuming that there is no rearrangement of the individual peaks in the bands.

All of the calculated vibrational frequencies are expressed as the wavenumber in cm^{-1}, while the so-called integral absorption coefficient, which is directly proportional to the sum of the components of the squares of the transition dipole moments, is represented by the formula:

$$\overline{A} = \frac{N\pi}{3c^2}\left[\left(\frac{\partial\mu_x}{\partial Q}\right)^2 + \left(\frac{\partial\mu_y}{\partial Q}\right)^2 + \left(\frac{\partial\mu_z}{\partial Q}\right)^2\right], \quad (1)$$

where N—Avogadro constant, c—speed of light and $\frac{\partial\mu_i}{\partial Q}$—the change in the component of the dipole moment with respect to the normal coordinates.

For a more precise analysis of the spectra, it was necessary to calculate the components of the transition dipole moment. Information about the components of the transition dipole moment for a specific vibration enables the parallel and perpendicular components of the spectral density to be calculated. The parallel component of the absorption coefficient was calculated as the square of the component of the transition dipole moment along the axis that coincided with the long axis of the dimer $|\mu_z|^2$. To determine the perpendicular component of the spectral density, the sum of the squares of the transition dipole moments along the perpendicular directions was used $|\mu_x|^2 + |\mu_y|^2$. The direction of the transition dipole moment was determined according to the molecular reference system (see Figure S1 in Supplementary Materials).

In order to compare the theoretical bands with the experimental results, the discrete spectrum lines were extended using the Gaussian function with a half-width of 7 cm^{-1}. Then, the frequencies were rescaled using the scaling coefficient that was determined for the base that was used, which is 0.967 ± 0.021, according to Computational Chemistry Comparison and Benchmark Database. The intensities of the bands were divided by the intensity of the highest band in the theoretical spectrum for a given molecule (or the conformation in the case of molecules for which two stable conformers were calculated). The results of the calculations were compared with the experimental data from the nematic phase in order to identify the most important bands in the spectrum.

2.3. Vibrational Spectra and Their Assignments

It is practically impossible to obtain perfect order in an experiment, and additionally, the dichroism of the bands is influenced by many factors, mainly the intermolecular interactions. Therefore, based only on the analysis of the experimental spectra, it is difficult to say with absolute certainty which band and which of the cyanobiphenyl para axes will be characterized by the behavior of the long axis of a dimer, i.e., to independently describe the direction of the dimer arms. In the case of the theoretical spectra for symmetric dimers, for a given band of the cyanobiphenyl, two characteristic vibrations with a similar frequency but with different intensities should always be obtained. These correspond to the simultaneous vibrations in both arms, one of which is in phase and the other is out of phase. These vibrations, while coupling with each other, can ultimately cause a transition dipole moment to be along the long axis of the dimer or across it. Based on the analysis of the band dichroism for the theoretical spectra, which illustrate the theoretical, "ideal" order, it can be said that if the so-called infinite dichroism can be obtained, i.e., when one of the components is maximum and the other is close to zero, then such a band will then describe the behavior of the long axis of a dimer. On the other hand, in any other case, when the band dichroism is intermediate, there is a lack of vibration coupling for both arms and this band will describe the behavior of the para axis of the mesogens.

The experimental spectra in the entire range of wavenumbers (500–3500 cm^{-1}) for all of the studied dimers in the nematic phase are presented in Figure S2 in the Supplementary Materials.

Overall, the spectra of the cyanobiphenyl dimers can be divided into the following frequency ranges:

- 500–600 cm^{-1} and 700–900 cm^{-1} ranges, which cover the deformational vibrations of the carbon atoms (C-C) and hydrogen atoms (C-H) out of the benzene plane;
- 900–1650 cm^{-1} range, which includes both the characteristic deformation vibrations in the benzene plane as well as the deformation vibrations of the methylene groups of the alkyl chain of the linker in a dimer;
- 2100–2400 cm^{-1} range, which includes the stretching vibrations of the cyan group (C≡N) that were observed as a sharp and very intense peak in the experimental spectrum;
- 2800–2950 cm^{-1} range, which includes the C-H stretching vibrations of the methylene groups. In this range, the vibrations were not well reproduced by the theoretical spectra because the calculations did not take into account the anharmonic effect;
- 2900–3100 cm^{-1} range, which represents the stretching vibrations of the hydrogen atoms (C-H) in the aromatic ring. These vibrations were also not well reproduced by the theoretical frequencies. The bands in this range corresponded to the mixed vibrations, which were strongly overlapping, and the vibrations stretching the C-H hydrogen atoms, significantly disturbed by the Fermi resonance effect.

Most of the fundamentals in the range of 500–2300 cm^{-1} were very well reproduced by the vibrations in the experimental spectra. Most of the observed changes, relative to the transition from the nematic phase to the twist-bend and with the conformational changes that were observed for the thioether dimers, were in the range from 500 to 1200 cm^{-1}.

In this section, we will present the spectra analysis for the CBS7SCB and CBS7OCB dimers in more detail. A comparison of the experimental and theoretical spectra for the dimers, CBC9CB and CBOC7OCB, is summarized in the Supplementary materials (Figures S3–S5).

2.3.1. The CBSC7SCB Dimer

The most important information in comparing the conformers is polarity of the prominent bands. The simulated spectra correspond to the perfect order. On the contrary, the experimental spectra are measured in the nematic phase, and corresponding orientational order is rather weak (S~0.5) due to the bending shape of the molecules. In order to make them comparable, we converted experimental spectra to those of the perfect order (we used the Equation (2a,2b)) [46]. Then, we analyzed differences in the theoretical spectra between the two conformations (flat and upright), with comparison to the experimental spectra in the nematic phase.

Figure 3 shows the comparison between the polarized experimental and theoretical spectra for the CBSC7SCB dimer in the region of 500–1150 cm^{-1}. In the low frequency range, two medium-intensity bands at 520 and 560 cm^{-1}, which were assigned to the deformational vibration of the carbon atoms out of the benzene plane (γCC op CB; op–out of benzene plane CB–cyanobiphenyl), and a low-intensity band on the slope of the 560 band at the wavenumber of 551 cm^{-1}, which referred to the deformational vibration of the C≡N group (δCN), were observed. In the experimental spectra, the band at the wavenumber of 520 cm^{-1} had a perpendicular direction of the transition dipole moment. This band also involved the thioether bridge and made a significant contribution to the deformation vibration of the sulfur atom (γCC op CB + δCS). In the theoretical spectrum for the planar conformation (F), as a result of the coupling of the vibrations of both arms, a band at the wavenumber of 514 cm^{-1} was observed with the maximum dichroism, and the transition dipole moment was perpendicular to the long axis of the dimer. In the case of the second conformation, which was called upright (U), two bands of comparable intensities were observed, one representing the vibration in the phase of both arms, which indicated the perpendicular direction of the transition dipole moment, while the antiphase vibration of both arms indicated the direction of the transition dipole moment parallel to the long axis dimer. Therefore, this band did not show dichroism for the U conformation. Additionally, for the upright conformation (U) in the theoretical spectrum, for the band at 650 cm^{-1}, we observed a splitting into two maxima, which was not observed in the experimental spectra.

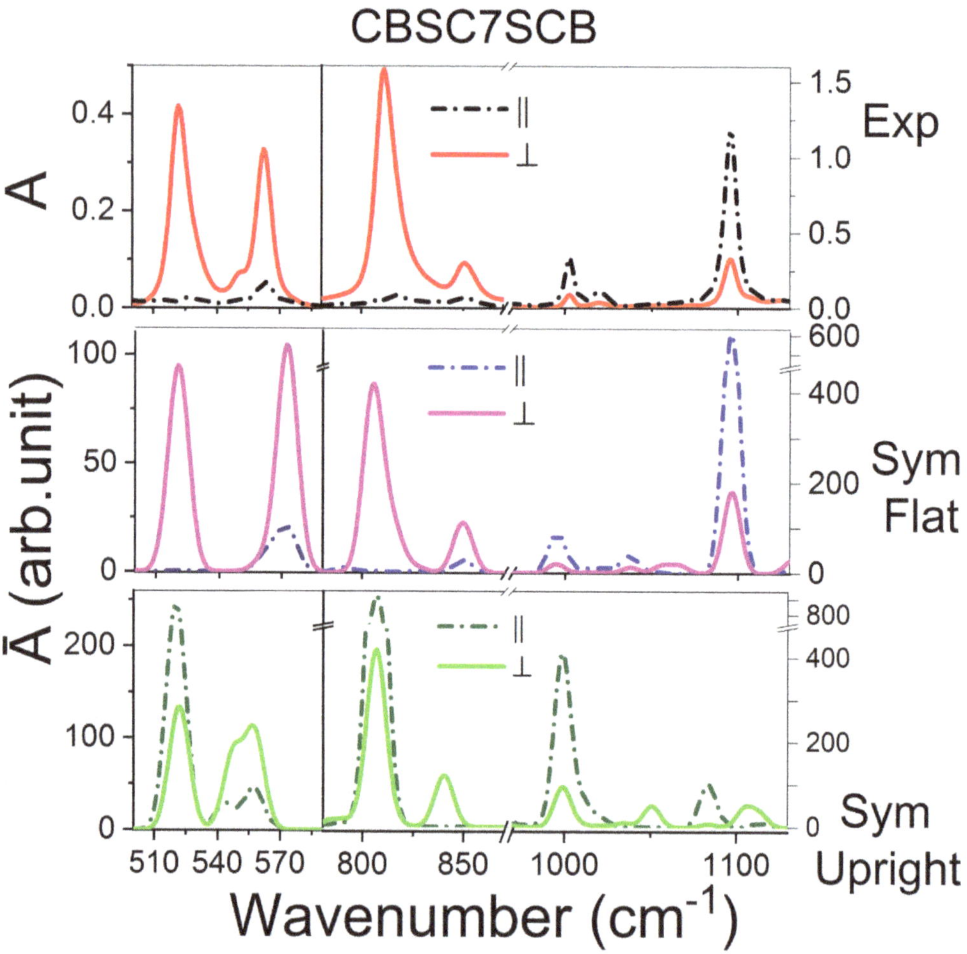

Figure 3. The comparison of the polarized experimental spectrum with the theoretical spectra (B3-LYP/6-311G (d,p)) for the CBSC7SCB dimer in the region of 500–1150 cm^{-1}. A_{II}–the parallel absorbance component in the z-axis direction of the molecular system, which coincided with the ordering axis of the sample. $A_{\perp}$–the perpendicular absorbance component was perpendicular to the rubbing direction. Top Figures–experimental spectra of the nematic phase (370 K). Middle Figures–calculated spectra for a planar conformer (dihedral angles CAl-S-CAr-CAr = 0°). Bottom Figures–calculated spectra for an upright conformation (dihedral angles C_{Al}-S-C_{Ar}-C_{Ar} = 90°).

In the range of 800 to 1120 cm^{-1}, the biggest differences between the conformers were associated with the out-of-plane deformation vibrations of the C-H groups (811 cm^{-1}; γCH op CB) and the deformation C-H vibrations in the benzene plane (1100 cm^{-1}). In the case of the upright conformation (U), there were two vibrations with similar wavenumbers and comparable intensities, which were similar to the 520 cm^{-1} band. In the experimental spectra, the 811 cm^{-1} band indicated the perpendicular direction of the transition dipole moment with a high dichroism, which corresponded to the flat conformer (F). The band at 1100 cm^{-1} was a complexed vibration and involved a sulfide bridge and was assigned to the asymmetric stretching vibration of the C_{Ar}-S group (βCH ip CB + ν_{as} C_Ar-S; Ar–aromatic ring; as–asymmetric, ip–in the benzene plane); therefore, this band was quite sensitive to the dimer conformational changes.

Another indication that the flat conformation was more favored was the absence of a band at 1100 cm^{-1} for the U conformer. In this range, there was also a band of low intensity at 1000 cm^{-1}, which was attributed to the breathing deformation vibration of the carbon atoms in the benzene ring (βCC).

The next spectrum range that is discussed is the 1200–1650 cm^{-1} range, where several high-intensity bands (1460, 1594, 1600 cm^{-1}) were observed, which were assigned to the benzene ring deformations (νCC br; br–benzene ring). They corresponded to the longitudinal transition dipole, which was similar to the bands at 1000, 1100 and 2300 cm^{-1} wavenumbers (2300 cm^{-1}–assigned to the stretching vibrations of the cyan group, νCN).

In this range, a low-intensity band with a wavenumber of about 1395 cm^{-1} was also observed, the direction of which was the dipole transition perpendicular to the long axis of the dimer. This band was assigned to the in-plane deformation of the C-H groups (βCH ip CB). Differences between the two conformations for the benzene ring vibrations were also observed at 1600 cm^{-1}. In the experimental spectrum, this peak was split into two maxima with similar wavenumbers (1594 and 1604 cm^{-1}), for which the intensity shares changed with temperature. In the nematic phase, both peaks had a similar intensity, while in the twist-bend phase, an increase in peak intensity was observed at a lower wavenumber (1594 cm^{-1}). For the F conformer, this vibration was observed at a lower wavenumber, i.e., about 1595 cm^{-1}, while for the U conformer, it was observed at 1598 cm^{-1} wavenumber. The above observations could lead to the conclusion that in the high temperature phases (nematic phase), the coexistence of both conformations and/or some intermediate conformers are probable, while when the temperature is lower, the planar conformation predominates, which is more preferred energetically.

In the Supplementary materials, a comparison of the unpolarized experimental and theoretical spectra for both conformations is additionally summarized in Figure S6. This comparison was intended to show that the band frequencies in the experimental spectrum are remarkably well reproduced by the theoretical ones for the flat conformation.

2.3.2. The CBSC7OCB Dimer

Figure 4 shows the comparison between the polarized experimental and theoretical spectra for the asymmetric CBSC7OCB dimer in the region of 500–1150 cm^{-1}. In the low frequency range, three medium-intensity bands were observed at 520, 530 and 560 cm^{-1}. It was observed that the band at the wavenumber 520 cm^{-1}, which characterized the vibration of the C-C groups out of plane, together with the deformation vibration of the C-S group, had a perpendicular direction of the transition dipole moment in the experimental spectrum and was split into two maxima. An additional maximum appeared at the wavenumber 530 cm^{-1}, which was not observed for the CBSC7SCB dimer. This new band was assigned to the deformation vibration of the C-C group out of the benzene plane along with the deformation vibration of the C_{Ar}-O-C_{Al} group (γCC op CB + $\delta C_{Ar}OC_{Al}$; Al–alkyl chain). In the theoretical spectrum of the F conformer, a split band was also obtained and the transition dipole moment was directed perpendicular to the long axis of the dimer.

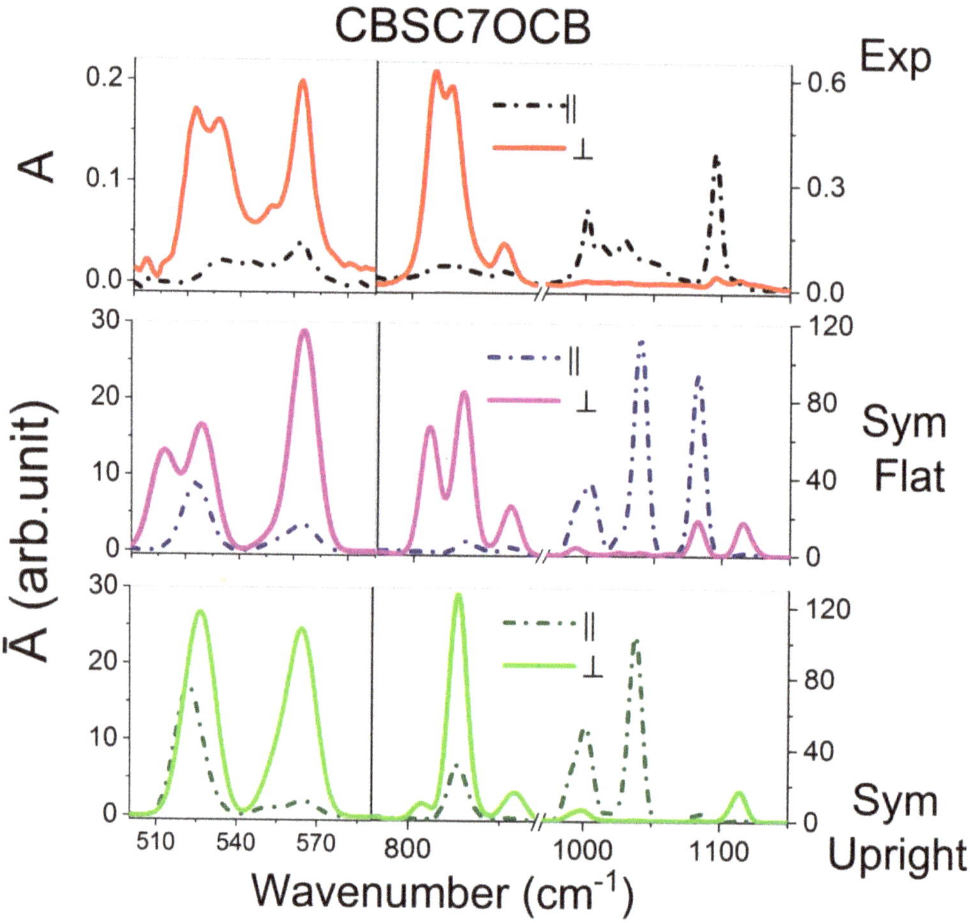

Figure 4. The comparison of the polarized experimental spectrum with the theoretical spectra (B3-LYP/6-311G (d,p)) for the CBSC7OCB dimer in the region of 500–1150 cm^{-1}. A_{II}–the parallel absorbance component in the z-axis direction of the molecular system, which coincided with the ordering axis of the sample. $A_{\perp}$–the perpendicular absorbance component was perpendicular to the rubbing direction. Top Figures–experimental spectra of the nematic phase (375 K). Middle Figures–calculated spectra for a flat conformer (dihedral angles CAl-S-CAr-CAr = 0°, CAl-O-CAr-CAr = 0°). Bottom Figures—calculated spectra for a mixed conformation (dihedral angles C_{Al}-S-C_{Ar}-C_{Ar} = 90°, C_{Al}-O-C_{Ar}-C_{Ar} = 0°).

In contrast to the mixed conformation (M), only one undivided band was observed. Additionally, in the experimental spectrum, three bands were observed, which also corresponded to the transversal transition dipole moment: wavenumbers 813, 821 and 850 cm^{-1}. Compared to the CBS7SCB dimer, a new maximum appeared at 821 cm^{-1}, which was a complex band and, in addition to involving the out-of-plane deformation vibrations of the C-H group, also described the symmetrical stretching vibration of the C_{Ar}-O group (γCH op CB + ν_s C_{Ar}O; s–symmetric). In the spectrum of the CBSC7OCB dimer, other new bands that were characteristic for the oxygen bridge were observed: asymmetric stretching vibrations of the C_{Al}-O group (ν_{as} C_{Al}O) at 1029 and 1050 cm^{-1} and strong vibrations that were assigned to the asymmetric stretching of the C-O group (νas $C_{Ar}O$) at 1249 and 1266 cm^{-1}.

In the 800–1120 cm^{-1} range, the greatest differences between the conformers were associated with the out-of-plane vibration of the C-H group (813, 821 cm^{-1}). In the case of the mixed conformation, only one maximum of high intensity at 816 cm^{-1} was observed. In the case of the F conformer, two maxima were observed, which was in good agreement with the experiment. Another indicator that a planar conformation is more likely was the absence of a maximum at 1100 cm^{-1} in the spectra of M conformer, which was the same as that for the symmetric dimer.

Figure S7 in the Supplementary Materials shows a comparison of the unpolarized experimental and theoretical spectra for both conformations.

3. Discussion

The spontaneous emergence of nanoscale helices in non-chiral materials is one of the most fascinating topics in liquid crystal research with implications for soft matter research in general. An example is the nematic twist bend (N_{TB}) phase, where typically dimeric molecules assemble in helices of the 6–16 nm scale. The helix pitch has mainly been identified by anomalous (k-edge) scattering–a technique globally available at a very limited number of synchrotron stations. Based on the information gathered there, models for molecular assembly have been developed, essentially packing bend and twisted molecules into helices. In this contribution, we show the powerful tool of IR spectroscopy, a much more widespread and in many ways more democratic technique, as it is much cheaper and does not require access to scarce and massively oversubscribed beamlines to provide additional information on the assembly in the N_{TB} phase.

To clarify matters, we investigated a set of systems where the helical structure had been unambiguously identified, and we measured and analyzed the impact of different moieties using detailed polarized infrared measurements (IR) of critical groups. Additionally we employed molecular modelling methods, mainly the density functional theory (DFT) to analyze structural and conformational changes of the dimers. The analysis of the conformational changes of the molecules in the transition from the nematic to the twist-bend nematic phase based on a comparison of simulated and experimental vibrational spectra was performed for the first time for such complex systems.

In summary, it can be seen that the frequencies of the most important bands in the experimental spectra are well reproduced by the theoretical frequencies, while the differences in band intensities are due to the lack of consideration of the intermolecular interactions. The torsional barrier of the S-bridge is quite small (~2 kJ/mol). Therefore, the nearest neighbor's molecules may significantly influence the vibration of the particular molecule. In the paper [74], we show the effect of the vibrational coupling between nearest neighbors. As a result, transition dipoles as well as absorbances of the particular vibrations significantly differ from those of a single molecule.

Table 2 presents a comparison of the main experimental bands and their assignments to the vibrations of the appropriate functional groups for all of the studied dimers. More-detailed and raw data on the theoretical and experimental frequencies, dichroism values, relative intensity, direction of the transition dipole moment and approximate band assignments for all of the investigated dimers are included in the Supplementary Materials (Tables S1–S4).

Table 2. Vibrational frequencies, IR intensities and assignments of the dimers.

CBC9CB		CBSC7SCB		CBSC7OCB		CBOC7OCB		Assignments
$\bar{\nu}$	I_r	$\bar{\nu}$	I_r	$\bar{\nu}$	I_r	$\bar{\nu}$	I_r	
520	vvw	522	m	523	w	–	–	γCC op CB + δCS
–	–	–	–	532	w	532	w/m	γCC op CB + δCO
554	w	562	w	562	w	562	w	γCC op CB + δCN
816	vs	811	vs	813	vs	–		γCH op CB
–	–	–	–	821	s, sh	821	vs	ν_sCOC + γCH op CB
836 850	w	852	w	850	w	850	w	γCH op CB + νCCC sk + $\delta_{as}CH_2$ rocking
1007	w	999	w	999	w	1000	w	βCC ip CB, breathable
1026	vw	1020	vw	1013	vw	1013	vw	νCCC sk + βCH ip CB
–	–	–	–	1029 1051	w	1032	w	$\nu_{as}C_{Al}O$ + βCH ip CB
–	–	1097	m	1095	m	–	–	$\nu_{as}C_{Ar}S$ + βCH ip CB
1112	vw	–	–	–	–	1120	w	βCH ip CB
1185	m	1185	s	1180	s	1178	s	βCH ip CB
–	–	–	–	1249 1266	vs m, sh	1249 1266	vs m,sh	$\nu_{as}C_{Ar}O$ + βCH ip CB
1284 1315 1360	vw	1279 1315	vw	1290 1311	vw	1290 1313	vw	γ_sCH_2 wagging γ_sCH_2 twisting
1397	w	1395	w	1392	w	1390	w	βCH ip CB
1460	w	1437 1462	w	1435 1472	w	1472	w	β_sCH_2 scissoring
– 1493	s	1484 –	vs	1485 1494	vs	– 1493	vs	νCC br
–	–	–	–	1522 1577	m vw	1523 1580	m w	νCC br + β_sCH_2 + $\nu_{as}C_{Ar}O$
– 1605	s	1594 1604	vs	– 1603	vs	– 1602	vs	νCC br
2224	vs	2223	vs	2223	vs	2223	vs	νCN

Key: ip—in plane vibration; op—out of plane deformation, br—stretching and deformation vibrations of the ring (benzene ring), s—symmetrical, as—asymmetric, Al —alkyl chain, Ar—aromatic ring, sk—skeletal, ν—stretching, γ—deforming out of plane, β—deforming in plane, δ—deforming, vs—very strong, s—strong, m—medium, w—weak, vw—very weak, sh—shoulder. Ir—relative intensity of the bands.

4. Materials and Methods

4.1. Materials

The symmetrical and asymmetrical liquid crystal dimers with the cyanobiphenyl (CB) mesogenic groups were investigated. We present three symmetric dimers, which had the general acronym CBXC7XCB (X = C or S or O), that contain nine functional groups in the chain linking the two mesogenic cores (methylene, thioether, ether); the CBC9CB that contains nine methylene groups in the linker and the CBSC7SCB and CBOC7OCB in which the alkyl chains contain seven methylene groups are connected to the cyanobiphenyls by two thioether or two ether bridges, respectively. In the asymmetric dimers with the acronym CBSCnOCB (n = 5, 7), the mesogens are linked to an alkyl chain with five or seven methylene groups on one side by a thioether bridge and on the other by an ether bridge. The details for the materials CBC9CB were reported earlier [75–77]. Synthesis details concerning the thioether/ether compounds were published recently [59,73]. Figure 5 shows the chemical structures of the investigated compounds.

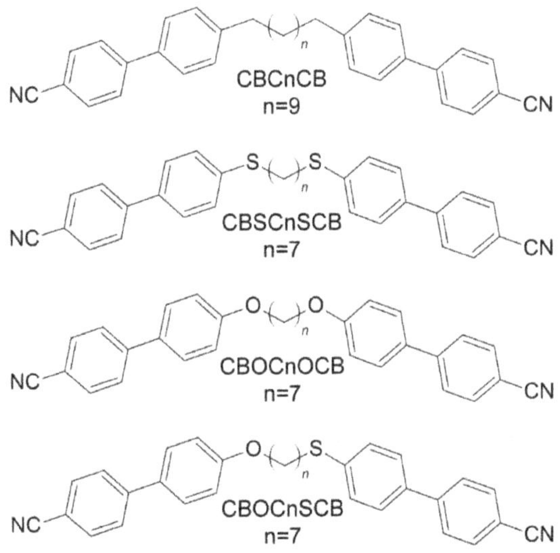

Figure 5. Chemical structures of the investigated dimers.

4.2. Infrared Spectroscopy

The planarly aligned cells were prepared between two optically polished zinc selenide (ZnSe) discs. The thickness of the fabricated cells was determined to be within the range of 5.1–5.6 μm by the measurements of the interference fringes using a spectrometer that was interfaced with a PC (Avaspec-2048). The infrared spectra were acquired using a Fourier infrared spectrometer (Agilent Cary 670 FTIR). The experiment was conducted using the transmission method with a polarized IR beam. An IR-KRS5 grid polarizer was used to polarize the IR beam. The IR spectra were measured as a function of the polarizer rotation angle in the range 500–4000 cm^{-1} of wavenumbers. The details of the sample preparation and absorbance measurement are published in Kocot et al. [46]. These measurements enabled the orientation of the transition dipole moment of the bands to be determined with respect to the long molecular axis and the temperature dependencies of the absorbance of the samples. To determine all three components of absorbance (A_x, A_y, A_z), it is necessary to measure two samples with different orientations: planar (homogeneous) and homeotropic. Unfortunately, in the case of the tested materials, i.e., for the cyanobiphenyl dimers, it was extremely difficult to obtain a good homeotropic alignment. Therefore, in order to calculate the mean absorbance of the sample and assuming that the material was uniaxial, it was assumed that $A_x = A_y$ and, therefore, the mean absorbance was determined as $A_0 = (2A_y + A_z)/3$. The absorbance components were determined as the area that was bound by the contour of a given band using Bio-Rad Win-IR Pro version 2.96e. In the case of complex bands that contained more vibrations, they were separated using Origin Pro 2021 software with the Pearson VII fit.

4.3. Density Functional Theory Calculations

In this work, the calculations of the electronic structure of the molecules were performed using the Gaussian09 program (version E.01) [78]. The molecular structures, intermolecular binding energy, harmonic vibrational force constants, absolute IR intensities and components of the transition dipole moment were calculated using the density functional theory (DFT) with Becke's three-parameter exchange functional in combination with the Lee, Yang and Parr correlation functional B3-LYP method with the basis set: 6-311(d,p) [79,80]. The results were visualized using GaussView 5.0.8.

All of the DFT optimizations were performed with the following convergence criteria, which were used with the Berny algorithm (all values in atomic units): the maximum component of the force was set to 0.00045, the root-mean square (RMS) of the forces that were calculated for the next step—smaller than 0.0003; the computed displacement for the next step—smaller than 0.0018 and the RMS of the displacement below 0.0012. These criteria restricted the dependence of the final geometry parameters on the initial starting geometry.

5. Conclusions

The extensive research on the structure of the N_{TB} phase, to date, permits specific conclusions to be drawn, namely, that a molecular curvature is fundamental for the formation of the phase and that the stability of this phase increases with a decrease in the molecular bending angle.

Using the DFT simulations of the electron structure of molecules, the rotational barriers were determined and the optimization of the most probable conformers that could appear in real samples was performed. Two energetically stable conformations were found for the molecules that contained sulfide bridges, and one was found for the molecules CBC9CB and CBOC7OCB. The opening angles of the tested dimers were also determined. The simulated theoretical spectra turned out to be an invaluable tool for identifying the most important bands in the experimental spectra as well as in determining the directions of the dipole transition moments of the molecules. Combined with the observations of the deflections in the normal modes of the atoms from the equilibrium positions, bands were assigned to the appropriate fragments of the molecules, and thus information about the geometry of the system was obtained.

However, due to the torsional barrier of the S-bridge that is very small (~2 kJ/mol) the nearest neighbor's molecules may significantly influence the vibration of the particular molecule.

Supplementary Materials: The following supporting information can be downloaded at: https://www.mdpi.com/article/10.3390/ijms23148005/s1.

Author Contributions: Conceptualization: K.M. and A.K.; Synthesis: G.H.M. and Y.A.; Methodology: K.M. and A.K.; Investigation: K.M. and B.L.; Formal Analysis: B.L. and K.M.; Writing—original draft: K.M.; Writing—review and editing: K.M. and A.K.; Visualization: B.L. and K.M.; Supervision: K.M. and A.K. All authors have read and agreed to the published version of the manuscript.

Funding: This research was funded by National Science Centre, Poland through Grant No. 2018/31/B/ST3/03609.

Institutional Review Board Statement: Not applicable.

Informed Consent Statement: Not applicable.

Data Availability Statement: The data will be made publicly available when all research results are published. All data stored on the RUJ will have a DOI number, with descriptions including project name, data of the introductory person and project manager, and dates indicating project duration. https://ruj.uj.edu.pl/xmlui/ (accessed on 18 July 2022).

Acknowledgments: K.M. and A.K. thank the National Science Centre for funding through the Grant No. 2018/31/B/ST3/03609. All DFT calculations were carried out with the Gaussian09 program using the PL-Grid Infrastructure on the ZEUS and Prometheus cluster.

Conflicts of Interest: The authors declare no conflict of interest.

References

1. de Gennes, P.G. Soft Matter. *Science* **1992**, *256*, 495–497. [CrossRef] [PubMed]
2. Chandrasekhar, S.; Madhusudana, N.V. Liquid Crystals. *Annu. Rev. Mater. Sci.* **1980**, *10*, 133. [CrossRef]

3. Alfutimie, A.; Curtis, R.T.G.J. *Tiddy in Handbook of Liquid Crystals*, 2nd ed.; Goodby, J.W., Collings, P.J., Kato, T., Tschierske, C., Gleeson, H., Raynes, P., Eds.; Wiley-VCH: Weinheim, Germany, 2014; Volume 6, Chapter 12.
4. Kirov, N.; Simova, P. Infrared Absorption Spectra of Liquid Crystals. *Mol. Cryst. Liq. Cryst.* **1975**, *30*, 59–71.
5. Fringeli, U.P.; Schadt, M.; Riĥak, P.; Günthard, H.H. Hydrocarbon Chain Ordering in Liquid Crystals Investigated by Means of Infrared Attenuated Total Reflection (IR-ATR) Spectroscopy. *Z. Naturforsch.* **1976**, *31*, 1098–1107. [CrossRef]
6. Kruk, G.; Kocot, A.; Wrzalik, R.; Vij, J.K.; Karthaus, O.; Ringsdorf, H. Infrared absorption study of hexapentyloxytriphenylene a discotic liquid crystal. *Liq. Cryst.* **1993**, *14*, 807–819. [CrossRef]
7. Sakagami, S.; Nonaka, K.; Koga, T.; Takase, A. Infrared Spectroscopic Study of Liquid Crystalline N-[4-(4-n-alkoxybenzoyloxy)-2-hydroxybenzylidene]-3-cyanoanilines. *Mol. Cryst. Liq. Cryst.* **1998**, *312*, 23–32. [CrossRef]
8. Orgasińska, B.; Perova, T.S.; Merkel, K.; Kocot, A.; Vij, J.K. Surface phenomena in discotic liquid crystals investigated using polarized FTIR transmission spectroscopy. *Mat. Sci. Eng. C* **1999**, *8–9*, 283–289. [CrossRef]
9. Jang, W.G.; Park, C.S.; Kim, K.H.; Glaser, M.A.; Clark, N.A. Infrared spectroscopic study of molecular hydrogen bonding in chiral smectic liquid crystals. *Phys. Rev. E* **2000**, *62*, 5027–5035. [CrossRef]
10. Merkel, K.; Wrzalik, R.; Kococt, A. Calculation of vibrational spectra for cyanobiphenyl liquid crystals. *J. Mol. Struct.* **2001**, *563–564*, 477–490. [CrossRef]
11. Wrzalik, R.; Merkel, K.; Kocot, A.; Cieplak, B. Analysis of experimental and simulated vibrational spectra for the antiferroelectric liquid crystal 12OBBB1M6. *J. Chem. Phys.* **2002**, *117*, 4889–4895. [CrossRef]
12. Merkel, K.; Kocot, A.; Vij, J.K.; Korlacki, R.; Mehl, G.H.; Meyer, T. Thermotropic Biaxial Nematic Phase in Liquid Crystalline Organo-Siloxane Tetrapodes. *Phys. Rev. Lett.* **2004**, *93*, 237801–237805. [CrossRef]
13. Merkel, K.; Kocot, A.; Vij, J.K.; Stevenson, P.J.; Panov, A.; Rodriguez, D. Anomalous temperature dependence of layer spacing of de Vries liquid crystals: Compensation model. *Appl. Phys. Lett.* **2016**, *108*, 243301–243306. [CrossRef]
14. Imrie, C.T.; Luckhurst, G.R. *Liquid Crystal Dimers and Oligomers in Handbook of Liquid Crystals Vol. 2B: Low Molecular Weight Liquid Crystals*; Demus, D., Goodby, J.W., Gray, G.W., Spiess, H.W., Vill, V., Eds.; Wiley-VCH: Weinheim, Germany, 1998; Chapter X; p. 801.
15. Imrie, C.T.; Henderson, P.A. Liquid crystal dimers and higher oligomers: Between monomers and polymers. *Chem. Soc. Rev.* **2007**, *36*, 2096–2124. [CrossRef]
16. Mandle, R.J. The dependency of twist-bend nematic liquid crystals on molecular structure: A progression from dimers to trimers, oligomers and polymers. *Soft Matter* **2016**, *12*, 7883–7901. [CrossRef]
17. Morris, S.M.; Clarke, M.J.; Blatch, A.E.; Coles, H.J. Structure-flexoelastic properties of bimesogenic liquid crystals. *Phys. Rev. E* **2007**, *75*, 041701–041710. [CrossRef]
18. Shamid, S.M.; Dhakal, S.; Selinger, J.V. Statistical mechanics of bend flexoelectricity and the twist-bend phase in bent-core liquid crystals. *Phys. Rev. E* **2013**, *87*, 052503. [CrossRef]
19. Meyer, C.; Luckhurst, G.R.; Dozov, I. Flexoelectrically driven electroclinic effect in the twist-bend nematic phase of achiral molecules with bent shapes. *Phys. Rev. Lett.* **2013**, *111*, 067801–067806. [CrossRef]
20. Panov, V.P.; Balachandran, R.; Vij, J.K.; Tamba, M.G.; Kohlmeier, A.; Mehl, G.H. Microsecond linear optical response in the unusual nematic phase of achiral bimesogens. *Appl. Phys. Lett.* **2011**, *99*, 261903–261906. [CrossRef]
21. Chen, D.; Porada, J.H.; Hooper, J.B.; Klittnick, A.; Shen, Y.; Tuchband, M.R.; Korblova, E.; Bedrov, D.; Walba, D.M.; Glaser, M.A.; et al. Chiral heliconical ground state of nanoscale pitch in a nematic liquid crystal of achiral molecular dimers. *Proc. Natl. Acad. Sci. USA* **2013**, *110*, 15931–15936. [CrossRef]
22. Panov, V.P.; Vij, J.K.; Mehl, G.H. Twist-bend nematic phase in cyanobiphenyls and difluoroterphenyls bimesogens. *Liq. Cryst.* **2017**, *44*, 147–159.
23. Meyer, C. Nematic twist-bend phase under external constraints. *Liq. Cryst.* **2016**, *43*, 2144–2162. [CrossRef]
24. Panov, V.P.; Nagaraj, M.; Vij, J.K.; Panarin, Y.P.; Kohlmeier, A.; Tamba, M.G.; Lewis, R.A.; Mehl, G.H. Spontaneous Periodic Deformations in Nonchiral Planar-Aligned Bimesogens with a Nematic-Nematic Transition and a Negative Elastic Constant. *Phys. Rev. Lett.* **2010**, *105*, 167801. [CrossRef] [PubMed]
25. Borshch, V.; Kim, Y.K.; Xiang, J.; Gao, M.; Jákli, A.; Panov, V.P.; Vij, J.K.; Imrie, C.T.; Tamba, M.G.; Mehl, G.H.; et al. Nematic twist-bend phase with nanoscale modulation of molecular orientation. *Nat. Commun.* **2013**, *4*, 2635–2638. [CrossRef] [PubMed]
26. Parsouzi, Z.; Shamid, S.M.; Borshch, V.; Challa, P.K.; Baldwin, A.R.; Tamba, M.G.; Welch, C.; Mehl, G.H.; Gleeson, J.T.; Jakli, O.D.A.; et al. Fluctuation Modes of a Twist-Bend Nematic Liquid Crystal. *Phys. Rev. X* **2016**, *6*, 021041-16. [CrossRef]
27. Chen, D.; Nakata, M.; Shao, R.; Tuchband, M.R.; Shuai, M.; Baumeister, U.; Weissflog, W.; Walba, D.M.; Glaser, M.A.; Maclennan, J.E.; et al. Twist-bend heliconical chiral nematic liquid crystal phase of an achiral rigid bent-core mesogen. *Phys. Rev. E* **2014**, *89*, 022506. [CrossRef]
28. Babakhanova, G.; Parsouzi, Z.; Paladugu, S.; Wang, H.; Nastishin, Y.A.; Shiyanovskii, S.V.; Sprunt, S.; Lavrentovich, O.D. Elastic and viscous properties of the nematic dimer CB7CB. *Phys. Rev. E* **2017**, *96*, 062704. [CrossRef]
29. Connor, P.L.M.; Mandle, R.J. Chemically induced splay nematic phase with micron scale periodicity. *Soft Matter* **2020**, *16*, 324–329. [CrossRef]
30. Merkel, K.; Kocot, A.; Welch, C.; Mehl, G.H. Soft modes of the dielectric response in the twist–bend nematic phase and identification of the transition to a nematic splay bend phase in the CBC7CB dimer. *Phys. Chem. Chem. Phys.* **2019**, *21*, 22839–22848. [CrossRef]

31. Meyer, C.; Blanc, C.; Luckhurst, G.R.; Davidson, P.; Dozov, I. Biaxiality-driven twist-bend to splay-bend nematic phase transition induced by an electric field. *Sci. Adv.* **2020**, *6*, eabb8212. [CrossRef]
32. Zhu, C.; Tuchband, M.R.; Young, A.; Shuai, M.; Scarbrough, A.; Walba, D.M.; Maclennan, J.E.; Wang, C.; Hexemer, A.; Clark, N.A. Resonant carbon k-edge soft x-ray scattering from lattice-free heliconical molecular ordering: Soft dilative elasticity of the twist-bend liquid crystal phase. *Phys. Rev. Lett.* **2016**, *116*, 147803–147809. [CrossRef]
33. Cruickshank, E.; Salamończyk, M.; Pociecha, D.; Strachan, G.J.; Storey, J.M.D.; Wang, C.; Feng, J.; Zhu, C.; Gorecka, E.; Imrie, C.T. Sulfur-linked cyanobiphenyl-based liquid crystal dimers and the twist-bend nematic phase. *Liq. Cryst.* **2019**, *46*, 1595–1609. [CrossRef]
34. Cao, Y.; Feng, J.; Nallapaneni, A.; Arakawa, Y.; Zhao, K.; Zhang, H.; Mehl, G.H.; Zhu, C.; Liu, F. Deciphering helix assembly in the heliconical nematic phase via tender resonant X-ray scattering. *J. Mater. Chem. C* **2021**, *9*, 10020–10028. [CrossRef]
35. Cestari, M.; Diez-Berart, S.; Dunmur, D.A.; Ferrarini, A.; de la Fuente, M.R.; Jackson, D.J.B.; Lopez, D.O.; Luckhurst, G.R.; Perez-Jubindo, M.A.; Richardson, R.M.; et al. Phase behavior and properties of the liquid-crystal dimer 1,7-bis(4-cyanobiphenyl-4-yl) heptane: A twist-bend nematic liquid crystal. *Phys. Rev. E* **2011**, *84*, 031704. [CrossRef] [PubMed]
36. Adlem, K.; Čopič, M.; Luckhurst, G.R.; Mertelj, A.; Parri, O.; Richardson, R.M.; Snow, B.D.; Timimi, B.A.; Tuffin, R.P.; Wilkes, D. Chemically induced twist-bend nematic liquid crystals, liquid crystal dimers, and negative elastic constants. *Phys. Rev. E* **2013**, *88*, 022503. [CrossRef]
37. Stevenson, W.D.; Zou, H.; Zeng, X.; Welch, C.; Ungar, G.; Mehl, G.H. Dynamic calorimetry and XRD studies of the nematic and twist-bend nematic phase transitions in a series of dimers with increasing spacer length. *Phys. Chem. Chem. Phys.* **2018**, *20*, 25268–25274. [CrossRef]
38. Salamończyk, M.; Vaupotič, N.; Pociecha, D.; Wang, C.; Zhu, C.; Gorecka, E. Structure of nanoscale-pitch helical phases: Blue phase and twist-bend nematic phase resolved by resonant soft X-ray scattering. *Soft Matter* **2017**, *13*, 6694–6699. [CrossRef]
39. Stevenson, W.; Ahmed, Z.; Zeng, X.; Welch, C.; Ungar, G.; Mehl, G. Molecular organization in the twist–bend nematic phase by resonant X-ray scattering at the Se K-edge and by SAXS, WAXS and GIXRD. *Phys. Chem. Chem. Phys.* **2017**, *19*, 13449–13454. [CrossRef]
40. Tuchband, M.R.; Paterson, D.A.; Salamo ´nczyk, M.; Norman, V.A.; Scarbrough, A.N.; Forsyth, E.; Garcia, E.; Wang, C.; Storey, J.M.; Walba, D.M.; et al. Distinct differences in the nanoscale behaviors of the twist–bend liquid crystal phase of a flexible linear trimer and homologous dimer. *Proc. Natl. Acad. Sci. USA* **2019**, *116*, 10698–10704. [CrossRef]
41. Saha, R.; Feng, C.; Welch, C.; Mehl, G.H.; Feng, J.; Zhu, C.; Gleeson, J.; Sprunt, S.; Jákli, A. The interplay between spatial and heliconical orientational order in twist-bend nematic materials. *Phys. Chem. Chem. Phys.* **2021**, *23*, 4055–4063. [CrossRef]
42. Zhang, Z.; Panov, V.P.; Nagaraj, M.; Mandle, R.J.; Goodby, J.W.; Luckhurst, G.R.; Jones, J.C.; Gleeson, H.F. Raman scattering studies of order parameters in liquid crystalline dimers exhibiting the nematic and twist-bend nematic phases. *J. Mater. Chem. C* **2015**, *3*, 10007–10016. [CrossRef]
43. Knežević, A.; Dokli, I.; Sapunar, M.; Sĕgota, S.; Baumeister, U.; Lesac, A. Induced smectic phase in binary mixtures of twist-bend nematogens. *Beilstein J. Nanotechnol.* **2018**, *9*, 1297–1307. [CrossRef] [PubMed]
44. Merkel, K.; Loska, B.; Welch, C.; Mehl, G.H.; Kocot, A. Molecular biaxiality determines the helical structure—Infrared measurements of the molecular order in the nematic twist-bend phase of difluoro terphenyl dimer. *Phys. Chem. Chem. Phys.* **2021**, *23*, 4151–4160. [CrossRef] [PubMed]
45. Merkel, K.; Loska, B.; Welch, C.; Mehl, G.H.; Kocot, A. The role of intermolecular interactions in stabilising the structure of the nematic twist-bend phase. *RSC Adv.* **2021**, *11*, 2917–2925. [CrossRef] [PubMed]
46. Kocot, A.; Loska, B.; Arakawa, Y.; Merkel, K. Structure of the twist-bend nematic phase with respect to the orientational molecular order of the thioether-linked dimers. *Phys. Rev. E* **2022**, *105*, 044701. [CrossRef]
47. Beguin, L.; Emsley, J.W.; Lelli, M.; Lesage, A.; Luckhurst, G.R.; Timimi, B.A.; Zimmermann, H. The chirality of a twist–bend nematic phase identified by NMR spectroscopy. *J. Phys. Chem. B* **2012**, *116*, 7940–7951. [CrossRef] [PubMed]
48. Emsley, W.; Lesot, P.; Luckhurst, G.R.; Meddour, A.; Merlet, D. Chiral solutes can seed the formation of enantiomorphic domains in a twist-bend nematic liquid crystal. *Phys. Rev. E* **2013**, *87*, 040501. [CrossRef] [PubMed]
49. Emsley, J.W.; Lelli, M.; Lesage, A.; Luckhurst, G.R. A comparison of the conformational distributions of the achiral symmetric liquid crystal dimer cb7cb in the achiral nematic and chiral twist-bend nematic phases. *J. Phys. Chem. B* **2013**, *117*, 6547–6557. [CrossRef]
50. Jokisaari, J.P.; Luckhurst, G.R.; Timimi, B.A.; Zhu, J.F.; Zimmermann, H. Twist-bend nematic phase of the liquid crystal dimer cb7cb: Orientational order and conical angle determined by xe-129 and h-2 nmr spectroscopy. *Liq. Cryst.* **2015**, *42*, 708–721.
51. Ivšić, T.; Vinković, M.; Baumeister, U.; Mikleušević, A.; Lesac, A. Towards understanding the NTB phase: A combined experimental, computational and spectroscopic study. *RSC Adv.* **2016**, *6*, 5000–5007. [CrossRef]
52. Forsyth, E.; Paterson, D.A.; Cruickshank, E.; Strachan, G.J.; Gorecka, E.; Walker, R.; Storey, J.M.D.; Imrie, C.T. Liquid crystal dimers and the twist-bend nematic phase: On the role of spacers and terminal alkyl chains. *J. Mol. Liq.* **2020**, *320*, 114391. [CrossRef]
53. Mandle, R.J.; Goodby, J.W. Molecular Flexibility and Bend in Semi-Rigid Liquid Crystals: Implications for the Heliconical Nematic Ground State. *Chem. Eur. J.* **2019**, *25*, 14454–14459. [CrossRef] [PubMed]
54. Archbold, C.T.; Mandle, R.J.; Andrews, J.L.; Cowling, S.J.; Goodby, J.W. Conformational landscapes of bimesogenic compounds and their implications for the formation of modulated nematic phases. *Liq. Cryst.* **2017**, *44*, 2079–2088. [CrossRef]

55. Pocock, E.E.; Mandle, R.J.; Goodby, J.W. Molecular shape as a means to control the incidence of the nanostructured twist bend phase. *Soft Matter* **2018**, *14*, 2508–2514. [CrossRef] [PubMed]
56. Pocock, E.E.; Mandle, R.J.; Goodby, J.W. Experimental and Computational Study of a Liquid Crystalline Dimesogen Exhibiting Nematic, Twist-Bend Nematic, Intercalated Smectic, and Soft Crystalline Mesophases. *Molecules* **2021**, *26*, 532. [CrossRef] [PubMed]
57. Longa, L.; Tomczyk, W. Twist−Bend Nematic Phase from the Landau−de Gennes Perspective. *J. Phys. Chem. C* **2020**, *124*, 22761–22775. [CrossRef]
58. Paterson, D.A.; Gao, M.; Kim, Y.-K.; Jamali, A.; Finley, K.L.; Robles-Herna´ndez, B.; Diez-Berart, S.; Salud, J.; de la Fuente, M.R.; Timimi, B.A.; et al. Understanding the twist-bend nematic phase: The characterisation of 1-(4-cyanobiphenyl-40-yloxy)-6-(4-cyanobiphenyl-40-yl)hexane (CB6OCB) and comparison with CB7CB. *Soft Matter* **2016**, *12*, 6827–6840. [CrossRef]
59. Arakawa, Y.; Komatsu, K.; Tsuji, H. Twist-bend nematic liquid crystals based on thioether linkage. *New J. Chem.* **2019**, *43*, 6786–6793. [CrossRef]
60. Kumar, A. Dependency of the twist-bend nematic phase formation on the molecular shape of liquid crystal dimers: A view through the lens of DFT. *J. Mol. Liq.* **2022**, *354*, 118858. [CrossRef]
61. Lesac, A.; Baumeister, U.; Dokli, I.; Hameršak, Z.; Kontrec, I.D.; Viskić, M.; Knežević, A.; Mandle, R.J. Geometric aspects influencing N-N TB transition—Implication of intramolecular torsion. *Liq. Cryst.* **2018**, *45*, 1101–1110. [CrossRef]
62. Memmer, R. Liquid crystal phases of achiral banana-shaped molecules: A computer simulation study. *Liq. Cryst.* **2002**, *29*, 483–496. [CrossRef]
63. Wilson, M.R.; Yu, G.; Potter, T.D.; Walker, M.; Gray, S.J.; Li, J.; Boyd, N.J. Molecular Simulation Approaches to the Study of Thermotropic and Lyotropic Liquid Crystals. *Crystals* **2022**, *12*, 685. [CrossRef]
64. Yu, G.; Wilson, M.R. All-atom simulations of bent liquid crystal dimers: The twist-bend nematic phase and insights into conformational chirality. *Soft Matter* **2022**, *18*, 3087–3096. [CrossRef] [PubMed]
65. Jansze, S.M.; Martínez-Felipe, A.; Storey, J.M.D.; Marcelis, A.T.M.; Imrie, C.T. A Twist-Bend Nematic Phase Driven by Hydrogen Bonding. *Angew. Chem. Int. Ed.* **2015**, *54*, 643–646. [CrossRef]
66. Martinez-Felipe, A.; Cook, A.G.; Abberley, J.P.; Walker, R.; Storey, J.M.D.; Imrie, C.T. An FT-IR spectroscopic study of the role of hydrogen bonding in the formation of liquid crystallinity for mixtures containing bipyridines and 4-pentoxybenzoic acid. *RSC Adv.* **2016**, *6*, 108164–108179. [CrossRef]
67. Walker, R.; Pociecha, D.; Abberley, J.P.; Martinez-Felipec, A.; Paterson, D.A.; Forsyth, E.; Lawrence, G.B.; Henderson, P.A.; Storey, J.M.D.; Gorecka, E.; et al. Spontaneous chirality through mixing achiral components: A twist-bend nematic phase driven by hydrogen-bonding between unlike components. *Chem. Commun.* **2018**, *54*, 3383–3386. [CrossRef]
68. Alaasar, M.; Tschierske, C. Nematic phases driven by hydrogen-bonding in liquid crystalline nonsymmetric dimers. *Liq. Cryst.* **2019**, *49*, 124–130. [CrossRef]
69. Kima, J.; Koo, J.; Park, J.; Jeong, K.; Lee, J. Enhancement of flexoelastic ratio of nematic liquid crystal doped with hydrogen-bonded bimesogen molecules. *J. Mol. Liq.* **2019**, *277*, 541–545. [CrossRef]
70. Kato, T.; Gupta, M.; Yamaguchi, D.; Gan, K.P.; Nakayama, M. Supramolecular Association and Nanostructure Formation of Liquid Crystals and Polymers for New Functional Materials. *Bull. Chem. Soc. Jpn.* **2021**, *94*, 357–376. [CrossRef]
71. Knežević, A.; Sapunar, M.; Buljan, A.; Dokli, I.; Hameršak, Z.; Kontrec, D.; Lesac, A. Fine-tuning the effect of p–p interactions on the stability of the NTB phase. *Soft Matter* **2018**, *14*, 8466–8474. [CrossRef]
72. Emsley, J.W.; de Luca, G.; Lesage, A.; Merlet, D.; Pileio, G. The structure and conformation of a mesogenic compound between almost zero and almost complete orientational order. *Liq Cryst.* **2007**, *34*, 1071–1093. [CrossRef]
73. Arakawa, Y.; Ishida, Y.; Tsuji, H. Ether- and Thioether-Linked Naphthalene-Based Liquid-Crystal Dimers: Influence of Chalcogen Linkage and Mesogenic-Arm Symmetry on the Incidence and Stability of the Twist–Bend Nematic Phase. *Chem. Eur. J.* **2020**, *26*, 3767–3775. [CrossRef]
74. Merkel, K.; Loska, B.; Arakawa, Y.; Mehl, G.H.; Karcz, J.; Kocota, A. Increased intermolecular interactions and cluster formation at the onset of the twist-bend nematic phase in thioether cyanobiphenyl-based liquid crystal dimers. *arXiv* **2022**, arXiv:2206.07437. [CrossRef]
75. Barnes, P.J.; Douglass, A.G.; Heeks, S.K.; Luckhurst, G.R. An enhanced odd-even effect of liquid crystal dimers Orientational order in the α,ω-bis(4′-cyanobiphenyl-4-yl)alkanes. *Liq. Cryst.* **1993**, *13*, 603–613. [CrossRef]
76. Carvalho, J.; Cruz, C.; Figueirinhas, J.L.; Tamba, M.G.; Kohlmeier, A.; Mehl, G.H. Proton and deuterium NMR study of the CBC9CB dimer system. *J. Phys. Chem. B* **2019**, *123*, 1442–1451. [CrossRef] [PubMed]
77. Aluculesei, A.; Cachitas, H.; Carvalho, J.; Chavez, F.V.; Figueirinhas, J.L.; Sebastião, P.J.; Cruz, C.; Tamba, M.G.; Kohlemeir, A.; Mehl, G.H. ^{1}H NMR study of molecular order and dynamics in the liquid crystal CB-C9-CB. *Phys. Chem. Chem. Phys.* **2019**, *21*, 4523–4537. [CrossRef]
78. Frisch, M.J.; Trucks, G.W.; Schlegel, H.B.; Scuseria, G.E.; Robb, M.A.; Cheeseman, J.R.; Scalmani, G.; Barone, V.; Mennucci, B.; Petersson, G.A.; et al. *Gaussian 09, Revision E.01*; Gaussian, Inc.: Wallingford CT, UK, 2009.
79. Hertwig, R.H.; Koch, W. On the parameterization of the local correlation functional. What is Becke-3-LYP? *Chem. Phys. Lett.* **1997**, *268*, 345. [CrossRef]
80. Becke, A.D. Density-functional exchange-energy approximation with correct asymptotic behavior. *Phys. Rev. A* **1988**, *38*, 3098. [CrossRef]

International Journal of
Molecular Sciences

MDPI

Article

Rashba Splitting and Electronic Valley Characteristics of Janus Sb and Bi Topological Monolayers

Qi Gong and Guiling Zhang *

School of Material Science and Chemical Engineering, Harbin University of Science and Technology, Harbin 150040, China; gongqi.hrbust@gmail.com
* Correspondence: guiling-002@163.com

Abstract: Janus Sb and Bi monolayers as a new class of 2D topological insulator materials, which could be fulfilled by asymmetrical functionalizations with methyl or hydroxyl, are demonstrated by first-principles spin–orbit coupling (SOC) electronic structure calculations to conflate nontrivial topology, Rashba splitting and valley-contrast circular dichroism. Cohesive energies and phonon frequency dispersion spectra indicate that all Janus Sb and Bi monolayers possess a structural stability in energetic statics but represent virtual acoustic phonon vibrations of the hydrogen atoms passivating on monolayer surfaces. Band structures of Janus Sb and Bi monolayers and their nanoribbons demonstrate they are nontrivial topological insulators. Rashba spin splitting at G point in Brillouin zone of Janus Bi monolayers arises from the strong SOC p_x and p_y orbitals of Bi bonding atoms together with the internal out-of-plane electric field caused by asymmetrical functionalization. Janus Sb and Bi monolayers render direct and indirect giant bandgaps, respectively, which are derived from the strong SOC p_x and p_y orbitals at band-valley Brillouin points K and K′ where valley-selective circular dichroism of spin valley Hall insulators is also exhibited.

Keywords: two-dimensional Janus structure; topological insulator; quantum spin Hall effect; valley electronics

1. Introduction

Two-dimensional (2D) topology insulators and nanomaterials have gained more attention since graphene was exfoliated successfully, and the recently focused Rashba effects and valleytronics promote emergence of various practical spintronic devices, especially for operating at room temperature [1,2]. Janus polar-structures have attracted intensive attention due to their specific intrinsic properties and potential applications of exploiting Janus anisotropy in spintronic devices, catalysts, drug delivery, biosensors and antibacterials [3–5]. Two-dimensional Janus discs have been demonstrated both in theory and practice to provide new schemes of modifying electronic structure and surface functionality of graphene-like 2D materials [6–8]. Spin–orbit coupling (SOC) plays an important role together with internal built-in electric fields in Janus quantum systems with a 2D honeycomb atomic structure to cause various physical effects, such as Rashba spin-splitting and electronic valley freedom from Dirac-point electronic states at Brillouin zone boundary [9,10].

Surface functionalizations of physical or chemical decorations on 2D graphene-like materials provides a new routine of manipulating electronic properties by band structure engineering. Physical adsorption of gas or small organic molecules provides a rather feasible method to achieve preferable electronic properties, which will generally cause substantial reconstruction in local atomic structures to engender great changes in electronic structures [11–13]. However, covalent functionalization is a better way of acquiring higher stability and homogeneity. Previous researches on the covalent functionalizations of elemental 2D topological materials incorporate hydrogenation and halogenation, oxidation, organic group decoration and cycloaddition reaction [14–19]. The plasma method of halogenation leads to a serious increase in lattice defects and thus even ruins topological properties, and the hydrogenated surface is easy to be oxidized at room temperature due to the poor chemical stability [20,21].

Extensive effort has been devoted to the search for new 2D topological insulators (TIs) with a large bulk band-gap. Some layered materials such as silicene, germanene and stanene have been proposed, and the bulk band-gap of 2D TIs has been promoted remarkably to 0.3 eV in chemically modified tin films [22–25]. Recently, ultrathin Bi films have drawn much attention as promising candidates for quantum spin Hall effect (QSH) insulators, and the 2D topological properties of ultrathin Bi(111) films have been reported [26]. To the best of our knowledge, neither 2D nor 3D TIs have achieved a bulk band-gap exceeding 0.7 eV [27]. Because Bi and Sb are well known for their strong SOC that can generate and stabilize topologically nontrivial electronic-states, it is worthwhile to explore large band-gap QSH insulators from Bi- or Sb-based 2D materials, such as Sb or Bi monolayers symmetrically functionalized by methyl, amino and hydroxyl groups [28]. Especially when the inversion symmetry of honeycomb lattice is broken, Bi and Sb monolayers become a quantum valley Hall insulator, whilst Rashba spin splitting and chiral optical selectivity of electronic valleys are expected to be feasibly achieved. To date, no research has been reported focusing on the asymmetrical functionalization of Sb or Bi monolayers to realize these versatile exotic phenomena from nontrivial topology in inversion asymmetry and fabricate new quantum devices operating at room temperature. Such novel structures will promote graphene-like elemental 2D topological materials to a controllable and practical level of fulfilling advanced applications in innovative 2D spintronic devices.

In the present study, we intensively investigate the asymmetrical (Janus) functionalizations of Sb and Bi monolayers by chemically attaching hydrogen atoms on one side and chemical groups (methyl or hydroxyl) on the other side according to the so-called "chair" prototypical geometry [29]. First-principles calculations of structure optimization, phonon dispersion and electronic structures are carried out to theoretically verify that Sb and Bi monolayers asymmetrically functionalized by methyl or hydroxyl are stable giant band-gap QSH insulators with a giant bulk band-gap. It is suggested that their giant bulk band-gaps originate from strong SOC near Fermi level that is dominantly contributed by Sb or Bi p_x and p_y orbitals in Janus monolayers, which makes them promising for topological insulator applications at ambient temperature. Meanwhile, when the inversion symmetry of Bi or Sb monolayers in honeycomb lattices is broken by asymmetrical functionalization of hydrogen and methyl or hydroxyl respectively on two surface sides, we obtain Janus 2D structures with evident Rashba splitting and chiral optical selectivity.

2. Results and Discussion

2.1. Atomic Geometry and Stability

Janus Sb or Bi monolayers (HSb_2XH_n and HBi_2XH_n, $XH_n = CH_3$, OH) are obtained by asymmetrical functionalizations (decorations) where the radical groups of methyl (CH_3) and hydroxyl (OH) are bonded to Sb or Bi atoms on one surface side of Sb or Bi monolayer with the other surface side being saturated by hydrogen (H), as schematically shown in Figure 1, which possess space group of P3M1 in a threefold rotational symmetry of 2D honeycomb lattices. Compared with Sb or Bi and their hydrogen-saturated monolayers, the inversion symmetry has been broken for Janus Sb or Bi monolayers due to asymmetrical decoration. Lattice constants, bond lengths of Sb or Bi with center atoms of CH_3 or OH, center buckle thickness, entire layer thickness, cohesive energy, Mulliken population charge of Janus Sb or Bi monolayers are listed in Table 1. We can see the appreciable buckle configurations of central Sb or Bi atomic layers for SbH or BiH monolayer and the quasi-planar geometry of Janus structures (for comparison). It has been demonstrated by preliminary calculations that the cohesive energy of the SbH or BiH monolayer (without chemical group decoration and no Janus) is higher than that of their purely elemental counterparts. Even though the bonding strength of Sb or Bi with C or O atoms is substantially higher than that with H atoms, the remarkably larger buckling of SbH and BiH monolayers leads to a comparable and even higher cohesive energy as that of Janus Sb and Bi monolayers, as shown in Table 1. The bonding strength, which is characterized by bond length and cohesive energy, increases with the increase in electronegativities of C and O elements. In particular, the cohesive energy approaches the highest and lowest values of 9.59 and 7.53 eV/unitcell for HSb_2OH and HBi_2CH_3, respectively. It is an energetic indication that Janus Sb or Bi monolayers can be realized by strongly bonding CH_3 or OH with Sb or Bi atomic layer into perfect crystalline structures. Thus, it is preferential to synthesize these asymmetrically functionalized monolayers with a high static stability in structure.

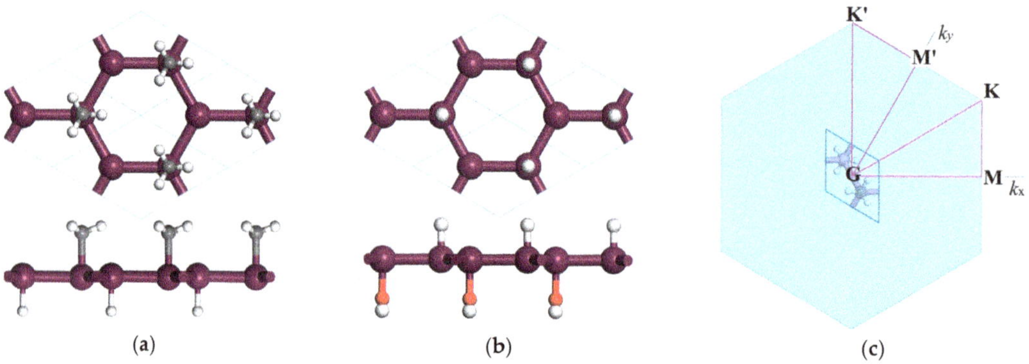

Figure 1. (**a**,**b**) Top-view (top) and side-view (bottom) schematics of Janus Sb or Bi monolayers with asymmetrical decorations of H and CH_3 or OH on two surface sides, respectively, where the violet, gray, red and white balls symbolize the bonding atoms of Sb or Bi, C, O and H, respectively; (**c**) high-symmetry points (M, K, M′, K′) specifying the dispersion paths of phonon frequency or electron energy in Brillouin zone.

Table 1. Lattice constants a, bond lengths of Sb or Bi with center atoms of CH_3 or OH (d_{MX}), center buckle thickness (h_M), entire layer thickness (h), cohesive energy (E_{coh}), Mulliken population charge (q_M) where M_1 and M_2 indicate two atoms of Sb or Bi and X denotes the functional-group center atom (C or O) or the H atom (for SbH and BiH only) of Janus Sb and Bi monolayers together with the SbH and BiH monolayers for comparison.

Monolayers	a/Å	d_{MX}/Å	h_M/Å	h/Å	E_{coh}/eV·unitcell^{-1}	q_M/e		
						M_1	M_2	X
SbH	5.218	1.719	0.092	3.346	9.409	0.13	0.13	−0.13
HSb_2CH_3	5.207	2.173	0.003	4.252	9.245	0.29	0.09	−0.53
HSb_2OH	5.219	1.912	0.003	4.596	9.591	0.71	−0.13	−0.71
BiH	5.467	1.851	0.175	3.526	7.771	0.73	0.73	−0.73
HBi_2CH_3	5.478	2.333	0.046	4.475	7.534	1.08	0.48	−0.93
HBi_2OH	5.494	2.120	0.034	4.656	7.619	1.18	0.46	−1.32

Kinetic stability under atomic oscillations is evaluated with frequency dispersion spectra of phonon modes, as shown in Figure 2. Preliminary phonon calculations by the linear response method [30] indicate that LO–TO frequencies at Brillouin zone center are degenerated for all these Janus and SbH or BiH monolayers, verifying that LO–TO splitting can be neglected. Phonon dispersion spectra of Janus Sb and Bi monolayers represent 12 branches comprised of three acoustic and nine optical branches. Acoustic phonon modes are classified into longitudinal acoustic waves, in-plane and out-of-plane transverse acoustic waves, while optical phonons contain an isolated out-of-plane transverse optical wave and a pair of G-point degenerated longitudinal and in-plane transverse optical waves. Both Janus Sb or Bi monolayers and SbH or BiH monolayers represent partially negative acoustic phonon frequencies in dispersion along G–M while approaching positive values near K point, which is attributed to virtual vibrations of H atoms passivating on Sb or Bi atomic layers. This result suggests that it is preferable to use other elements with a heavier mass and higher bonding strength rather than H atoms to saturate p_z orbitals of Sb or Bi monolayers for complete kinetic stability.

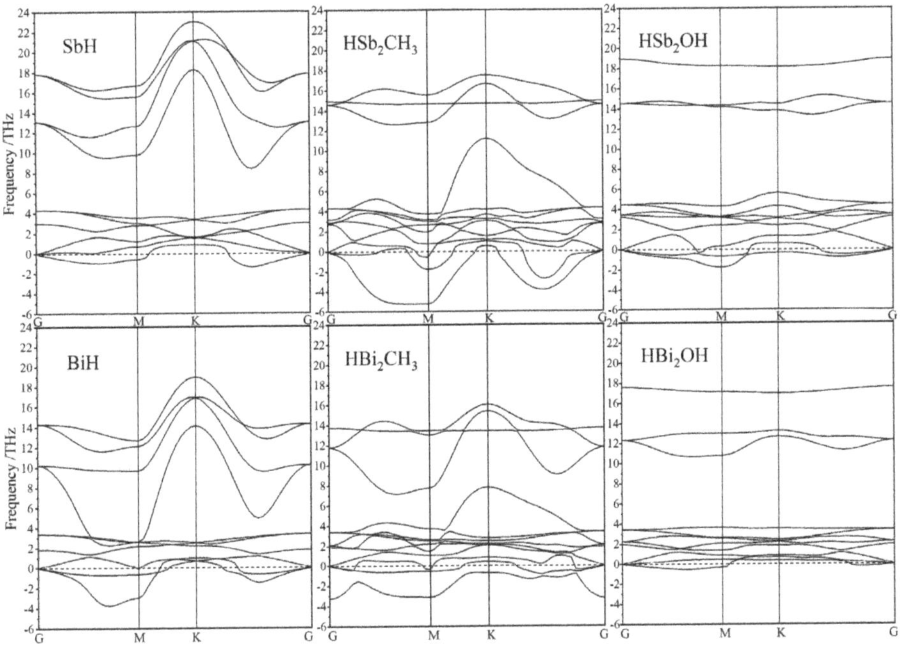

Figure 2. Phonon dispersion spectra of Janus Sb and Bi monolayers in comparison to SbH and BiH monolayers.

2.2. Internal Gradient of Electrostatic Potential

Mulliken population charges on the C and O bonding atoms of −0.5 and −0.7 e (or −0.9 and −1.3 e), respectively, due to their higher electronegativity than that of Sb (or Bi) are in contrast to −0.1 e (or 0.7 e) on hydrogen atoms directly bonding with Sb or Bi atoms, as listed in Table 1, implying substantial charge transfer and electronic cloud shifting out of central layer plane in Janus Sb and Bi monolayers. Accordingly, significant internal electric fields of Janus monolayers are created by the potential differences of $\Delta\phi$ = 5.586, 3.179, 5.262 and 1.734 eV for HSb_2CH_3, HSb_2OH, HBi_2CH_3 and HBi_2OH, as shown in Figure 3. Therefore, it is suggested that inversion symmetry of Sb and Bi monolayers has been broken in these Janus monolayers due to the asymmetric decoration of CH_3 or OH. For the Janus monolayers based on a heavy main group element, the electronegativity of center atoms in asymmetrically decorating chemical groups is a judgement giving rise to an intrinsic electric field.

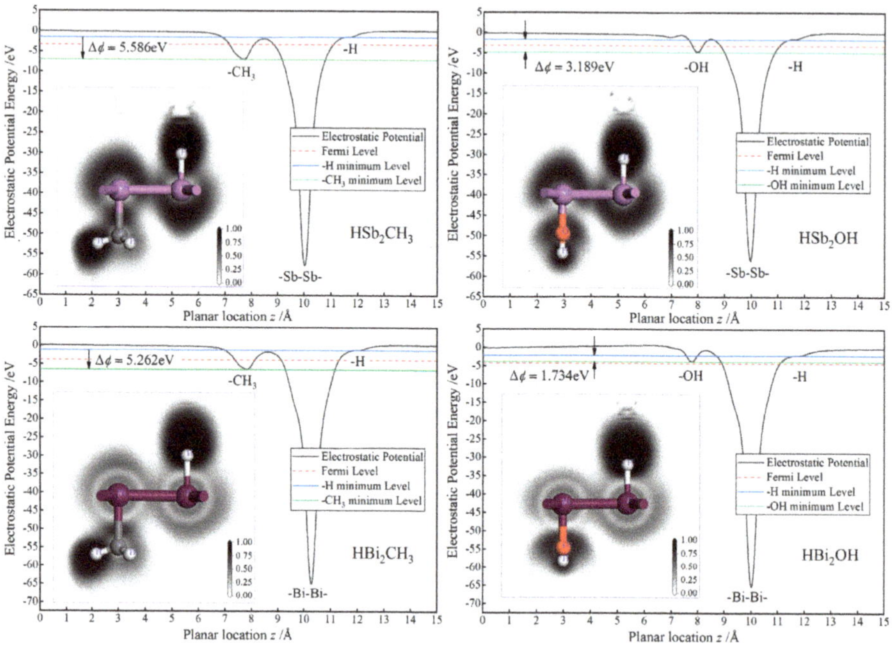

Figure 3. Internal potential gradient and bonding characteristics of the Janus Sb and Bi monolayers: mean electrostatic potential as a function of the planar position, with the inserted panels showing the electron localization function contoured on the normal plane crossing both Sb or Bi and C or O atomic centers.

Bonding characteristics of Sb or Bi with C or O atoms are evaluated by electron localization function (ELF), as depicted with the white–black contoured maps on the normal plane through atomic centers in the inserted panels of Figure 3. For all these Janus Sb and Bi monolayers, the highest ELF appears around the bonding middle between C atom center of CH_3 and Sb or Bi atoms as a manifestation of covalent bonding in consistence with the minimal electron transfer according to Mulliken population charges. By contrast, for H or O atoms bonding with Sb or Bi atoms, the high ELF region resides around H or O center and promptly decreases at the middle of bonds, which implies ionic character, accounting for an internal electric field internal Janus Sb or Bi monolayers. ELF distributions reveal the chemical bonding attributes of the internal electric fields caused by asymmetrical functionalization with CH_3 or OH.

2.3. Band Structure

All the band structures of Janus Sb and Bi monolayers from our calculations are merged in Figure 4, where the SOC effect could be distinguished by red and black dispersion curves. According to partial and atomic projections, it is determined that band-edge electronic-states near Fermi level are primarily composed of p_x and p_y orbitals from Sb or Bi bonding atoms [31]. Even without SOC, the highest valence band and the lowest conduction band crossing linearly at K and K′ points imply the Dirac-cone-like characteristics of these 2D honeycomb systems. When SOC is included, the degeneracy at Dirac-cone point splits off with the highest valence band and the lowest conduction band shifted towards high and low energy from Fermi level, respectively, thereby opening a notable band-gap by SOC for both the symmetrically hydrogenated and Janus monolayers. By contrast, at G point, the valence and conduction bands are upshifted and downshifted respectively, which accounts for a global indirect band-gap in the functionalized Bi monolayers with a remarkably larger K-point splitting caused by a much stronger SOC of Bi systems than that of Sb systems. Global band-gaps of Janus Sb and Bi monolayers range in 0.2~0.4 eV and 0.8~0.9 eV, respectively, as listed in Table 2.

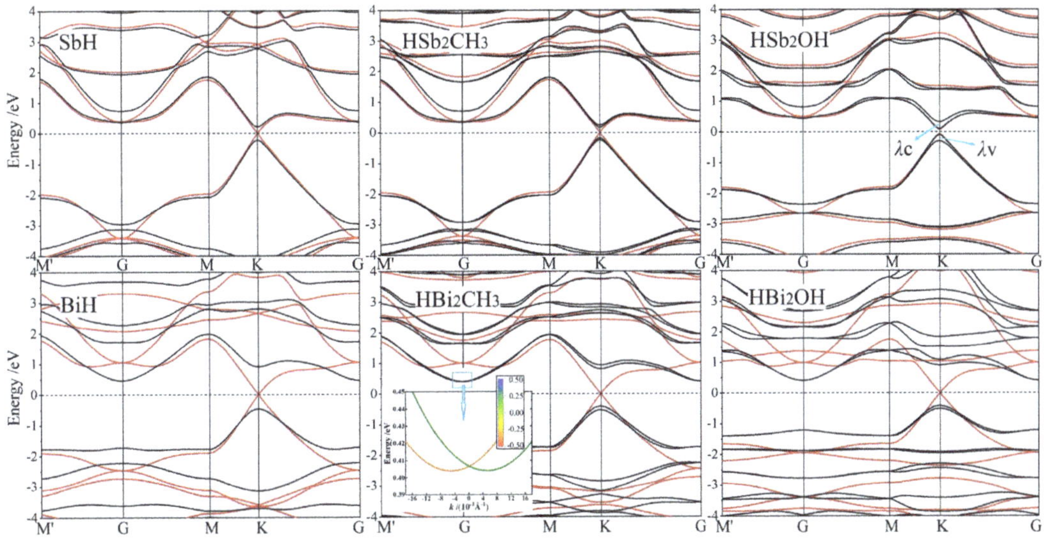

Figure 4. Band structures without SOC (red curves) and with SOC (black curves) of Janus Sb (top panels) and Bi (bottom panels) monolayers, and SbH and BiH monolayers with inversion symmetry are presented for comparison, in which Fermi level is referenced as energy zero (horizontal dash line).

Table 2. Rashba spin-splitting R_G of the highest valence band at G point, SOC-induced splitting λv and λc at K point for valence and conduction band-edges respectively, SOC-induced Dirac-cone splitting gaps at K point E_D(K), global band-gaps E_g(global) of Janus Sb and Bi monolayers compared with SbH and BiH monolayers.

Monolayers	R_G/eV·Å	λv/meV	λc/meV	E_D(K)/eV	E_g(global)/eV
SbH	0	0	0	0.426	0.426
HSb_2CH_3	0	0.060	0.065	0.351	0.351
HSb_2OH	0	0.194	0.229	0.194	0.194
BiH	0	0	0	1.372	0.918
HBi_2CH_3	0.224	0.124	0.123	1.219	0.809
HBi_2OH	0	0.093	0.177	1.323	0.824

It has been reported by ab-initio calculations that Rashba spin-splitting occurs in all the Janus Mo and W dichalcogenide monolayers due to the simultaneous existences of strong SOC and internal electric field. Consistently, Rashba spin-splitting arises in HBi_2CH_3 monolayers with a built-in potential gradient as shown in Figure 3, leading to a perceptible Rashba parameter R_G of 0.224 eV·Å due to the strong SOC from p_x and p_y orbitals of Bi bonding atoms, as manifested by comparing non-SOC and SOC band structures in Figure 4. It is suggested by the lower built-in electric field in Rashba splitting $HBiCH_3$ than in HSb_2CH_3 with a lower SOC that strong SOC dominates the generation of Rashba splitting in Janus Sb and Bi monolayers. Since SOC and angular momentum cannot be simultaneous represented in quantum mechanics, the spin-up and spin-down states cannot be completely determined in SOC band structures, as shown by the color-mapped spin components near Fermi level for Rashba $HBiCH_3$ monolayer in the inserted panel of Figure 4. Meanwhile,

SOC also causes remarkable band-edge splittings at K point for Janus Sb and Bi monolayers, as indicated by λv and λc for the upmost valence band and the downmost conduction band, respectively, with the values listed in Table 2. For HBi_2CH_3 and HBi_2OH monolayers, the valley shape of the upmost valence band is preferred over the flat shape of the downmost conduction band at K point for valleytronics applications.

Topological properties of SbH and BiH monolayers are characterized by the topological Z_2 invariant evaluated from first-principles electronic Bloch wave-functions, where $Z_2 = 1$ and $Z_2 = 0$ represent the nontrivial and trivial topology, respectively. Since the atomic structures of SbH and BiH monolayers have space inversion symmetry, the Z_2 topological invariant is directly calculated by Bloch wave-function parities of valence electrons at time-reversal-invariant points (TRIPs) in Brillouin zone. Electronic structures of SbH and BiH monolayers have four TRIPs, one at G point and three at M point, so the topological index P can be calculated as follows:

$$P(K_i) = \prod_{m=1}^{N} \delta_{2m}^{i}, \quad (-1)^{\nu} = \prod_{i=1}^{4} P(K_i) = P(\mathrm{G})P(\mathrm{M})^3 \tag{1}$$

where P denotes the parity product of Bloch wave-functions on TRIPs, $\delta = \pm 1$ indicates parity, and N is the number of valence bands. The 12 valence electrons in primitive cells of SbH or BiH monolayer constitute six spin-degenerate electronic-states at a TRIP, and the parities of spin-degenerate electron eigenfunctions of SbH and BiH monolayers on all TRIPs are identical. As shown in Table 3, the values of P at G and M points are −1 and +1 respectively, which leads to the nontrivial $Z_2 = 1$, demonstrating that SbH and BiH monolayers are nontrivial topological insulators.

Table 3. Parities δ and their product P of spin-degenerate states at TRIPs for SbH and BiH monolayers.

TRIP			Parity δ				P
G	+1	+1	−1	+1	−1	+1	−1
3 M	+1	−1	+1	−1	+1	−1	+1

In contrast, it is invalid of evaluating Z_2 invariant by the parity of valence electronic-states to determine whether Janus Sb and Bi monolayers are topologically trivial or nontrivial, because they are lacking of space inversion symmetry due to the asymmetrical functionalization of chemical groups. Nevertheless, nontrivial topology of Janus Sb and Bi monolayers could be represented in the odd number of conductive channels from edge electronic-states without band-gap, where Dirac-cone point appears to be crossed by the lowest conduction band and the highest valence band of nanoribbon at Fermi level. To investigate this important topological property, the zigzag nanoribbons of Janus Sb and Bi monolayers are constructed, in which atomic structures on both edges are mirror-symmetrical to each other, and the unsaturated edge bonds are passivated by hydrogen atoms. Nanoribbon width is specified about 35 Å to avoid overlapping between two edge states of nanoribbon. After geometrical optimization, the band structures of Janus Sb and Bi monolayer nanoribbons are calculated, with the results shown in Figure 5. The energy dispersion of edge electronic-states of these nanoribbons intersects into a Dirac-cone point of Fermi level at boundary X point of one-dimensional Brillouin zone. The odd number of edge-state conductance channels is derived from nontrivial topology that will also present edge-state Dirac-cones in band structures of armchair nanoribbons. Therefore, this Dirac-cone signature of edge states proves these Janus Sb and Bi monolayers are (nontrivial) topological insulators to realize QSH effect at ambient temperature supported by their giant band-gaps.

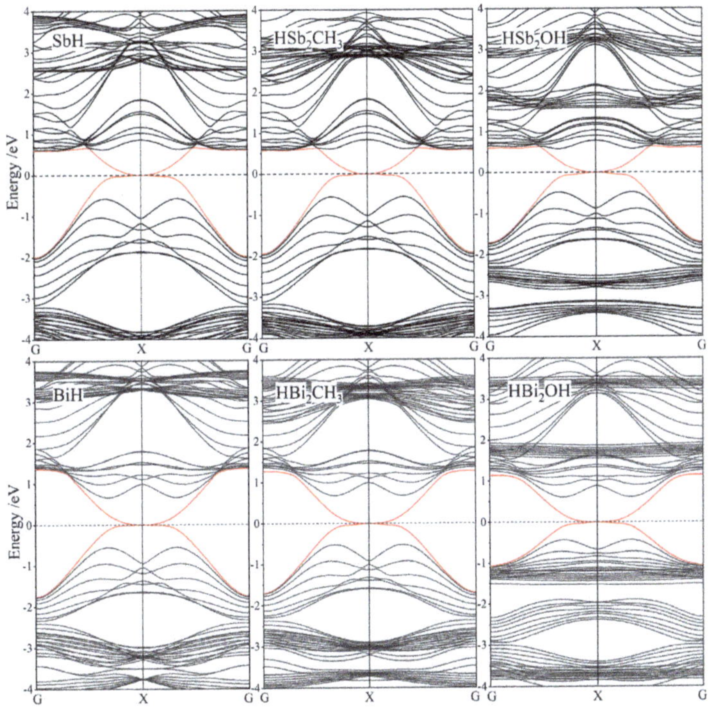

Figure 5. SOC-included band structures of zigzag nanoribbons of Janus Sb and Bi monolayers (functionalized asymmetrically by H atom and CH_3 or OH), and SbH and BiH nanoribbons are presented for comparison, with red color indicating edge states and Fermi level being referenced as energy zero.

2.4. Valley-Contrast Circular Dichroism

There is a valley degree of freedom in Janus Sb and Bi monolayers, and the strong SOC opens a large electronic energy gap at the corners of Brillouin zone. Inversion symmetry is broken in Janus Sb and Bi monolayers due to the asymmetrical functionalizations with CH_3 or OH, leading to valley-contrast Berry curvature and spin that yield valley spin Hall effect and valley-contrast circular dichroism. With a giant bulk band-gap and nontrivial topology, these Janus monolayers can afford to be a quantum spin Hall insulator and a quantum valley Hall insulator at ambient temperature. Valley-contrast Berry curvature and spin can be described by valley polarization $\eta(\boldsymbol{k})$ in Brillouin zone as follows:

$$\eta(\boldsymbol{k}) = \frac{|P_+(\boldsymbol{k})|^2 - |P_-(\boldsymbol{k})|^2}{|P_+(\boldsymbol{k})|^2 + |P_-(\boldsymbol{k})|^2}, \quad P_\pm = P_x \pm iP_y \tag{2}$$

where P_α ($\alpha = x, y$) denotes momentum matrix element between the lowest conduction band and the highest valence band. Prototypically, the calculated $\eta(\boldsymbol{k})$ of HBi_2OH monolayer along the high-symmetry path through valley K and K′ points in Brillouin zone is shown in Figure 6, illustrating the contrast of optical selection rule between two valley counterparts. The left-hand and right-hand circular polarized photons are almost purely absorbed at K and K′ valleys, respectively. Therefore, circular polarized photons can distinguish electronic excitation and imbalance electron occupation between the two inversely distributed nonequivalent electronic-valleys in Janus Sb and Bi monolayers without external electric field.

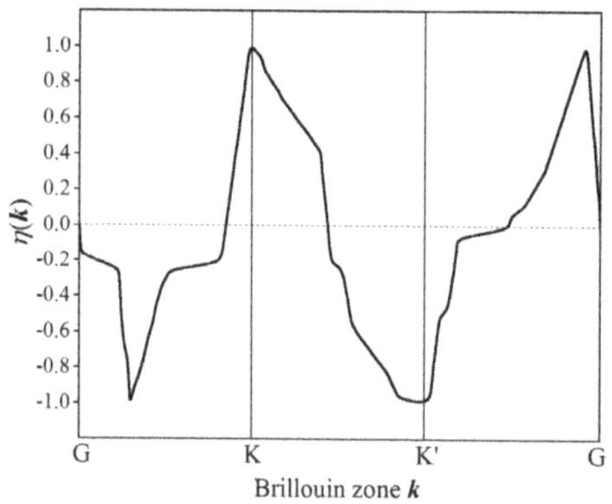

Figure 6. Valley polarization of HBi_2OH monolayer along high-symmetry path through valley K and K′ points.

According to a low-power model, the valley magnetic moments in Janus Sb and Bi monolayers are related to both valley pseudo-spin and real spin which are substantially coupled due to giant band-gaps induced by strong SOC. Valley pseudo-spin can be rendered by strong SOC under the internal built-in electric field in Janus Sb and Bi monolayers, which is different from inversion symmetry systems that need an external electric field. Notably, valleytronics can hardly be achieved due to minimal SOC in graphene or silicene, whilst requires a high external electric field to break inversion symmetry in SbH and BiH monolayers.

3. Theoretical Methodology

Employing the fully relativistic pseudopotential planewave first-principles method based on spin density functional theory, the atomic structure, phonon dispersion, SOC electronic structures of Janus Sb and Bi monolayers are calculated as implemented by CASTEP module of Materials Studio 2020 package (Accelrys Inc., Materials Studio version 2020.08, San Diego, CA, USA) [32]. Gradient-corrected exchange–correlation functional Wu and Cohen (WC) is adopted to perform the electronic state and energy calculations [33]. WC functional has been shown to achieve a significant improvement over LDA, PBE and TPSS functionals for predicting atomic structures of solids including ionic compounds and describing covalent bonds of molecules, especially for group I, group IV and group III–V semiconductors [34]. SOC is incorporated into the calculations of energy, atomic force and geometry, electronic structure and optoelectronic matrix elements. Band structures without SOC are also calculated for comparison to reveal the SOC-induced source of Rashba splitting, valley freedom and giant band-gaps.

Interactions between atomic cores and valence electrons are presented by norm-conserving pseudopotential, and the relativistic effect is treated with Koelling–Harmon schemes [35]. Electron wave functions are expanded by plane-wave basis-set under 720 eV energy cutoff, whilst finite basis-set correction is used in energy and stress calculations by estimating energy derivatives from three-points numerical differentiation [36]. Pulay density mixing scheme at 0.5 charge magnitude is utilized to carry out self-consistent field (SCF) iterations with a convergence tolerance of 5×10^{-7} eV/atom and on an FFT grid of $48 \times 48 \times 200$ for electron density calculations [37,38]. Monkhorst–Pack $3 \times 3 \times 1$ grid of k-sampling is adopted for integration in Brillouin zone [39].

Two-dimensional atomic structures are modeled in virtual three-dimensional space by setting a sufficient vacuum layer (15 Å) between the periodic images of monolayers. Geometry optimization is fulfilled by minimizing energy functional with BFGS algorithm in delocalized internal coordinates instead of Cartesian coordinates, so as to reach the energy, atomic force and stress convergences of 5.0×10^{-6} eV/atom, 0.01 eV/Å and 0.02 GPa, respectively [40]. Phonon structures are calculated by finite-displacement method using several small supercells without SOC-induced bonding force and ignoring frequency splitting between the longitude and transverse optical vibration modes (LO–TO splitting) at G point, in which the convergence tolerance of mechanical constant is specified as 1.0×10^{-5} eV/Å^2.

4. Conclusions

Spin–orbit coupling electronic structures of Janus (asymmetrically functionalized by CH_3 or OH) Sb and Bi monolayers (HSb_2XH_n and HBi_2XH_n) with nontrivial topology, Rashba splitting and valley-contrast circular dichroism are first proposed and demonstrated in the present study. Cohesive energies and phonon dispersion indicate that all these Janus systems have innate static stability, whilst undergoing kinetic instability of H atoms capping another surface of Sb or Bi atomic layer. Under SOC effects, Janus Sb and Bi monolayers show direct and indirect band-gap characteristics of band structures, respectively. It is identified by means of first-principles calculations that HSb_2XH_n and HBi_2XH_n are a new family of giant band-gap 2D TIs with the largest bulk band-gap of ~0.8 eV, by far exceeding the current band-gaps of 2D TI materials realized in experiments. These giant band-gaps are entirely attributed to strong SOC of Sb or Bi p_x and p_y atomic orbitals around K and K′ valleys in Brillouin zone of honeycomb lattice, which suffice for practical applications at ambient temperature. Rashba splitting originates dominantly from strong SOC together with built-in out-of-plane electric field, as manifested by the unique Rashba HBi_2CH_3 monolayer possessing strong SOC but with a minimal out-of-plane potential gradient. Janus Sb and Bi monolayers are identified to be a spin valley Hall insulator by valley-selective circular dichroism. Due to strong SOC, the valley magnetic moments could be modified by intrinsic out-of-plane electric field to produce a substantial correlation between real and valley-pseudo spins. These results represent a significant advance in 2D TIs to stimulate further work of developing these new 2D TIs for Rashba spintronic devices and quantum valley Hall insulators.

Author Contributions: Data curation, formal analysis, writing—original draft preparation, writing—review and editing, Q.G.; conceptualization, investigation, project administration, G.Z. All authors have read and agreed to the published version of the manuscript.

Funding: This research was funded by the National Natural Science Foundation of China (Grant No. 51973046).

Institutional Review Board Statement: Not applicable.

Informed Consent Statement: Not applicable.

Data Availability Statement: Theoretical methods and results are available from all authors.

Conflicts of Interest: The authors declare no conflict of interest.

References

1. Salem, A.K.; Searson, P.C.; Leong, K.W. Multifunctional nanorods for gene delivery. *Nat. Mater.* **2003**, *2*, 668–671. [CrossRef] [PubMed]
2. Roh, K.H.; Martin, D.C.; Lahann, J. Biphasic Janus particles with nanoscale anisotropy. *Nat. Mater.* **2005**, *4*, 759–763. [CrossRef] [PubMed]
3. Nisisako, T.; Hatsuzawa, T. A microfluidic cross-flowing emulsion generator for producing biphasic droplets and anisotropically shaped polymer particles. *Microfluid. Nanofluid.* **2010**, *9*, 427–437. [CrossRef]
4. Zhang, L.; Luo, Q.; Zhang, F.; Zhang, D.M.; Wang, Y.S.; Sun, Y.L.; Dong, W.F.; Liu, J.Q.; Huo, Q.S.; Sun, H.B. High-performance magnetic antimicrobial Janus nanorods decorated with Ag nanoparticles. *J. Mater. Chem.* **2012**, *22*, 23741–23744. [CrossRef]
5. Kirillova, A.; Schliebe, C.; Stoychev, G.; Jakob, A.; Lang, H.; Synytska, A. Hybrid hairy Janus particles decorated with metallic nanoparticles for catalytic applications. *ACS Appl. Mater. Interfaces* **2015**, *7*, 21218–21225. [CrossRef]
6. Zhang, L.; Yu, J.; Yang, M.; Xie, Q.; Peng, H.; Liu, Z. Janus graphene from asymmetric two-dimensional chemistry. *Nat. Commun.* **2013**, *4*, 1443. [CrossRef]
7. Bissett, M.A.; Takesaki, Y.; Tsuji, M.; Ago, H. Increased chemical reactivity achieved by asymmetrical 'Janus' functionalisation of graphene. *RSC Adv.* **2014**, *4*, 52215–52219. [CrossRef]

8. Li, F.; Li, Y.F. Band gap modulation of Janus graphene nanosheets by interlayer hydrogen bonding and the external electric field: A computational study. *J. Mater. Chem. C* **2015**, *3*, 3416–3421. [CrossRef]
9. Chico, L.; Latgé, A.; Brey, L. Symmetries of quantum transport with Rashba spin-orbit: Graphene spintronics. *Phys. Chem. Chem. Phys.* **2015**, *17*, 16469–16475. [CrossRef] [PubMed]
10. Lv, M.H.; Li, C.M.; Sun, W.F. Spin-orbit coupling and spin-polarized electronic structures of Janus vanadium-dichalcogenide monolayers: First-principles calculations. *Nanomaterials* **2022**, *12*, 382. [CrossRef]
11. Hu, W.; Xia, N.; Wu, X.J.; Li, Z.Y.; Yang, J.L. Silicene as a highly sensitive molecule sensor for NH_3, NO and NO_2. *Phys. Chem. Chem. Phys.* **2014**, *16*, 6957–6962. [CrossRef]
12. Kaloni, T.P.; Schreckenbach, G.; Freund, M.S. Large enhancement and tunable band gap in silicene by small organic molecule adsorption. *J. Phys. Chem. C* **2014**, *118*, 23361–23367. [CrossRef]
13. Özçelik, V.O.; Kecik, D.; Durgun, E.; Ciraci, S. Adsorption of group IV elements on graphene, silicene, germanene, and stanene: Dumbbell formation. *J. Phys. Chem. C* **2015**, *119*, 845–853. [CrossRef]
14. Denis, P.A. Stacked functionalized silicene: A powerful system to adjust the electronic structure of silicene. *Phys. Chem. Chem. Phys.* **2015**, *17*, 5393–5402. [CrossRef] [PubMed]
15. Yu, W.Z.; Yan, J.A.; Gao, S.P. Band gap characters and ferromagnetic/antiferromagnetic coupling in group-IV monolayers tuned by chemical species and hydrogen adsorption configurations. *Nanoscale Res. Lett.* **2015**, *10*, 351. [CrossRef] [PubMed]
16. Qiu, J.; Fu, H.; Xu, Y.; Zhou, Q.; Meng, S.; Li, H.; Chen, L.; Wu, K. From silicene to half-silicane by hydrogenation. *ACS Nano* **2015**, *9*, 11192–11199. [CrossRef] [PubMed]
17. Zhang, W.B.; Song, Z.B.; Dou, L.M. The tunable electronic structure and mechanical properties of halogenated silicene: A first-principles study. *J. Mater. Chem. C* **2015**, *3*, 3087–3094. [CrossRef]
18. Özçelik, V.O.; Cahangirov, S.; Ciraci, S. Stable single-layer honeycomblike structure of silica. *Phys. Rev. Lett.* **2014**, *112*, 246803. [CrossRef]
19. Stephan, R.; Hanf, M.C.; Sonnet, P. Opening the way to molecular cycloaddition of large molecules on supported silicene. *J. Chem. Phys.* **2015**, *143*, 154706. [CrossRef]
20. Wu, J.; Xie, L.M.; Li, Y.; Wang, H.; Ouyang, Y.; Guo, J.; Dai, H.G. Controlled chlorine plasma reaction for noninvasive graphene doping. *J. Am. Chem. Soc.* **2011**, *133*, 19668–19671. [CrossRef]
21. Freitas, R.R.Q.; Rivelino, R.; de Brito Mota, F.; de Castilho, M.C. Topological insulating phases in two-dimensional bismuth containing single layers preserved by hydrogenation. *J. Phys. Chem. C* **2015**, *119*, 23599–23606. [CrossRef]
22. Liu, C.C.; Jiang, H.; Yao, Y.G. Low-energy effective Hamiltonian involving spin-orbit coupling in silicene and two-dimensional germanium and tin. *Phys. Rev. B* **2011**, *84*, 195430. [CrossRef]
23. Liu, C.C.; Feng, W.X.; Yao, Y.G. Quantum spin Hall effect in silicene and two-dimensional germanium. *Phys. Rev. Lett.* **2011**, *107*, 076802. [CrossRef] [PubMed]
24. Wang, Z.F.; Liu, Z.; Liu, F. Organic topological insulators in organometallic lattices. *Nat. Commun.* **2013**, *4*, 1471. [CrossRef] [PubMed]
25. Xu, Y.; Yan, B.; Zhang, H.J.; Wang, J.; Xu, G.; Tang, P.; Duan, W.; Zhang, S.C. Large-gap quantum spin Hall insulators in tin films. *Phys. Rev. Lett.* **2013**, *111*, 136804. [CrossRef]
26. Yang, F.; Miao, L.; Wang, Z.F.; Yao, M.Y.; Zhu, F.; Song, Y.R.; Wang, M.X.; Xu, J.P.; Fedorov, A.V.; Sun, Z.; et al. Spatial and energy distribution of topological edge states in single Bi(111) bilayer. *Phys. Rev. Lett.* **2012**, *109*, 016801. [CrossRef]
27. Yan, B.H.; Jansen, M.; Felser, C. A large-energy-gap oxide topological insulator based on the superconductor $BaBiO_3$. *Nat. Phys.* **2013**, *9*, 709–711. [CrossRef]
28. Gong, Q.; Zhang, G. Spin-orbit coupling electronic structures of organic-group functionalized Sb and Bi topological monolayers. *Nanomaterials* **2022**, *12*, 2041. [CrossRef]
29. Klintenberg, M.; Lebegue, S.; Katsnelson, M.; Eriksson, O. Theoretical analysis of the chemical bonding and electronic structure of graphene interacting with group IA and group VIIA elements. *Phys. Rev. B* **2010**, *81*, 085433. [CrossRef]
30. Baroni, S.; de Gironcoli, S.; dal Corso, A.; Giannozzi, P. Phonons and related crystal properties from density-functional perturbation theory. *Rev. Mod. Phys.* **2001**, *73*, 515–562. [CrossRef]
31. Ma, Y.D.; Dai, Y.; Kou, L.Z.; Frauenheim, T.; Heine, T. Robust two-dimensional topological insulators in methyl functionalized bismuth, antimony, and lead bilayer films. *Nano Lett.* **2015**, *15*, 1083–1089. [CrossRef] [PubMed]
32. Clark, S.J.; Segall, M.D.; Pickard, C.J.; Hasnip, P.J.; Probert, M.I.J.; Refson, K.; Payne, M.C. First principles methods using CASTEP. *Z. Krist.* **2005**, *220*, 567–570. [CrossRef]
33. Wu, Z.; Cohen, R.E. More accurate generalized gradient approximation for solids. *Phys. Rev. B* **2006**, *73*, 235116. [CrossRef]
34. Tran, F.; Laskowski, R.; Blaha, P.; Schwarz, K. Performance on molecules, surfaces, and solids of the Wu-Cohen GGA exchange-correlation energy functional. *Phys. Rev. B* **2007**, *75*, 115131. [CrossRef]
35. Lin, J.S.; Qteish, A.; Payne, M.C.; Heine, V. Optimized and transferable nonlocal separable ab initio pseudopotentials. *Phys. Rev. B* **1993**, *47*, 4174–4180. [CrossRef] [PubMed]
36. Milman, V.; Lee, M.H.; Payne, M.C. Ground-state properties of $CoSi_2$ determined by a total-energy pseudopotential method. *Phys. Rev. B* **1994**, *49*, 16300–16308. [CrossRef]
37. Payne, M.C.; Teter, M.P.; Allan, D.C.; Arias, T.A.; Joannopoulos, J.D. Iterative minimization techniques for ab initio total energy calculations: Molecular dynamics and conjugate gradients. *Rev. Mod. Phys.* **1992**, *64*, 1045–1097. [CrossRef]

38. Kresse, G.; Furthmuller, J. Efficient iterative schemes for ab initio total-energy calculations using a plane-wave basis set. *Phys. Rev. B* **1996**, *54*, 11169–11186. [CrossRef]
39. Monkhorst, H.J.; Pack, J.D. Special points for Brillouin-zone integrations—A reply. *Phys. Rev. B* **1977**, *16*, 1748.
40. Packwood, D.; Kermode, J.; Mones, L.; Bernstein, N.; Woolley, J.; Gould, N.; Ortner, C.; Csanyi, G. A universal preconditioner for simulating condensed phase materials. *J. Chem. Phys.* **2016**, *144*, 164109. [CrossRef]

International Journal of
Molecular Sciences

MDPI

Article

DFT Study of Molecular Structure, Electronic and Vibrational Spectra of Tetrapyrazinoporphyrazine, Its Perchlorinated Derivative and Their Al, Ga and In Complexes

Igor V. Ryzhov, Alexey V. Eroshin, Yuriy A. Zhabanov *, Daniil N. Finogenov and Pavel A. Stuzhin

Research Institute of Chemistry of Macroheterocyclic Compounds, Ivanovo State University of Chemistry and Technology, Sheremetievskiy Av. 7, 153000 Ivanovo, Russia; ryzhoff.ihor@yandex.ru (I.V.R.); alexey.yeroshin@gmail.com (A.V.E.); dan.finogenof@gmail.com (D.N.F.); stuzhin@isuct.ru (P.A.S.)

* Correspondence: zhabanov@isuct.ru; Tel.: +7-4932-35-98-74

Abstract: Electronic and geometric structures of metal-free, Al, Ga and In complexes with tetrapyrazinoporphyrazine (TPyzPA) and octachlorotetrapyrazinoporphyrazine ($TPyzPACl_8$) were investigated by density functional theory (DFT) calculations and compared in order to study the effect of chlorination on the structure and properties of these macrocycles. The nature of the bonds between metal atoms and nitrogen atoms was described using the NBO-analysis. Simulation and interpretation of electronic spectra were performed with the use of time-dependent density functional theory (TDDFT). A description of calculated IR spectra was carried out based on the analysis of the distribution of the potential energy of normal vibrational coordinates.

Keywords: tetrapyrazinoporphyrazine; octachlorotetrapyrazinoporphyrazine; DFT study; molecular and electronic structure; electronic spectra; vibrational spectra

1. Introduction

Tetrapyrrole macroheterocycles are essential for a wide range of technologies, including optoelectronics and information storage devices [1–7]. The complexes of tetrapyrazinoporphyrazine ($H_2TPyzPA$) with metals possessing π-deficient heterocyclic fragments are promising compounds for organic electronics [8–11]. The results of the recent electrochemical study [8] of the perchlorinated complexes $M(Cl)TPyzPACl_8$ (M = Al, Ga, In) allow them to be considered as potential acceptor materials. Recently this was also demonstrated for perchlorinated tripyrazinosubporphyrazines [12].

Fine tuning of the physico-chemical properties of the metal complexes with macroheterocyclic ligands requires a deep knowledge of their molecular and electronic structures. However, only a little attention has, of yet, been paid to structural investigations of the metal complexes of TPyzPA [13]. Solid-state X-ray studies of Fe(II) [14] and Sn(IV) complexes [15] revealed that the $TPyzPACl_8$ macrocycle is nearly planar. The possible planarity of the structures of $MTPyzPACl_8$ complexes gives rise to the extended π-electron conjugation, which in turn results in an increased packing density.

The present contribution is devoted to the combined computational investigation of the geometry structures, features of the metal–ligand bonding and spectral properties of M(Cl)TPyzPA (M = Al, Ga, In) complexes and their perchlorinated analogues. It also extends our recent study [16] of the tetrabenzoporphyrin (TBP) complexes with the same set of metals. Density functional theory (DFT) [17,18] was used as a theoretical framework since it was earlier established to provide a good description of the features of the geometry and electronic structures of the analogous macroheterocycles and their metal complexes [16,19–25].

2. Results

2.1. Molecular Structures

According to the performed quantum chemical calculations, the equilibrium structures of the Al(III), Ga(III) and In(III) complexes of tetrapyrazinoporphyrazine (M(Cl)TPyzPA) and their perchlorinated analogues are doming-distorted and possess C_{4v} symmetry (Figure 1). The distortion is caused by a metal atom as the metal-free H_2TPyzPA and H_2TPyzPACl_8 molecules are planar (D_{2h} symmetry). The H_2TPyzPA internuclear distances obtained by the authors of [26] are systematically higher compared to our data. This discrepancy can be explained by the use of corrections for the dispersion interaction in this paper and pseudopotentials in [26]. The H_2TPyzPA bond lengths calculated in [27] are also systematically higher than those obtained in this work. The main geometric parameters of compounds under consideration are given in Table 1.

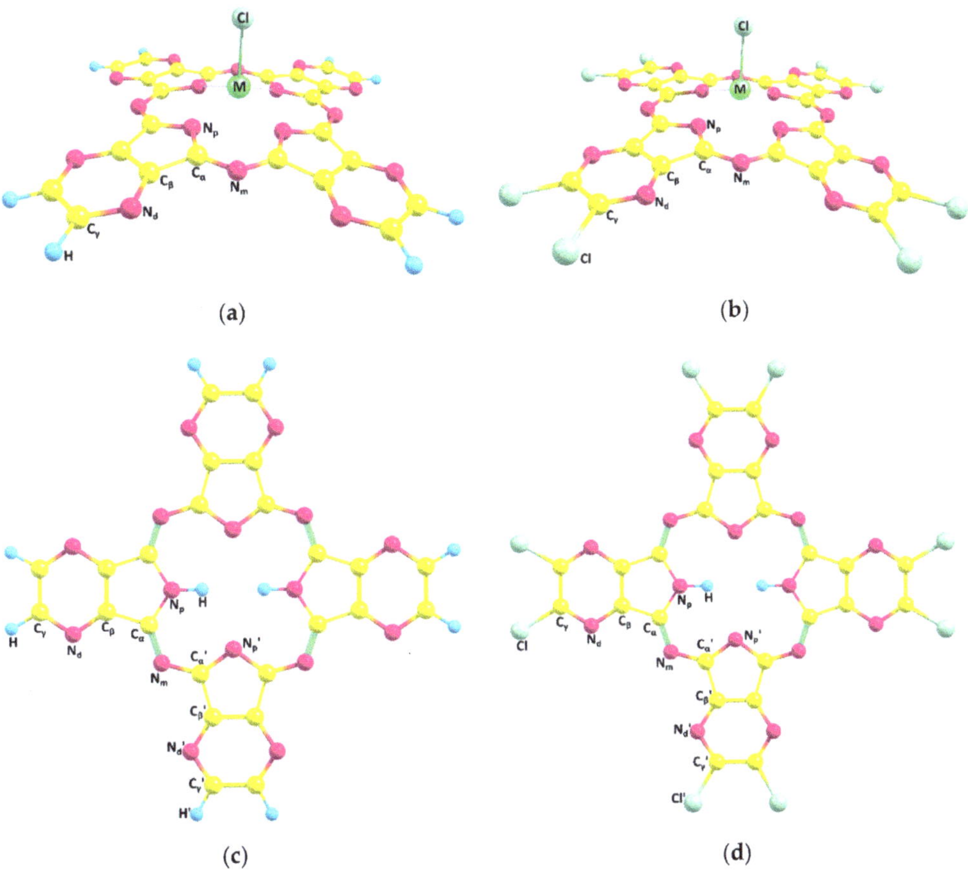

Figure 1. Molecular structures of the Al(III), Ga(III) and In(III) complexes with pyrazinoporphyrazine (M(Cl)TPyzPA) (**a**), octachloropyrazinoporphyrazines (M(Cl)TPyzPACl_8) (**b**) and metal-free molecules ((**c**) and (**d**), respectively) with atom labeling.

Table 1. Internuclear distances (r_e, in Å) and valence angles (∠, in deg.) of the equilibrium structures by PBE0-D3/def2-TZVP calculations.

	Al(Cl)TPyzPA	Ga(Cl)TPyzPA	In(Cl)TPyzPA	H_2TPyzPA
Symmetry	C_{4v}	C_{4v}	C_{4v}	D_{2h}
		Distances		
$M\text{-}N_p/M\text{-}N_p'$	1.981	2.022	2.161	-
$M\text{-}Cl_{ax}$	2.145	2.187	2.336	-
$N_p\text{-}C_\alpha/N_p'\text{-}C_\alpha'$	1.371	1.367	1.364	1.368/1.355
$C_\alpha\text{-}N_m/C_\alpha'\text{-}N_m$	1.308	1.310	1.316	1.304/1.321
$C_\alpha\text{-}C_\beta/C_\alpha'\text{-}C_\beta'$	1.450	1.453	1.458	1.450/1.464
$C_\beta\text{-}C_\beta/C_\beta'\text{-}C_\beta'$	1.391	1.393	1.399	1.401/1.393
$C_\beta\text{-}N_d/C_\beta'\text{-}N_d'$	1.328	1.328	1.326	1.329/1.324
$N_d\text{-}C_\gamma/N_d'\text{-}C_\gamma'$	1.322	1.322	1.323	1.320/1.327
$C_\gamma\text{-}C_\gamma/C_\gamma'\text{-}C_\gamma'$	1.408	1.407	1.405	1.408/1.401
$C_\gamma\text{-}H/C_\gamma'\text{-}H'$	1.086	1.086	1.086	1.086/1.086
$(N_p \ldots N_p)_{opp}$	3.859	3.918	4.037	4.054/3.899
$(N_p \ldots N_p)_{adj}$	2.728	2.770	2.854	2.812
P [1]	21.432	21.416	21.440	21.376
		Bond angles		
$\angle(Cl_{ax}MN_p)$	103.1	104.3	110.9	-
$\angle(MN_pC_\alpha)$	125.2	124.5	123.6	123.2
$\angle(N_pC_\alpha N_m)/(N_p'C_\alpha'N_m)$	127.9	128.0	128.0	128.5/128.1
$\angle(C_\alpha N_m C_\alpha')$	122.5	123.4	125.5	124.2
$\angle(C_\alpha C_\beta N_d)/(C_\alpha'C_\beta'N_d')$	129.8	129.8	129.8	129.0/130.6
$\angle(C_\beta N_d C_\gamma)/(C_\beta'N_d'C_\gamma')$	113.2	113.3	113.6	113.6/113.3
M-X [2]	0.448	0.499	0.772	-
α [3]	175.5	174.3	164.7	180
	Al(Cl)TPyzPACl$_8$	Ga(Cl)TPyzPACl$_8$	In(Cl)TPyzPACl$_8$	H_2TPyzPACl$_8$
Symmetry	C_{4v}	C_{4v}	C_{4v}	D_{2h}
		Distances		
$M\text{-}N_p/M\text{-}N_p'$	1.980	2.021	2.162	-
$M\text{-}Cl_{ax}$	2.142	2.183	2.332	-
$N_p\text{-}C_\alpha/N_p'\text{-}C_\alpha'$	1.372	1.368	1.364	1.368/1.356
$C_\alpha\text{-}N_m/C_\alpha'\text{-}N_m$	1.308	1.310	1.316	1.304/1.321
$C_\alpha\text{-}C_\beta/C_\alpha'\text{-}C_\beta'$	1.447	1.450	1.455	1.448/1.461
$C_\beta\text{-}C_\beta/C_\beta'\text{-}C_\beta'$	1.384	1.387	1.392	1.394/1.386
$C_\beta\text{-}N_d/C_\beta'\text{-}N_d'$	1.330	1.329	1.328	1.330/1.325
$N_d\text{-}C_\gamma/N_d'\text{-}C_\gamma'$	1.305	1.306	1.307	1.304/1.310
$C_\gamma\text{-}C_\gamma/C_\gamma'\text{-}C_\gamma'$	1.433	1.432	1.430	1.433/1.424
$C_\gamma\text{-}Cl/C_\gamma'\text{-}Cl'$	1.709	1.709	1.709	1.709/1.712
$(N_p \ldots N_p)_{opp}$	3.856	3.915	4.031	4.051/3.892
$(N_p \ldots N_p)_{adj}$	2.726	2.768	2.850	2.809
P [1]	21.440	21.424	21.440	21.376
		Bond angles		
$\angle(Cl_{ax}MN_p)$	103.2	104.5	111.2	-
$\angle(MN_pC_\alpha)$	125.2	124.5	123.5	123.3
$\angle(N_pC_\alpha N_m)/(N_p'C_\alpha'N_m)$	128.0	128.1	128.1	128.5/128.2
$\angle(C_\alpha N_m C_\alpha')$	122.3	123.3	125.3	124.0
$\angle(C_\alpha C_\beta N_d)/(C_\alpha'C_\beta'N_d')$	130.3	130.3	130.3	129.5/131.1
$\angle(C_\beta N_d C_\gamma)/(C_\beta'N_d'C_\gamma')$	114.9	115.0	115.2	115.3/114.9
M-X [2]	0.451	0.504	0.782	-
A [3]	175.3	174.0	164.6	180

[1] P is the coordination cavity perimeter (in Å). [2] X is dummy atom located in center between N_p atoms. [3] α is the dihedral angle between planes of opposite pyrrole rings.

The degree of the doming-distortion can be described by the distance M-X between the metal atom and the center of the plane formed by four nitrogen atoms N_p and the dihedral angle between the planes of opposite pyrrole rings α. The distance M-X increases in line with the size of a metal atom. Furthermore, the dihedral angle between the opposite pyrrole fragments α decreases by 10 degrees from Ga to In, while a slight change in this parameter is noted from Al to Ga (r_{ionic}(Al) = 0.39, r_{ionic}(Ga) = 0.47, r_{ionic}(In) = 0.62) [28]. Structural parameters of the TPyzPA macrocyclic ligand are practically independent of the nature of a metal atom and peripheral –Cl substitution. The only noticeable change occurs for the peripheral C_γ-C_γ bond that is 0.025 Å longer in the $TPyzPACl_8$ metal complexes due to the electron-withdrawing effect of the –Cl substituents. A similar picture was previously observed for the complexes of tetrabenzoporphyrin with Al(Cl), Ga(Cl), In(Cl) [10] and porphyrazine, and tetrakis(1,2,5-thiadiazole)porphyrazine with Y(Cl), La(Cl) and Lu(Cl) [29]. The perimeter of the coordination cavity is not affected by the nature of a metal atom and increases by ca. 0.05 Å in the metal complexes as compared to the metal-free $H_2TPyzPA$ and $H_2TPyzPACl_8$; nevertheless, the distances between the nitrogen atoms of pyrrole fragments $(N_p \ldots N_p)_{opp}$ and $(N_p \ldots N_p)_{adj}$ are noticeably different (up to 0.2 Å). This can be explained by the fact that in the Al-Ga-In series, there is a consistent decrease in distances N_p-C_α along with valence angles $\angle(MN_pC_\alpha)$ and an increase in C_α-N_m distances and $\angle(C_\alpha N_m C_\alpha)$, which leads to the possibility of a significant difference in distances $(N_p \ldots N_p)_{opp}$ and $(N_p \ldots N_p)_{adj}$ without noticeable perimeter change.

2.2. NBO-Analysis

To gain a deeper insight into the electronic structures of M(Cl)TPyzPA and M(Cl)TPyz $PACl_8$ complexes and features of the metal–ligand chemical bonding, we performed the NBO-analysis of the electron density distribution. The obtained results suggest that the chemical bonding between a metal atom and $TPyzPACl_8$ macrocyclic ligand can be described in terms of the donor–acceptor interactions of the types: LP(N) → ns(M) and LP(N) → np(M), where n is the principal quantum number for the valence shell of the metal atom (n = 3 for Al, n = 4 for Ga, n = 5 for In). It should be noted that in the case of Al complexes, the interactions occur between LP(N) and two 3p(Al) orbitals, while for the complexes of Ga and In, all three valence p-orbitals provide a favorable overlap (Figure 2).

Comparing periphery-substituted $TPyzPACl_8$ complexes with their non-substituted analogues (Table 2), we found that peripheral –Cl substituents do not affect the characteristics of the metal–ligand bonding. The enhanced covalent contribution into the Ga–N bond is in line with the trend of the electronegativities of the metal atoms: Al (χ = 1.47), Ga (χ = 1.82) and In (χ = 1.49) [30,31]. According to Table 2, the NPA charge of Al is the most positive, in contrast to the NPA charge of Ga. More than that, an analogical situation is observed for the negativity of NPA charges of chlorine atoms. Therefore, these charges also correlate with the electronegativities of the metals.

Table 2. Selected parameters of M(Cl)PyzPA complexes from NBO calculations.

	Al(Cl) TPyzPA	Al(Cl) $TPyzPACl_8$	Ga(Cl) TPyzPA	Ga(Cl) $TPyzPACl_8$	In(Cl) TPyzPA	In(Cl) $TPyzPACl_8$
E(HOMO),eV	−6.329	−6.803	−6.362	−6.836	−6.411	−6.882
E(LUMO),eV	−3.815	−4.294	−3.829	−4.308	−3.864	−4.343
ΔE, eV	2.514	2.509	2.533	2.528	2.547	2.539
q(M) NPA, *e*	1.718	1.716	1.637	1.635	1.694	1.692
q(N) NPA, *e*	−0.651	−0.651	−0.631	−0.631	−0.623	−0.623
q(Cl) NPA, *e*	−0.550	−0.545	−0.523	−0.518	−0.528	−0.521
configuration	$3s^{0.42}3p^{0.83}$	$3s^{0.42}3p^{0.83}$	$4s^{0.53}4p^{0.81}$	$4s^{0.53}4p^{0.81}$	$5s^{0.54}5p^{0.76}$	$5s^{0.54}5p^{0.76}$
$\sum E$(d-a), kcal mol^{-1}	526.0	526.5	539.6	539.0	510.7	507.8
Q(M-N), *e*	0.335	0.330	0.343	0.342	0.327	0.325
r(M-N), Å	1.981	1.980	2.022	2.021	2.161	2.162

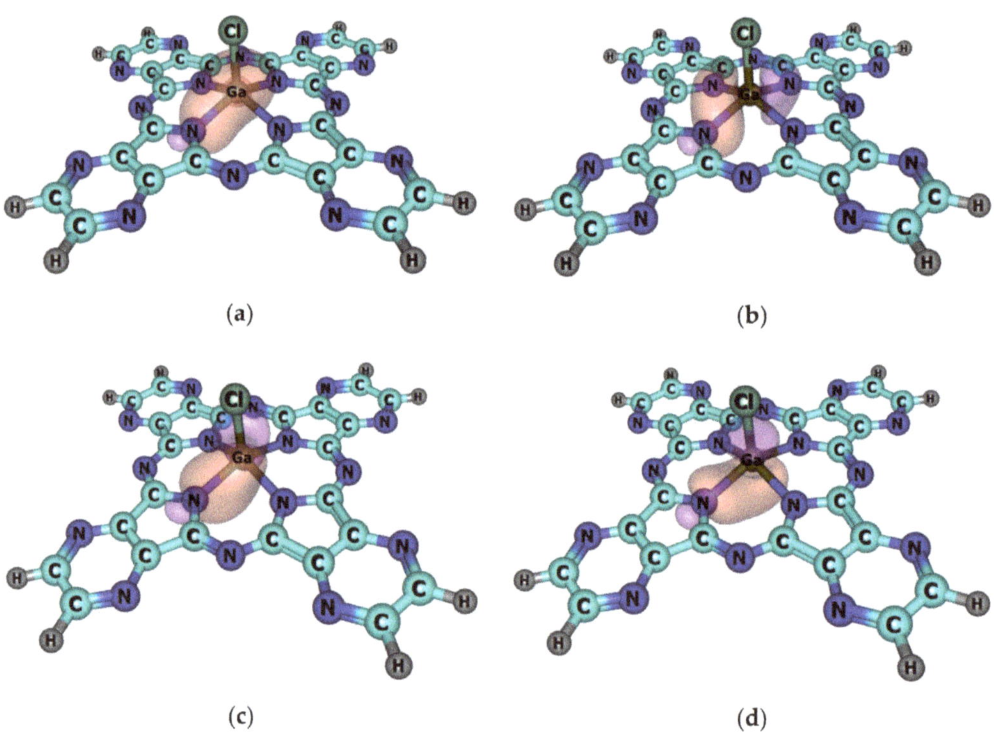

Figure 2. Schemes of the dominant donor–acceptor interactions between Ga and TPyzPA ligand. (**a**) The result of the orbital interaction of the type LP(N) → 4s(Ga) ($E^{(2)}$ = 59.2 kcal mol^{-1}); (**b**) the result of the orbital interaction of the type LP(N) → $4p_x$(Ga) ($E^{(2)}$ = 30.7 kcal mol^{-1}); (**c**) the result of the orbital interaction of the type LP(N) → $4p_y$(Ga) ($E^{(2)}$ = 30.7 kcal mol^{-1}); (**d**) the result of the orbital interaction of the type LP(N) → $4p_z$(Ga) ($E^{(2)}$ = 14.3 kcal mol^{-1}). Only one of the four corresponding interactions is demonstrated.

2.3. Electronic Absorption Spectra

Analyzing the absorption spectra (Figure 3), we can note that three intense absorption bands are observed in the visible light and near UV in the case of M(Cl)TPyzPACl_8, in contrast to M(Cl)TPyzPA, of which spectra contain only two bands in the visible light and near UV range. The B_1-band in the latter has low oscillator strength, so it is practically imperceptible in the absorption spectrum, while this peak is clearly observed in the perchlorinated complexes. This can be related to the fact that the peripheral Cl atoms contribute significantly to $6a_1$ molecular orbital (MO) in M(Cl)TPyzPACl_8 in contrast to the hydrogen atoms in M(Cl)TPyzPA to $3a_1$ MO (Figure S1). The B_1-band is predominantly formed by the electron transitions from these orbitals to doubly-degenerated LUMOs (Table 3).

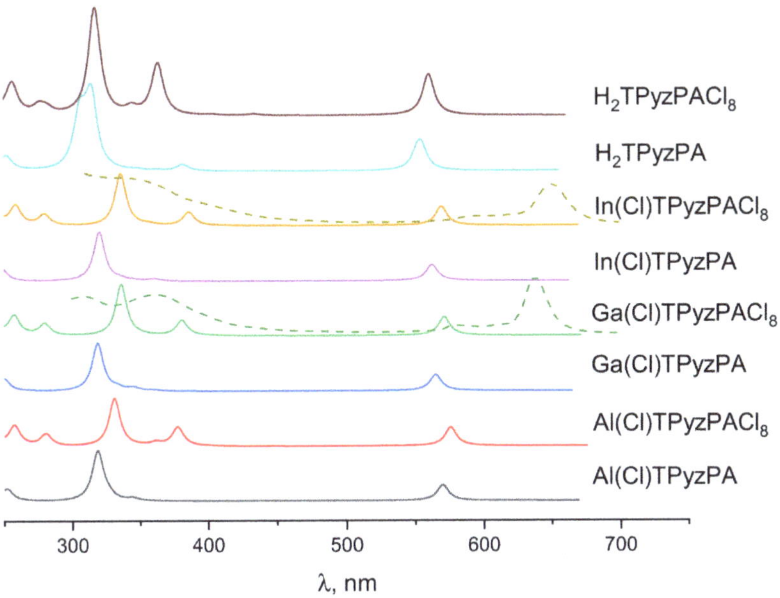

Figure 3. Simulated (solid lines) and experimental (dashed lines) electronic absorption spectra for M(Cl)TpyzPA and M(Cl)TPyzPACl$_8$ (M = Al, Ga, In) and its metal-free complexes.

It can be seen that the position of the Q-band maxima in the electronic absorption spectra shifted bathochromically in the series Ga < Al < In. The changes in the theoretical and experimental spectra are similar to each other and therefore allow the nature of the results obtained to be explained. This indicates that the HOMO–LUMO gap is influenced not only by the degree of doming-distortion of the macrocycle, but also by electronegativity of the central metal, which is the highest for Ga. In addition, a minor bathochromic shift of the Q-band occurs for perchlorinated complexes M(Cl)TPyzPACl$_8$ as compared to M(Cl)TPyzPA. Noteworthy is the single absorption maximum (B-band) in the near UV Soret region (300–420 nm) that appears for M(Cl)TPyzPA, while B$_1$- and B$_2$-bands are present for M(Cl)TPyzPACl$_8$. Experimental spectra of Ga(OH)TPyzPACl$_8$ and In(OH)TPyzPACl$_8$ can be found in ESI (Figures S2 and S3).

In addition, attention should be paid to the pronounced hyperchromic effect in metal-free complexes caused by the superposition of several absorption bands with close wavelengths. The composition of the excited states forming the absorption bands is given in Table 3.

It can be seen that the excited states of the compounds under consideration are formed as a result of similar sets of electron transitions. The main composition of the excited states responsible for the Q-band consists of transitions between the frontier π-orbitals. Hence Q-band wavelengths correlate with the HOMO–LUMO gaps. The molecular orbitals level diagram is shown in Figure 4. Stabilization of LUMO upon peripheral chlorination of the TPyzPA macrocycle correlates with the electrochemical data [8], indicating that the reduction potentials for Cl$_8$TPyzPA complexes are shifted by 0.4–0.5 V in a less negative region as compared to the values typical for unsubstituted TPyzPA complexes [10]. It is also worth noting that the 1^1E states are formed mainly by Gouterman-type transitions $a_2 \rightarrow e^*$ for metal complexes and transitions of the $a_u \rightarrow b_{1g}$ and $a_u \rightarrow b_{3g}$ types for metal-free ones due to the lower symmetry of the latter [32–34]. The situation is typical for a number of macrocycles of similar structure [35–37]. Despite the fact that the higher energy states are formed by almost the same electronic transitions, they have a different composition, and therefore energy.

Table 3. Calculated composition of the lowest excited states and corresponding oscillator strengths for TPyzPA and TPyzPzCl$_8$ complexes.

Al(Cl)TPyzPA					**Al(Cl)TPyzPACl$_8$**				
State	Composition (%)	λ, nm	f	exp. λ, nm	State	Composition (%)	λ, nm	f	exp. λ, nm
1^1E	$4a_1 \rightarrow 1e^*$ (8) $3a_2 \rightarrow 1e^*$ (90)	570	0.31		1^1E	$5a_1 \rightarrow 1e^*$ (5) $3a_2 \rightarrow 1e^*$ (90)	576	0.37	643
7^1E	$3a_1 \rightarrow 1e^*$ (54) $2b_2 \rightarrow 1e^*$ (17) $4a_1 \rightarrow 1e^*$ (22)	327	0.07		4^1E	$2b_1 \rightarrow 1e^*$ (5) $6a_1 \rightarrow 1e^*$ (77) $3b_1 \rightarrow 1e^*$ (7) $3a_2 \rightarrow 2e^*$ (5)	377	0.37	
9^1E	$3a_1 \rightarrow 1e^*$ (27) $4a_1 \rightarrow 1e^*$ (58) $3a_2 \rightarrow 1e^*$ (6) $3a_2 \rightarrow 2e^*$ (6)	318	1.00		8^1E	$5a_1 \rightarrow 1e^*$ (70) $6a_1 \rightarrow 1e^*$ (7) $3a_2 \rightarrow 1e^*$ (7) $3a_2 \rightarrow 2e^*$ (9)	331	0.95	363
Ga(Cl)TPyzPA					Ga(Cl)TPyzPACl$_8$				
State	Composition (%)	λ, nm	f	exp. λ, nm	State	Composition (%)	λ, nm	f	exp. λ, nm
1^1E	$4a_1 \rightarrow 1e^*$ (8) $3a_2 \rightarrow 1e^*$ (90)	565	0.32		1^1E	$5a_1 \rightarrow 1e^*$ (6) $3a_2 \rightarrow 1e^*$ (90)	571	0.38	638
7^1E	$3a_1 \rightarrow 1e^*$ (26) $2b_2 \rightarrow 1e^*$ (34) $4a_1 \rightarrow 1e^*$ (32)	333	0.05		4^1E	$5a_1 \rightarrow 1e^*$ (6) $6a_1 \rightarrow 1e^*$ (75) $4b_2 \rightarrow 1e^*$ (13)	381	0.31	
9^1E	$3a_1 \rightarrow 1e^*$ (47) $4a_1 \rightarrow 1e^*$ (38) $3a_2 \rightarrow 1e^*$ (6) $3a_2 \rightarrow 2e^*$ (5)	319	0.95		8^1E	$5a_1 \rightarrow 1e^*$ (63) $6a_1 \rightarrow 1e^*$ (8) $3a_2 \rightarrow 1e^*$ (8) $3a_2 \rightarrow 2e^*$ (16)	337	1.03	360
In(Cl)TPyzPA					In(Cl)TPyzPACl$_8$				
State	Composition (%)	λ, nm	f	exp. λ, nm	State	Composition (%)	λ, nm	f	exp. λ, nm
1^1E	$4a_1 \rightarrow 1e^*$ (8) $3a_2 \rightarrow 1e^*$ (89)	562	0.32	589	1^1E	$3a_2 \rightarrow 1e^*$ (90) $5a_1 \rightarrow 1e^*$ (6)	569	0.38	649
7^1E	$3a_1 \rightarrow 1e^*$ (36) $2b_2 \rightarrow 1e^*$ (26) $4a_1 \rightarrow 1e^*$ (28) $3a_2 \rightarrow 2e^*$ (7)	336	0.04	335	4^1E	$6a_1 \rightarrow 1e^*$ (76) $5a_1 \rightarrow 1e^*$ (6) $3b_2 \rightarrow 1e^*$ (14)	386	0.26	
9^1E	$3a_1 \rightarrow 1e^*$ (43) $4a_1 \rightarrow 1e^*$ (39) $3a_2 \rightarrow 1e^*$ (7) $3a_2 \rightarrow 2e^*$ (7)	320	0.98		8^1E	$5a_1 \rightarrow 1e^*$ (55) $6a_1 \rightarrow 1e^*$ (6) $3a_2 \rightarrow 1e^*$ (8) $3a_2 \rightarrow 2e^*$ (22)	336	1.05	339
H_2TPyzPA					H_2 TPyzPACl$_8$				
State	Composition (%)	λ, nm	f	exp. λ, nm	State	Composition (%)	λ, nm	f	exp. λ, nm
1^1B_{1u}	$3a_u \rightarrow 1b^*_{1g}$ (90)	555	0.35	645	1^1B_{1u}	$3a_u \rightarrow 1b^*_{1g}$ (90)	560	0.43	656
1^1B_{3u}	$3a_u \rightarrow 1b^*_{3g}$ (86) $2b_{2u} \rightarrow 1b^*_{1g}$ (5) $3b_{2u} \rightarrow 1b^*_{1g}$ (8)	552	0.33	611	1^1B_{3u}	$1b_{2u} \rightarrow 1b^*_{1g}$ (5) $2b_{2u} \rightarrow 1b^*_{1g}$ (7) $3a_u \rightarrow 1b^*_{3g}$ (86)	560	0.39	626
2^1B_{3u}	$2b_{2u} \rightarrow 1b^*_{1g}$ (12) $3b_{2u} \rightarrow 1b^*_{1g}$ (80)	382	0.12		3^1B_{3u}	$1b_{2u} \rightarrow 1b^*_{1g}$ (5) $2b_{2u} \rightarrow 1b^*_{1g}$ (85) $3a_u \rightarrow 2b^*_{3g}$ (5)	364	0.78	358
4^1B_{3u}	$1b_{2u} \rightarrow 1b^*_{1g}$ (7) $2b_{2u} \rightarrow 1b^*_{1g}$ (66) $3b_{2u} \rightarrow 1b^*_{1g}$ (9) $3a_u \rightarrow 1b^*_{3g}$ (9) $3a_u \rightarrow 2b^*_{3g}$ (6)	315	1.32	330	4^1B_{1u}	$1b_{2u} \rightarrow 1b^*_{3g}$ (18) $2b_{2u} \rightarrow 1b^*_{3g}$ (78)	362	0.26	358
4^1B_{1u}	$1b_{2u} \rightarrow 1b^*_{3g}$ (62) $2b_{2u} \rightarrow 1b^*_{3g}$ (33)	311	0.29		4^1B_{3u}	$1b_{2u} \rightarrow 1b^*_{1g}$ (6) $3a_u \rightarrow 2b^*_{3g}$ (91)	345	0.13	
5^1B_{1u}	$1b_{2u} \rightarrow 1b^*_{3g}$ (33) $2b_{2u} \rightarrow 1b^*_{3g}$ (42) $3b_{2u} \rightarrow 1b^*_{3g}$ (10) $3a_u \rightarrow 1b^*_{1g}$ (6) $3a_u \rightarrow 2b^*_{1g}$ (5)	306	0.94		5^1B_{3u}	$1b_{2u} \rightarrow 1b^*_{1g}$ (79) $3a_u \rightarrow 1b^*_{3g}$ (8)	319	0.88	
					5^1B_{1u}	$1b_{2u} \rightarrow 1b^*_{3g}$ (68) $2b_{2u} \rightarrow 1b^*_{3g}$ (12)	316	1.34	

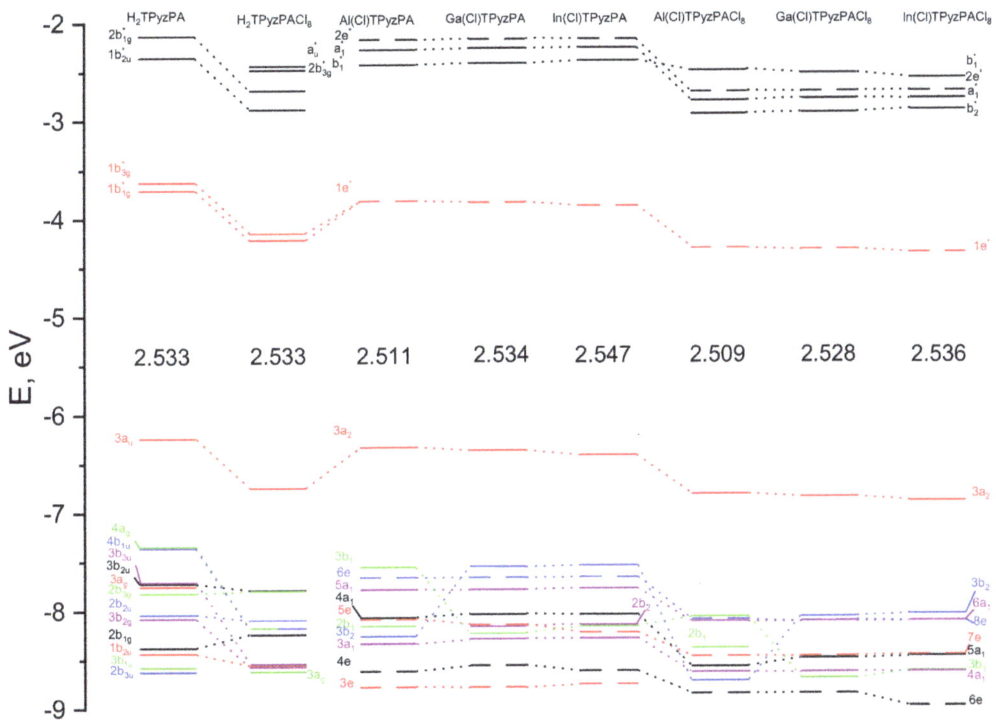

Figure 4. Molecular orbitals (MO) level diagram for $H_2TPyzPA$, $H_2TPyzPACl_8$, M(Cl)TPyzPA and M(Cl)$TPyzPACl_8$ complexes. The values of higher occupied molecular orbital–lowest unoccupied molecular orbital (HOMO–LUMO) gaps are given in eV.

Considering the shapes of the molecular orbitals (MOs) involved in the most probable electronic transitions, we can conclude that the metal nature should not affect the Q-band position since the boundary MOs are the linear combination of atomic orbitals (LCAO) of the macrocyclic ligand. Nevertheless, the size of the coordination cavity depends on the metal nature: changing the geometric structure provides an indirect effect on the position of the Q-band. Moreover, the shape of the LUMOs depends on the metal nature (Figure S1). In the case of Al, it consists of the bonding π-orbitals lying along the C_α-C_β bond and antibonding π-orbitals located around the C_α-N_m and C_α-N_p bonds, while for Ga and In, the C_α-N_p bonds have a significant contribution to the bonding π-orbital, probably due to the influence of the d-sublevel of the metal. The HOMO consists of the AOs localized on carbon atoms of pyrrolic fragments, which is a typical picture for porphyrazines [10,19,25,38]. Note that peripheral Cl atoms contribute to the formation of the MOs involved in the electronic transitions corresponding to the B_1-band in the absorption spectrum. For the rest, the MO's form is similar for all compounds; therefore, it can be concluded that the ligand rather than metal has a decisive influence on the absorption spectrum.

2.4. IR Spectra

The IR spectra were simulated on the basis of the normal mode frequencies and band intensities, which have been calculated by the DFT (PBE0/def2-TZVP) method in a harmonic approximation.

Bands of weak intensity appear in the region up to 600 cm^{-1}, as well as the shift of bands in the range from 600 to 1000 cm^{-1} occurs with the substitution of hydrogen atoms by the metal. It should be noted that the most intensive vibrational transitions of the metal complexes are degenerate, while the splitting of the bands in the case of the metal-free H_2TPyzPA and H_2TPyzPACl_8 is observed due to the lower symmetry. It results in two peaks with close frequencies formed by the normal vibrations of the pyrrole and pyrrolenine fragments. A medium peak of the N_p-H stretching at 3552 cm^{-1} and 3554 cm^{-1} (H_2TPyzPA and H_2TPyzPACl_8, respectively) is typical for macrocycles such as porphyrazines [25]. According to the experimental data [8,26], these frequencies are 3286 and 3287 cm^{-1}, respectively.

In-plane vibrations contribute to the medium band at 1400 cm^{-1} in the spectra of M(Cl)TPyzPA. Noteworthy are the main differences in the M(Cl)TPyzPA IR spectra observed in the range of 300–600 cm^{-1} related to the weak bands with a strong contribution of the M-Cl stretching.

It is worth noting that a low-frequency band shift occurs in both M(Cl)TPyzPA and M(Cl)TPyzPACl_8 in the Al→Ga→In series. Among the most intense peaks, the only exception is the ω_{92}-ω_{93} band at the ~1140 cm^{-1} for MTPyzPA, which corresponds to in-plane vibrations of the macrocyclic core. Moreover, a significant decrease in these bands' intensity occurs with an increase in the ionic radius of a metal [28].

The IR spectra of M(Cl)TPyzPA are similar in the range of 600–3500 cm^{-1} where the influence of the metal is almost absent. The replacement of H by Cl leads to the vanishing of the bands at 3190 cm^{-1}, corresponding to the C-H stretching vibrations, and an increase in the relative intensity of the peaks in the 800–1300 cm^{-1} region. Simulated spectra are shown in Figure 5. The most intense bands at ~1250 cm^{-1} of all investigated compounds are predominantly stretching vibrations of the bonds of pyrazine rings. Unlike M(Cl)TPyzPA spectra with one peak with a relative intensity >50%, the M(Cl)TPyzPACl_8 spectra also contain the strong bands ω_{117}-ω_{118}, composed by N_d-C_γ and C_β-N_d stretching vibrations. Moreover, M(Cl)TPyzPACl_8 spectra include more relatively intense peaks compared to MTPyzPA. A description of the main vibrations is listed in Table 4 (full list of the most active vibrations can be found in Supplementary Materials, Table S1). It is important to mention that the calculated spectra of Ga(Cl)TPyzPACl_8 and In(Cl)TPyzPACl_8 are consistent with the experimental ones with a factor ~0.95. Actually, the main differences may be caused by another axial ligand (OH instead of Cl). Experimental spectra can be found in ESI (Figures S4 and S5).

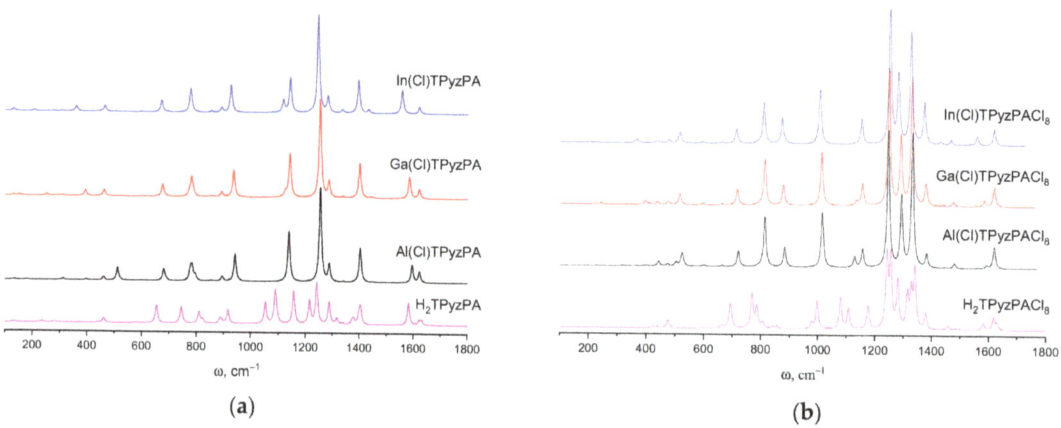

Figure 5. Calculated IR spectra of M(Cl)TPyzPA (**a**) and M(Cl)TPyzPACl_8 (**b**).

Table 4. Assignment of the IR vibrations of the M(Cl)TPyzPA and M(Cl)TPyzPACl$_8$ complexes.

Frequency, cm^{-1}	I_{rel}, %	Symmetry	Assignment [1]	Exp, cm^{-1}
H_2TPyzPA				KBr [26]
1091 (ω_{86})	78	B_{1u}	r(N_p-C_α) (29), r(N_m-C_α) (18), r(N_d-C_γ) (10), r(C_γ-C_γ) (12)	1040
1157 (ω_{90})	81	B_{2u}	r(N_p-C_α) (23), r(N_m-C_α) (18), r(C_α-C_β) (6), r(C_β-N_d) (8), r(N_d-C_γ) (6), φ(C_α-N_p-H_c) (8), φ(C_β-N_d-C_γ) (7)	1121
1242 (ω_{95})	100	B_{1u}	r(C_α-C_β) (14), r(C_β-C_β) (10), r(N_d-C_γ) (10), φ(N_d-C_γ-H_s) (7)r(C_β-N_d) (6), r(C_β-C_β) (42), r(C_γ-Cl) (26), φ(N_d-C_γ-Cl) (6)	1198
3193 (ω_{141})	24	B_{2u}	r(H_s-C_γ) (99)	3051
3552 (ω_{143})	45	B_{1u}	r(H_c-N_p) (99)	3286
H_2TPyzPACl$_8$				KBr [8]
1242 (ω_{107})	100	B_{1u}	r(C_β-N_d) (6), r(C_β-C_β) (42), r(C_γ-Cl) (26), φ(N_d-C_γ-Cl) (6)	1192
1255 (ω_{109})	98	B_{2u}	r(C_β-N_d) (8), r(C_γ-C_γ) (46), r(C_γ-Cl) (21), φ(N_d-C_γ-Cl) (6)	1235
1281 (ω_{111})	67	B_{1u}	r(C_α-C_β) (8), r(C_β-C_β) (19), r(N_d-C_γ) (13), r(C_γ-C_γ) (13)	
1328 (ω_{116})	52	B_{1u}	r(N_p-C_α) (7), r(N_m-C_α) (11), r(C_β-N_d) (12), r(N_d-C_γ) (45)	
1340 (ω_{117})	83	B_{2u}	r(N_p-C_α) (8), r(C_α-C_β) (10), r(C_β-C_β) (8), r(C_β-N_d) (19), φ(C_α-N_p-H_c) (28)	
3554 (ω_{143})	27	B_{1u}	r(H_c-N_p) (99)	3287

Table 4. *Cont.*

Frequency, cm^{-1}	I_{rel}, %	Symmetry	Assignment [1]	Exp, cm^{-1}
Al(Cl)TPyzPA				
1140 (ω_{92}-ω_{93})	53	E	r(N_p-C_α) (24), r(N_m-C_α) (14), r(C_α-C_β) (10), r(C_β-N_d) (12), r(N_d-C_γ) (9)	
1255 (ω_{98}-ω_{99})	100	E	r(N_p-C_α) (5), r(C_α-C_β) (15), r(C_β-C_β) (14), r(N_d-C_γ) (20), φ(C_β-N_d-C_γ) (9)	
1404(ω_{113}-ω_{114})	36	E	r(C_β-C_β) (12), r(C_γ-C_γ) (13), φ(N_d-C_γ-H_s) (36), φ(C_γ-C_γ-H_s) (23)	
3189 (ω_{142}-ω_{143})	18	E	r(H_s-N_p) (99)	
Al(Cl)TPyzPACl_8				KBr [8]
1246 (ω_{111}-ω_{112})	100	E	r(C_γ-C_γ) (43), r(C_γ-Cl) (25), φ(N_d-C_γ-Cl) (6)	1168
1293 (ω_{115}-ω_{116})	52	E	r(N_p-C_α) (8), r(C_β-C_β) (15), r(N_d-C_γ) (25), r(C_γ-C_γ) (13)	1260
1331 (ω_{117}-ω_{118})	97	E	r(C_α-C_β) (8), r(C_β-N_d) (23), r(N_d-C_γ) (49)	1323
Ga(Cl)TPyzPA				
1142 (ω_{92}-ω_{93})	44	E	r(N_p-C_α) (49), r(N_m-C_α) (11), r(C_α-C_β) (9), r(N_d-C_γ) (9)	
1254 (ω_{98}-ω_{99})	100	E	r(N_p-C_α) (6), r(C_α-C_β) (12), r(C_β-C_β) (11), r(C_β-N_d) (9), r(N_d-C_γ) (18)	
1401 (ω_{113}-ω_{114})	35	E	r(C_β-C_β) (11), r(C_γ-C_γ) (12), φ(N_d-C_γ-H_s) (36), φ(C_γ-C_γ-H_s) (23)	
3194 (ω_{142}-ω_{143})	17	E	r(H_s-N_p) (99)	
Ga(Cl)TPyzPACl_8				this work
1248 (ω_{111}-ω_{112})	100	E	r(C_γ-C_γ) (41), r(C_γ-Cl) (24), φ(N_d-C_γ-Cl) (6)	1232
1290 (ω_{115}-ω_{116})	50	E	r(N_p-C_α) (6), r(C_α-C_β) (7), r(C_β-C_β) (17), r(N_d-C_γ) (26), r(C_γ-C_γ) (13)	
1328 (ω_{117}-ω_{118})	94	E	r(C_α-C_β) (8), r(C_β-N_d) (22), r(N_d-C_γ) (50)	1349

Table 4. *Cont.*

Frequency, cm^{-1}	I_{rel}, %	Symmetry	Assignment [1]	Exp, cm^{-1}
In(Cl)TPyzPA				KBr [39]
1143 (ω_{92}-ω_{93})	36	E	r(N_p-C_α) (49), r(N_m-C_α) (10), r(C_α-C_β) (6), r(N_d-C_γ) (10)	1100
1247 (ω_{98}-ω_{99})	100	E	r(N_p-C_α) (5), r(C_α-C_β) (15), r(C_β-C_β) (14), r(N_d-C_γ) (20)	1213
1396 (ω_{113}-ω_{114})	34	E	r(C_β-C_β) (10), r(C_γ-C_γ) (10), φ(N_d-C_γ-H_s) (34), φ(C_γ-C_γ-H_s) (22)	1364
3194 (ω_{142}-ω_{143})	17	E	r(H_s-N_p) (99)	3316
In(Cl)TPyzPACl$_8$				this work
1250 (ω_{111}-ω_{112})	100	E	r(C_γ-C_γ) (44), r(C_γ-Cl) (26), φ(N_d-C_γ-Cl) (6)	1264
1280 (ω_{114}-ω_{115})	50	E	r(N_p-C_α) (5), r(C_α-C_β) (7), r(C_β-C_β) (19), r(N_d-C_γ) (26)	1323
1325 (ω_{117}-ω_{118})	83	E	r(C_α-C_β) (8), r(C_β-N_d) (23), r(N_d-C_γ) (49)	1364

[1] Coordinates are listed provided that their contributions (shown in parentheses) are greater than ~5%. Assignment of vibrational modes based on potential energy distribution. The following designations of the coordinates are used: r—stretching of the bond; φ—bending, a change in the angle; OPB—out-of-plane bending; θ—a change in the dihedral angle. Experimental frequencies are given for metal complexes with an axial–OH ligand.

3. Materials and Methods

3.1. Synthesis

Octachlorotetrapyrazinoporphyrazinatogallium(III) hydroxide, [Ga(OH)TPyzPACl$_8$]. A mixture of 5,6-dichloropyrazine-2,3-dicarbonitrile (200 mg, 1 mmol) and gallium(III) hydroxydiacetate (50 mg, 0.24 mmol) was melted at 200 °C for 10 min. The obtained solid was powdered, washed with CH_2Cl_2, then dissolved in conc. H_2SO_4, precipitated by pouring into ice-water, centrifuged and washed with MeOH. After drying, the complex was obtained as dark green hydrated material.

Octachlorotetrapyrazinoporphyrazinatoindium(III) hydroxide, [In(OH)TPyzPACl$_8$]. A mixture of 5,6-dichloropyrazine-2,3-dicarbonitrile (1) (200 mg, 1 mmol) and indium(III) hydroxydiacetate (50 mg, 0.2 mmol) was melted at 200 °C for 10 min. The obtained solid was powdered, washed with CH_2Cl_2, then dissolved in THF and chromatographed on silica with THF as eluent. The eluted product was precipitated by pouring into water, centrifuged and washed with MeOH. After drying, the complex was obtained as dark green hydrated material [8].

The IR spectra were measured on an IR-spectrometer Cary 630 FT-IR using KBr pellets. UV–Vis spectra were recorded using a Cary UV-Vis spectrophotometer in THF solution and can be found in ESI.

3.2. Computational Details

The DFT study of M(Cl)TPyzPA and M(Cl)TPyzPACl$_8$ included geometry optimization and calculations of the harmonic vibrations, followed by calculation of the electronic absorption spectrum by the TDDFT method. The number of the calculated excited states was 30. The calculations were performed using the PBE0 functional with the density functional dispersion correction D3 provided by Grimme [40] with the def2-TZVP basis set [41] taken from the EMSL BSE library [42–44]. Firefly QC package [45], which is partially based on the GAMESS (US) [46] source code, was used to obtain the optimized geometry, electronic absorption spectra and NBO-analysis. The optimized Cartesian coordinates of H_2TPyzPA, H_2TPyzPACl_8 and their metal complexes with Al, Ga, In are available in the Supplementary Materials.

The calculations of IR spectra were carried out with use of Gaussian09 [47] software package due to the fact it applies analytical functions in the second derivatives computing. The molecular models and orbitals demonstrated in the paper were visualized by means of the Chemcraft program [48].

4. Conclusions

The geometry and electronic structure of TPyzPA were investigated using PBE0-D3 functional with basis set def2-TZVP. The distance M-X between the metal atom and the center of the plane formed by four nitrogen atoms N_p increases in line with the size of a metal atom. Structural parameters of the TPyzPA macrocyclic ligand are practically independent of the nature of a metal atom and peripheral −Cl substitution. The fact that $(N_p \ldots N_p)_{opp}$ and $(N_p \ldots N_p)_{adj}$ are substantially different for Al, Ga and In complexes, while the perimeter of the coordination cavity almost remains invariant for both M(Cl)TPyzPA and M(Cl)TPyzPACl_8 complexes, indicates that the degree of distortion increases from Al to In. At the same time, the substitution of H-atoms for Cl mainly affects the internuclear distances of pyrazine rings and has a negligible effect on the N_p-C_α and C_α-N_m.

According to the NBO-analysis of electron density distribution, the strong donor–acceptor interactions of the types: LP(N) $\rightarrow$ ns(M) and LP(N) $\rightarrow$ np(M), where n is the principal quantum number for the valence shell of the metal atom, stabilize the complexes.

It was found that the HOMO–LUMO gap is slightly affected by the nature of the metal and increases in the Al-Ga-In series. The substitution of H-atoms for Cl leads to a decrease in the energy of the frontier MOs of M(Cl)TPyzPACl_8 as compared to M(Cl)TPyzPA. Furthermore, the LUMO is stabilized much more than the HOMO, resulting in a smaller HOMO–LUMO gap.

Simulated electronic absorption spectra show that the metal nature slightly influences the position of the band's position. In the series Al$\rightarrow$Ga$\rightarrow$In$\rightarrow H_2$, the Q-band is shifted to a shorter wavelength. In addition, a minor bathochromic shift occurs in M(Cl)TPyzPA compared to M(Cl)TPyzPACl_8. The excited states corresponding to the Q-bands are composed mainly of Gouterman-type transitions. The electronic absorption spectra of TPyzPACl_8 contain three intense maxima, while only two peaks were found for TPyzPA. The additional B_1 band in the M(Cl)TPyzPACl_8 spectra is caused by the transitions from MOs with a contribution of AOs of the peripheral Cl atoms.

The calculated IR spectra of TPyzPACl_8 have a significant difference in comparison with TPyzPA due to heavy Cl atoms being involved in the vibrations that form medium and strong bands in the region of 1000–1500 cm^{-1}.

Supplementary Materials: The following supporting information can be downloaded at: https://www.mdpi.com/article/10.3390/ijms23105379/s1.

Author Contributions: Conceptualization, P.A.S. and Y.A.Z.; methodology, P.A.S. and Y.A.Z.; investigation, I.V.R., A.V.E. and D.N.F.; resources, Y.A.Z.; data curation, A.V.E. and I.V.R.; writing—original draft preparation, A.V.E. and I.V.R. All authors have read and agreed to the published version of the manuscript.

Funding: This work is supported by the Russian Science Foundation (grant No. 21-73-10126).

Institutional Review Board Statement: Not applicable.

Informed Consent Statement: Not applicable.

Data Availability Statement: The data presented in this study are available on request from the corresponding author.

Acknowledgments: The research was carried out using the resources of the Center for Shared Use of Scientific Equipment of the ISUCT (with the support of the Ministry of Science and Higher Education of Russia, grant No. 075-15-2021-671). The authors are grateful to Arseniy A. Otlyotov for helpful suggestions and discussions.

Conflicts of Interest: The authors declare no conflict of interest.

References

1. De la Torre, G.; Vázquez, P.; Aguíló-López, F.; Torres, T. Role of structural factors in the nonlinear optical properties of phthalocyanines and related compounds. *Chem. Rev.* **2004**, *104*, 3723–3750. [CrossRef] [PubMed]
2. Drobizhev, M.; Makarov, N.S.; Rebane, A.; De La Torre, G.; Torres, T. Strong two-photon absorption in Push-Pull phthalocyanines: Role of resonance enhancement and permanent dipole moment change upon excitation. *J. Phys. Chem. C* **2008**, *112*, 848–859. [CrossRef]
3. Senge, M.O.; Fazekas, M.; Notaras, E.G.A.; Blau, W.J.; Zawadzka, M.; Locos, O.B.; Mhuircheartaigh, E.M.N. Nonlinear optical properties of porphyrins. *Adv. Mater.* **2007**, *19*, 2737–2774. [CrossRef]
4. McEwan, K.; Lewis, K.; Yang, G.Y.; Chng, L.L.; Lee, Y.W.; Lau, W.P.; Lai, K.S. Synthesis, Characterization, and Nonlinear Optical Study of Metalloporphyrins. *Adv. Funct. Mater.* **2003**, *13*, 863–867. [CrossRef]
5. Marder, S.R. Organic nonlinear optical materials: Where we have been and where we are going. *Chem. Commun.* **2006**, *2*, 131–134. [CrossRef]
6. Liu, F.; Qin, G.; Li, Z.; Wang, Z.; Peng, M.; Wu, S.; Li, C.; Yang, Y. The design and synthesis of nonlinear optical chromophores containing two short chromophores for an enhanced electro-optic activity†. *Mater. Adv.* **2021**, *2*, 728. [CrossRef]
7. Koifman, O.I.; Ageeva, T.A.; Beletskaya, I.P.; Averin, A.D.; Yakushev, A.A.; Tomilova, L.G.; Dubinina, T.V.; Tsivadze, A.Y.; Gorbunova, Y.G.; Martynov, A.G.; et al. Macroheterocyclic compounds—A key building block in new functional materials and molecular devices. *Macroheterocycles* **2020**, *13*, 311–467. [CrossRef]
8. Hamdoush, M.; Ivanova, S.S.; Koifman, O.I.; Kos'kina, M.; Pakhomov, G.L.; Stuzhin, P.A. Synthesis, spectral and electrochemical study of perchlorinated tetrapyrazinoporphyrazine and its AlIII, GaIII and InIII complexes. *Inorg. Chim. Acta* **2016**, *444*, 81–86. [CrossRef]
9. Kosov, A.D.; Dubinina, T.V.; Borisova, N.E.; Ivanov, A.V.; Drozdov, K.A.; Trashin, S.A.; De Wael, K.; Kotova, M.S.; Tomilova, L.G. Novel phenyl-substituted pyrazinoporphyrazine complexes of rare-earth elements: Optimized synthetic protocols and physicochemical properties. *New J. Chem.* **2019**, *43*, 3153–3161. [CrossRef]
10. Novakova, V.; Donzello, M.P.; Ercolani, C.; Zimcik, P.; Stuzhin, P.A. Tetrapyrazinoporphyrazines and their metal derivatives. Part II: Electronic structure, electrochemical, spectral, photophysical and other application related properties. *Coord. Chem. Rev.* **2018**, *361*, 1–73. [CrossRef]
11. Konovalov, A.I.; Antipin, I.S.; Burilov, V.A.; Madzhidov, T.I.; Kurbangalieva, A.R.; Nemtarev, A.V.; Solovieva, S.E.; Stoikov, I.I.; Mamedov, V.A.; Zakharova, L.Y.; et al. Modern Trends of Organic Chemistry in Russian Universities. *Russ. J. Org. Chem.* **2018**, *54*, 157–371. [CrossRef]
12. Skvortsov, I.A.; Kovkova, U.P.; Zhabanov, Y.A.; Khodov, I.A.; Somov, N.V.; Pakhomov, G.L.; Stuzhin, P.A. Subphthalocyanine-type dye with enhanced electron affinity: Effect of combined azasubstitution and peripheral chlorination. *Dyes Pigments* **2021**, *185*, 108944. [CrossRef]
13. Donzello, M.P.; Ercolani, C.; Novakova, V.; Zimcik, P.; Stuzhin, P.A. Tetrapyrazinoporphyrazines and their metal derivatives. Part I: Synthesis and basic structural information. *Coord. Chem. Rev.* **2016**, *309*, 107–179. [CrossRef]
14. Ustimenko, K.A.; Konarev, D.V.; Khasanov, S.S.; Lyubovskaya, R.N. Synthesis of iron(II) octachlorotetrapyrazinoporphyrazine, molecular structure and optical properties of the (X-)2FeTPyzPACl8 dianions with two axial anionic ligands (X- = CN-, CL-). *Macroheterocycles* **2013**, *6*, 345–352. [CrossRef]
15. Faraonov, M.A.; Romanenko, N.R.; Kuzmin, A.V.; Konarev, D.V.; Khasanov, S.S.; Lyubovskaya, R.N. Crystalline salts of the ring-reduced tin(IV) dichloride hexadecachlorophthalocyanine and octachloro- and octacyanotetrapyrazinoporphyrazine macrocycles with strong electron-withdrawing ability. *Dyes Pigments* **2020**, *180*, 108429. [CrossRef]
16. Eroshin, A.V.; Otlyotov, A.A.; Kuzmin, I.A.; Stuzhin, P.A.; Zhabanov, Y.A. DFT Study of the Molecular and Electronic Structure of Metal-Free Tetrabenzoporphyrin and Its Metal Complexes with Zn, Cd, Al, Ga, In. *Int. J. Mol. Sci.* **2022**, *23*, 939. [CrossRef]
17. Hohenberg, P.; Kohn, W. Inhomogeneous electron gas. *Phys. Rev.* **1964**, *136*, B864. [CrossRef]
18. Kohn, W.; Sham, L.J. Self-consistent equations including exchange and correlation effects. *Phys. Rev.* **1965**, *140*, A1133–A1138. [CrossRef]
19. Zhabanov, Y.A.; Tverdova, N.V.; Giricheva, N.I.; Girichev, G.V.; Stuzhin, P.A. DFT Study of molecular and electronic structure of magnesium (II) tetra(1,2,5-chalcogenadiazolo) porphyrazines, [TXDPzMg] (X = O, S, Se, Te). *J. Porphyr. Phthalocyanines* **2017**, *21*, 439–452. [CrossRef]
20. Yablokova, L.I.A.; Ivanova, S.S.; Novakova, V.; Zhabanov, Y.A.; Stuzhin, P.A. Perfluorinated porphyrazines. 3. Synthesis, spectral-luminescence and electrochemical properties of perfluorinated octaphenylporphyrazinatozinc(II). *J. Fluor. Chem.* **2018**, *214*, 86–93.

21. Stuzhin, P.A.; Skvortsov, I.A.; Zhabanov, Y.A.; Somov, N.V.; Razgonyaev, O.V.; Nikitin, I.A.; Koifman, O.I. Subphthalocyanine azaanalogues—Boron(III) subporphyrazines with fused pyrazine fragments. *Dyes Pigments* **2019**, *162*, 888–897. [CrossRef]
22. Otlyotov, A.A.; Zhabanov, Y.A.; Pogonin, A.E.; Kuznetsova, A.S.; Islyaikin, M.K.; Girichev, G.V. Gas-phase structures of hemiporphyrazine and dicarbahemiporphyrazine: Key role of interactions inside coordination cavity. *J. Mol. Struct.* **2019**, *1184*, 576–582. [CrossRef]
23. Hamdoush, M.; Nikitin, K.; Skvortsov, I.; Somov, N.; Zhabanov, Y.; Stuzhin, P.A. Influence of heteroatom substitution in benzene rings on structural features and spectral properties of subphthalocyanine dyes. *Dyes Pigments* **2019**, *170*, 107584. [CrossRef]
24. Eroshin, A.V.; Otlyotov, A.A.; Zhabanov, Y.A.; Veretennikov, V.V.; Islyaikin, M.K. Complexes of Ca(II), Ni(II) and Zn(II) with hemi- and dicarbahemiporphyrazines: Molecular structure and features of metal-ligand bonding. *Macroheterocycles* **2021**, *14*, 119–129. [CrossRef]
25. Zhabanov, Y.A.; Eroshin, A.V.; Stuzhin, P.A.; Ryzhov, I.V.; Kuzmin, I.A.; Finogenov, D.N. Molecular structure, thermodynamic and spectral characteristics of metal-free and nickel complex of tetrakis(1,2,5-thiadiazolo)porphyrazine. *Molecules* **2021**, *26*, 2945. [CrossRef]
26. Konarev, D.V.; Faraonov, M.A.; Kuzmin, A.V.; Osipov, N.G.; Khasanov, S.S.; Otsuka, A.; Yamochi, H.; Kitagawa, H.; Lyubovskaya, R.N. Molecular structures, and optical and magnetic properties of free-base tetrapyrazinoporphyrazine in various reduction states. *New J. Chem.* **2019**, *43*, 19214–19222. [CrossRef]
27. Liu, Z.; Zhang, X.; Zhang, Y.; Jiang, J. The molecular, electronic structures and IR and Raman spectra of metal-free, N,N′-dideuterio, and magnesium tetra-2,3-pyrazino-porphyrazines: Density functional calculations. *Vib. Spectrosc.* **2007**, *43*, 447–459. [CrossRef]
28. Shannon, R.D. Revised effective ionic radii and systematic studies of interatomic distances in halides and chalcogenides. *Acta Crystallogr. Sect. A* **1976**, *32*, 751–767. [CrossRef]
29. Zhabanov, Y.A.; Ryzhov, I.V.; Kuzmin, I.A.; Eroshin, A.V.; Stuzhin, P.A. DFT Study of Molecular and Electronic Structure of Y, La and Lu Complexes with Porphyrazine and Tetrakis(1,2,5-thiadiazole)porphyrazine. *Molecules* **2020**, *26*, 113. [CrossRef]
30. Allred, A.L.; Rochow, E.G. A scale of electronegativity based on electrostatic force. *J. Inorg. Nucl. Chem.* **1958**, *5*, 264–268. [CrossRef]
31. Mann, J.B.; Meek, T.L.; Allen, L.C. Configuration energies of the main group elements. *J. Am. Chem. Soc.* **2000**, *122*, 2780–2783. [CrossRef]
32. Gouterman, M. Spectra of porphyrins. *J. Mol. Spectrosc.* **1961**, *6*, 138–163. [CrossRef]
33. Gouterman, M.; Wagnière, G.H.; Snyder, L.C. Spectra of porphyrins. Part II. Four orbital model. *J. Mol. Spectrosc.* **1963**, *11*, 108–127. [CrossRef]
34. Weiss, C.; Kobayashi, H.; Gouterman, M. Spectra of porphyrins. Part III. Self-consistent molecular orbital calculations of porphyrin and related ring systems. *J. Mol. Spectrosc.* **1965**, *16*, 415–450. [CrossRef]
35. Nemykin, V.N.; Hadt, R.G.; Belosludov, R.V.; Mizuseki, H.; Kawazoe, Y. Influence of molecular geometry, exchange-correlation functional, and solvent effects in the modeling of vertical excitation energies in phthalocyanines using time-dependent density functional theory (TDDFT) and polarized continuum model TDDFT methods: Ca. *J. Phys. Chem. A* **2007**, *111*, 12901–12913. [CrossRef]
36. Belosludov, V.R.; Nevonen, D.; Rhoda, M.H.; Sabin, R.J.; Nemykin, N.V. Simultaneous Prediction of the Energies of Qx and Qy Bands and Intramolecular Charge-Transfer Transitions in Benzoannulated and Non-Peripherally Substituted Metal-Free Phthalocyanines and Their Analogues: No Standard TDDFT Silver Bullet Yet. *J. Phys. Chem. A* **2018**, *123*, 132–152. [CrossRef]
37. Stillman, M.; Mack, J.; Kobayashi, N. Theoretical aspects of the spectroscopy of porphyrins and phthalocyanines. *J. Porphyr. Phthalocyanines* **2002**, *6*, 296–300. [CrossRef]
38. Otlyotov, A.A.; Ryzhov, I.V.; Kuzmin, I.A.; Zhabanov, Y.A.; Mikhailov, M.S.; Stuzhin, P.A. DFT Study of Molecular and Electronic Structure of Ca(II) and Zn(II) Complexes with Porphyrazine and tetrakis(1,2,5-thiadiazole)porphyrazine. *Int. J. Mol. Sci.* **2020**, *21*, 2923. [CrossRef]
39. Dini, D.; Hanack, M.; Meneghetti, M. Nonlinear optical properties of tetrapyrazinoporphyrazinato indium chloride complexes due to excited-state absorption processes. *J. Phys. Chem. B* **2005**, *109*, 12691–12696. [CrossRef]
40. Grimme, S.; Antony, J.; Ehrlich, S.; Krieg, H. A consistent and accurate ab initio parametrization of density functional dispersion correction (DFT-D) for the 94 elements H-Pu. *J. Chem. Phys.* **2010**, *132*, 154104. [CrossRef]
41. Weigend, F.; Ahlrichs, R. Balanced basis sets of split valence, triple zeta valence and quadruple zeta valence quality for H to Rn: Design and assessment of accuracy. *Phys. Chem. Chem. Phys.* **2005**, *7*, 3297–3305. [CrossRef] [PubMed]
42. Schuchardt, K.L.; Didier, B.T.; Elsethagen, T.; Sun, L.; Gurumoorthi, V.; Chase, J.; Li, J.; Windus, T.L. Basis set exchange: A community database for computational sciences. *J. Chem. Inf. Model.* **2007**, *47*, 1045–1052. [CrossRef]
43. Feller, D. The role of databases in support of computational chemistry calculations. *J. Comput. Chem.* **1996**, *17*, 1571–1586. [CrossRef]
44. Pritchard, B.P.; Altarawy, D.; Didier, B.; Gibson, T.D.; Windus, T.L. New Basis Set Exchange: An Open, Up-to-Date Resource for the Molecular Sciences Community. *J. Chem. Inf. Model.* **2019**, *59*, 4814–4820. [CrossRef] [PubMed]
45. Granovsky, A.A. Firefly Version 8. Available online: http://classic.chem.msu.su/gran/firefly/index.html (accessed on 15 April 2021).
46. Schmidt, M.W.; Baldridge, K.K.; Boatz, J.A.; Elbert, S.T.; Gordon, M.S.; Jensen, J.H.; Koseki, S.; Matsunaga, N.; Nguyen, K.A.; Su, S.; et al. General atomic and molecular electronic structure system. *J. Comput. Chem.* **1993**, *14*, 1347–1363. [CrossRef]

47. Frisch, M.J.; Trucks, G.W.; Schlegel, H.B.; Scuseria, G.E.; Robb, M.A.; Cheeseman, J.R.; Scalmani, G.; Barone, V.; Mennucci, B.; Petersson, G.A.; et al. *Gaussian 09*; Revision A.1; Gaussian Inc.: Wallingford, UK, 2009.
48. Zhurko, G.A.; Zhurko, D.A. ChemCraft, Version 1.6 (Build 312); Ed. Available online: http://www.chemcraftprog.com/index.html (accessed on 1 December 2021).

International Journal of
Molecular Sciences

MDPI

Article

Protonation of Borylated Carboxonium Derivative $[2,6\text{-}B_{10}H_8O_2CCH_3]^-$: Theoretical and Experimental Investigation

Ilya N. Klyukin [1,*], Anastasia V. Kolbunova [1,2], Alexander S. Novikov [3], Aleksey V. Nelyubin [1], Nikita A. Selivanov [1], Alexander Yu. Bykov [1], Alexandra A. Klyukina [4], Andrey P. Zhdanov [1], Konstantin Yu. Zhizhin [1,5] and Nikolay T. Kuznetsov [1]

1 Kurnakov Institute of General and Inorganic Chemistry, Russian Academy of Sciences, Leninskii pr. 31, 117907 Moscow, Russia; zdimchi391@mail.ru (A.V.K.); nelyubin.av@yandex.ru (A.V.N.); goovee@yandex.ru (N.A.S.); bykov@igic.ras.ru (A.Y.B.); zhdanov@igic.ras.ru (A.P.Z.); zhizhin@igic.ras.ru (K.Y.Z.); boron@igic.ras.ru (N.T.K.)

2 Faculty of Chemistry, Higher School of Economics, Myasnitskaya yl., 101000 Moscow, Russia

3 Institute of Chemistry, Saint Petersburg State University, Universitetskaya Nab. 7-9, 199034 Saint Petersburg, Russia; a.s.novikov@spbu.ru

4 Research Centre of Biotechnology, Winogradsky Institute of Microbiology, Russian Academy of Sciences, 60 let Oktjabrja pr-t, 7, bld. 2, 117312 Moscow, Russia; alexandra.a.popova@gmail.com

5 Lomonosov Institute of Fine Chemical Technologies, MIREA Russian Technological University, 78 Vernadskiy Avenue, 119571 Moscow, Russia

* Correspondence: klukinil@igic.ras.ru; Tel.: +7-916-954-2737

Abstract: The process of protonation of $[2,6\text{-}B_{10}H_8O_2CCH_3]^-$ was investigated both theoretically and experimentally. The most suitable conditions for protonation of the derivative $[2,6\text{-}B_{10}H_8O_2CCH_3]^-$ were found. The process of protonation was carried out in the presence of an excess of trifluoromethanesulfonic acid CF_3SO_3H at room temperature in dichloromethane solution. The structure of the resulting complex $[2,6\text{-}B_{10}H_8O_2CCH_3{*}H^{fac}]^0$ was established using NMR data and the results of DFT calculations. An additional proton atom H^{fac} was found to be localized on one of the facets that was opposite the boron atom in a substituted position, and which bonded mainly with one apical boron atom. The main descriptors of the B-H^{fac} bond were established theoretically using QTAIM and NBO approaches. In addition, the mechanism of $[2,6\text{-}B_{10}H_8O_2CCH_3]^-$ protonation was investigated.

Keywords: *closo*-borates; boron cluster; protonation; DFT calculation; NMR spectra; QTAIM; Fukui function

1. Introduction

The investigation of covalent and noncovalent interactions is one of the main tasks of modern inorganic chemistry [1,2]. Such studies provide an opportunity to better understand the structures of chemical substances and their properties. By examining chemical bonds, it is possible to establish factors that influence their breaking and formation [3,4]. This fact can be used to produce new substances with given properties. A combined theoretical and experimental approach is the best way to investigate such phenomena [5,6].

There are several driving forces behind bonding formation: orbital, electrostatic, and the van der Waals interactions [7–9]. Information about the energy characteristics of such interactions is one of the most important descriptors. Theoretical methods allow exploration of the nature of chemical interactions and estimation of their energies in a simple and intuitive way. The application of such popular and well-established methods as QTAIM (Quantum Theory of Atoms in Molecules), ELF (Electron Localization Function), and NBO (Natural Bond Orbitals) analysis provides the opportunity to find essential information about the chemical structure and bonding of target compounds [10–14].

Closo-borates and related compounds have always attracted attention due to their unusual structure. Their electron structures cannot be described in terms of the classic Lewis approach [11,15]. For a correct description of the structure of *closo*-borate compounds, the terms of 3D-aromacity are used. *Closo*-borates possess various interesting properties, such as thermodynamic and oxidative stability and biological activity [16–18]. These properties determine the potential application of *closo*-borate systems for the preparation of solid-state batteries [19], magnetic and hydrogen storage materials [20,21], and drugs for BNCT (boron neutron capture therapy) [22–24]. Such systems can form various covalent and noncovalent interactions. It is impossible not to mention the tremendous contributions of the pioneers of boron cluster chemistry, such as Lipscomb [25–28], Hoffman [29–31], Wade [32–34], Mingos [35–37], and Muetterties [38–41]. Their works have significantly expanded ideas about the structure and properties of such systems. Currently, there is a great deal of interest in the chemistry of *closo*-borate anions [42,43]. The phenomena of the *exo*-polyhedral B-X (X = H, C, N, O, F) bonds have been extensively studied [44–46], for example, research into the main characteristics of these bonds for hydroxy-derivatives $[B_nH_{n-1}OH]^{2-}$ (n = 6, 10, 12) was carried out [47]. For the *exo*-polyhedral B-O bonds, it has been established that, for the *closo*-borate anion, covalent interactions increase with an increase in the boron cluster size. Noncovalent contacts of *closo*-borate anions and related compounds have also been investigated [48]. This type of interaction is weaker than covalent bonds but also has a significant effect on the structure and properties of the systems under study. For example, the interactions of nitrile derivatives with various nucleophiles have previously been studied [34,49–52]. It was shown that the structure of the products obtained was the result of noncovalent interactions between protons of nucleophile and hydrogen atoms of the cluster anion.

Closo-borate anions possess an excess of electron density and can act as an electron donor for an electrophilic system. They can form a complex with different Lewis acids: protons, $AlCl_3$, ML_n, and others [42,53]. Previously, the structure of the $[B_{10}H_{11}]^-$ anion was studied both theoretically and experimentally [54,55]. It was shown that H^{fac} moves freely in the equatorial belt, with a very low energy barrier (approximately 1 kcal/mol) [20].

The present study focused on $[2,6\text{-}B_{10}H_8O_2CCH_3]^-$. This anion is a member of the class of borylated heterocycles. Analogous derivatives based on borylated boroxazole were also investigated for *closo*-decaborate and dodecaborate anions [56–59]. Several approaches for the preparation of this class of *closo*-borate derivatives have been explored previously [60,61]. In addition, the disclosure reaction of borylated heterocycle has been established [62]. $[2,6\text{-}B_{10}H_8O_2CCH_3]^-$ is a convenient model system for the investigation of the protonation process. This system is more symmetrical than mono-substituted derivatives and the number of possible isomers is reduced, compared to them. The protonation process of a given anion can be readily investigated. It is important to assess the differences and similarities between the protonation of the original anion $[B_{10}H_{10}]^{2-}$ and the derivatives based on it. In addition, the protonation of a given derivative enables the preparation of a derivative with a degree of substitution equal to three. Thus, in the present research, a combined theoretical and experimental study of the protonation of $[2,6\text{-}B_{10}H_8O_2CCH_3]^-$ was carried out.

2. Results and Discussion

2.1. Experimental Protonation

First, the protonation process of $[2,6\text{-}B_{10}H_8O_2CCH_3]^-$ was conducted experimentally. As described previously [63], systems of the general form $[B_{10}H_9L]^-$ have a lesser possibility of forming a protonated complex than the $[B_{10}H_{10}]^{2-}$ anion. A possible reason for this is that the introduction of a positively charged group reduces the ability of the cluster cage to donate its electron density. Anion $[B_{10}H_{10}]^{2-}$ can be easily transferred into $[B_{10}H_{11}]^-$ in the presence of trifluoroacetic acid CF_3COOH. For the preparation of a protonated complex of the general form $[B_{10}H_9OR_1R_2{*}H^{fac}]^0$, however, trifluoromethanesulfonic acid CF_3SO_3H was used. In the present work, trifluoroacetic acid CF_3COOH was also used as a proton donor, but the protonation process did not take place. As in the case of oxonium derivatives $[B_{10}H_9OR_1R_2]^-$, the protonation of the derivative $[2,6\text{-}B_{10}H_8O_2CCH_3]^-$ was carried out in the presence of CF_3SO_3H at room temperature (Scheme 1). Dichloromethane CH_2Cl_2 was used as a solvent. The application of another solvent was impossible due to the interaction of $[2,6\text{-}B_{10}H_8O_2CCH_3{*}H^{fac}]^0$ with the molecules of the solvents. $[2,6\text{-}B_{10}H_8O_2CCH_3{*}H^{fac}]^0$ was formed immediately after the addition of the acid. The distinctive feature of this process is that the protonation of the cluster anion requires an excess of CF_3SO_3H acid. For an investigation of these phenomena, the experiment with sequential addition of CF_3SO_3H acid to a dichloromethane solution of the initial $[2,6\text{-}B_{10}H_8O_2CCH_3]^-$ derivative was conducted. The process was monitored by ^{11}B-NMR spectroscopy (Figure S1). Detailed information about the NMR spectra of the initial and target substances and the correlation of all signals are given below in the NMR spectra analysis section. When one equivalent of acid was added to the system, the overall structure of the spectrum characteristic of the original derivative was preserved, but one of the signals was broadened. This signal corresponds to boron atoms in apical positions. The given pattern indicates the possibility of the formation of intermolecular contacts between the carboxonium derivative and the trifluoromethanesulfonic acid molecule. When one more acid equivalent is added to the system, a more significant change in the spectrum was observed, which may indicate the formation of molecular complexes between the $[2,6\text{-}B_{10}H_8O_2CCH_3]^-$ derivative and the acid molecule. The formation of the target substance was also observed, as indicated by the presence of a broadened signal in the region of 20 ppm. However, the yield of the target product did not exceed approximately 10 percent. A more accurate estimation of the target product yield could not be made due to the overlapping signals of the target product and the cluster-acid complex. At ratios $[2,6\text{-}B_{10}H_8O_2CCH_3]^-$:$CF_3SO_3H$ equal to 1:3 and 1:4, an increase in the yield of the target product was observed. In the case of the 1:4 ratio, the yield of the target product reached approximately 40 percent. Finally, in the presence of 5 equivalents of CF_3SO_3H, complete protonation of $[2,6\text{-}B_{10}H_8O_2CCH_3]^-$ with formation of $[2,6\text{-}B_{10}H_8O2CCH_3{*}H^{fac}]^0$ occurred. Further addition of the acid excess did not change the form of the spectrum, indicating the completeness of the protonation process. It is noteworthy that $[2,6\text{-}B_{10}H_8O_2CCH_3{*}H^{fac}]^0$ is stable only as a solvent at room temperature and without air.

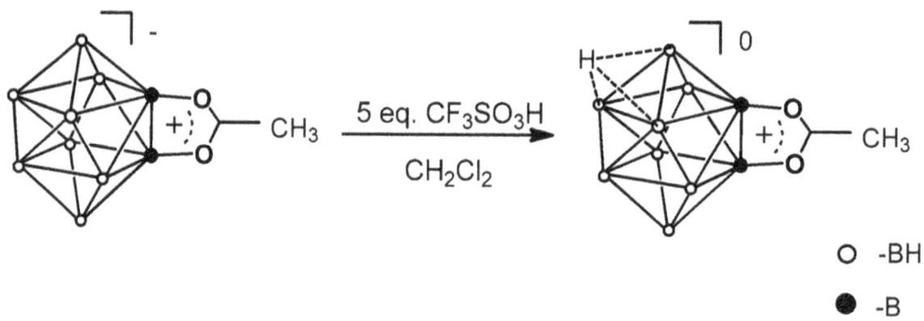

Scheme 1. Protonation of carboxonium derivative $[2,6\text{-}B_{10}H_8O_2CCH_3]^-$ anion.

2.2. Fukui Function

In the next step, the potential position in the boron cluster for the coordination of an additional proton atom was investigated. The protonation process of $[2,6\text{-}B_{10}H_8O_2CCH_3]^-$ can be considered to be an electrophilic attack on the boron cluster. The Fukui functions approach is the best tool for investigation of the most likely position in the molecule for an electrophilic or nucleophilic attack [64–66]. Previously, it has been shown, with the help of Fukui functions, that electrophilic attack on closo-borate anions is performed predominantly in the apical position [67,68]. In the present investigation, Fukui functions were used with different approaches employed for their calculation (Table S2). As in the case of $[B_nH_n]^{2-}$ systems [68], the Hirshfeld approach is the best for calculating the electrophilic attack positions. In the present case, electrophilic attack on closo-borate anions was performed predominantly in the apical position (the optimized structures of $[2,6\text{-}B_{10}H_8O_2CCH_3]^-$ and $[2,6\text{-}B_{10}H_8O_2CCH_3{}^*H^{fac}]^0$ are shown in Figure 1). Thus, the data from the Fukui functions analysis indicated that the H^{fac} was localized near apical positions (Table S3).

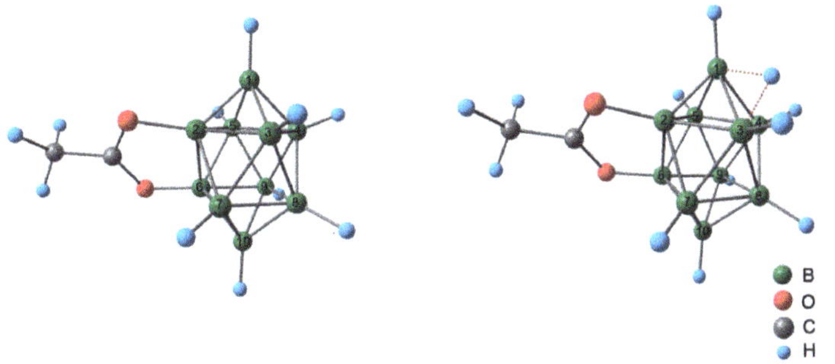

Figure 1. Optimized structures of $[2,6\text{-}B_{10}H_8O_2CCH_3]^-$ and $[2,6\text{-}B_{10}H_8O_2CCH_3{}^*H^{fac}]^0$.

2.3. Protonation Mechanism

The mechanism of $[2,6\text{-}B_{10}H_8O_2CCH_3]^-$ protonation was investigated. Initially, complex $[2,6\text{-}B_{10}H_8O_2CCH_3{}^*CF_3SO_3H]^-$ (Comp) formed endoergonically (by 23 $kJ \cdot mol^{-1}$, in terms of Gibbs free energies). The main driving force behind the formation of this complex is the dihydrogen bond between the proton atom from CF_3SO_3H and the equatorial hydrogen atom from the boron polyhedron. The contact length of H_{eq}–H was equal to 1.60 Å. The presence of the dihydrogen bond was proved with the help of QTAIM analysis of the Comp structure (Figure 2). In addition, the C–F···H–B contact was detected. According to the main topological descriptors of electron density C–F···H–B, contact can be characterized as being very weak.

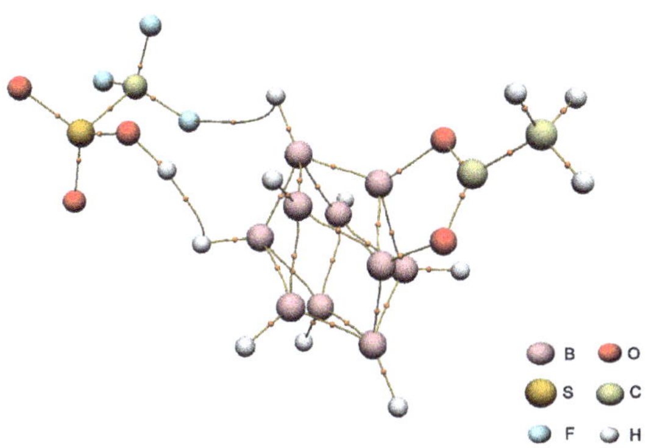

Figure 2. Molecular graph showing the results of the topological analysis of the electron density distribution in the model structure of the $[2,6\text{-}B_{10}H_8O_2CCH_3{}^*CF_3SO_3H]^-$ (Comp).

In the next step, the proton migrated to a boron polyhedral through the formation of the transition state (TS). The preferred location for the proton attack was the face of the polyhedron opposite the substituted position. The proton from CF_3SO_3H acid in TS connected mainly with the boron atom in the apical position, which was demonstrated with the help of QTAIM analysis (Figure 3). In the case of TS, the contact length between the apical boron atom and the proton atom from CF_3SO_3H was equal to 1.48 Å. The overall energy barrier of proton migration was 69 $kJ \cdot mol^{-1}$.

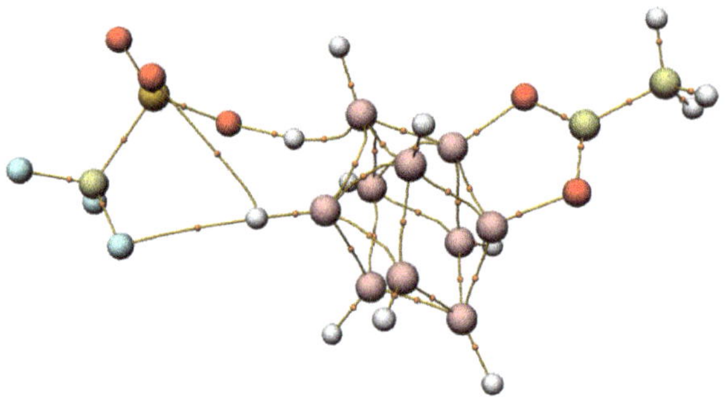

Figure 3. Molecular graph showing the results of the topological analysis of the electron density distribution in the model structure of TS. Atomic designations are given in Figure 2.

Finally, $[B_{10}H_8O_2CCH_3{*}H^{fac}]^0$ formed. The overall process of boron cluster protonation is endoergonical (by 18.5 kJ·mol^{-1}, in terms of Gibbs free energies) (Figure 4). Thus, due to $\Delta G > 0$ overall, a reaction did not occur in the presence of 1 equivalent of CF_3SO_3H. This finding correlates well with experimental data and the main reason it occurred is the weak stabilization of $CF_3SO_3^-$ in the dichloromethane solution. Additional proton donors are required to stabilize this anion. The use of an excess of trifluoromethanesulfonic acid can provide this. The $CF_3SO_3^-$ can be stabilized by hydrogen bonds between oxygen atoms of this anion and hydrogen atoms of CF_3SO_3H. For example, complex between $CF_3SO_3^-$ anion and CF_3SO_3H was observed (Figure S8). This complex possesses strong hydrogen bonds and the formation of a given system is an exergonic process. It is worth noting that the addition of the acid excess also significantly changed the properties of the medium, which in turn can change the solvation energy of the model system under consideration. This aspect may also contribute to the fact that the protonation process becomes exergonic when an excess of the acid is supplied.

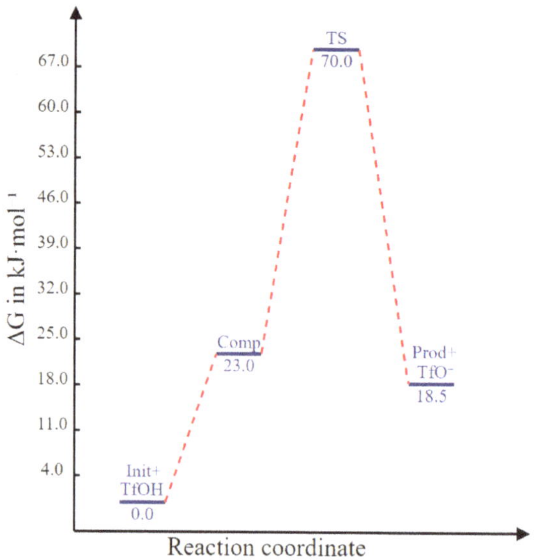

Figure 4. Energy profile of protonation of $[2,6\text{-}B_{10}H_8O_2CCH_3]^-$. TfOH–$CF_3SO_3H$, TfO$^-$–$CF_3SO_3^-$. The deciphering of other abbreviations is given in the main text.

2.4. NMR Spectra Analysis

^{11}B NMR spectra were used for the confirmation suggestion based on the Fukui index results and to establish the structure of the resulting complex. It is useful to consider the spectra of the $[2,6\text{-}B_{10}H_8O_2CCH_3]^-$ anion. Due to the high symmetry of the given system, the two boron atoms in the substituted positions and the two apical atoms are pairwise equivalent. Atoms in equatorial positions are split into two groups of signals (Figure 5). Thus, as described previously, the ^{11}B NMR spectra of the initial anion were characterized by four signals, as follows: a signal at 0.0 ppm with integral intensities I = 2, corresponding to the boron atoms from equivalent substituted positions. This signal did not split in the absence of broadband decoupling; a signal at −7.1 with integral intensities I = 2, corresponding to boron atoms from equivalent apical positions; and signals from equatorial positions appeared at −17.6 (I = 2) and −30.0 (I = 4). This interpretation of the spectrum is based on previously obtained spectral data for several *closo*-decaborate derivatives with B-O *exo*-polyhedral bonds. This approach does not, however, enable a complete correlation of signals and leaves some room for speculation.

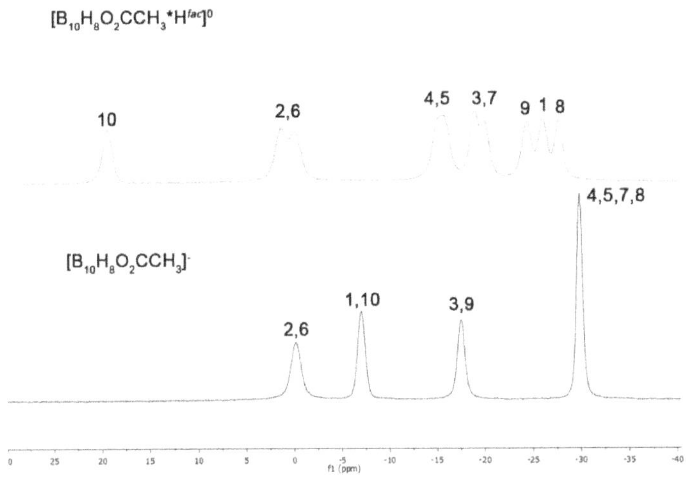

Figure 5. ^{11}B-^{1}H NMR spectra of $[B_{10}H_8O_2CCH_3]^-$ (**bottom**) and $[B_{10}H_8O_2CCH_3{*}H^{fac}]^0$ (**top**). The numbers are the positions of the boron atoms in the cluster polyhedral.

The theoretical calculation of chemical shielding enabled the full correlation of the signals in the ^{11}B NMR spectra with the boron positions in a cluster cage. Several computational levels for calculating NMR shifts were tested: wB97X-D3, B3LYP, and B97 DFT-functionals with 6–31++G(d,p), IGLO-III, and EPR-III basis sets (Table S5). To evaluate the difference between the theoretical and experimental data, the values of the root-mean-square deviation (RMSD) were used. The wB97X-D3 level of theory provided the worst results, with all basis sets. The values of RMSD were quite large (the range of RMSD values lay in the range 46–47) and applying this method did not enable the deciphering of the ^{11}B spectra of given compounds. Improved results were obtained with the application of the hybrid functionals B3LYP and B97. In the case of the B97 functional, the best values of RMSD were obtained (Table S5). The RMSD value for the B97/IGLO-III level of theory was equal to 3.7. Thus, applying the results of the theoretical calculation, the assumption about the chemical shifts of atoms in the substituted and apical positions proved to be correct. In addition, the signal at −17.6 ppm corresponded to boron atoms from the equatorial belt in the B3 and B9 positions. The signal from −30.0 ppm corresponded to atoms in the B4, B5, B7, and B8 positions.

The ^{1}H NMR spectra of $[2,6\text{-}B_{10}H_8O_2CCH_3]^-$ were considered. For a simple view of the spectrum and the correct correlation of signals, it is best to use ^{1}H-^{11}B NMR spectra (Figure S2). Some of the signals appeared, however, as broadened lines, making it difficult to identify them. Signals from the methyl substituent appeared at 2.18 ppm, and signals from apical positions at 3.16 ppm were well represented. Signals from equatorial positions appeared at 1.79, 0.39 and 0.12 ppm.

The addition of a proton reduced the symmetry of the resulting system and the spectra of $[2,6\text{-}B_{10}H_8O2CCH_3{*}H^{fac}]^0$ were more complicated than those of $[2,6\text{-}B_{10}H_8O_2CCH_3]^-$. Boron atoms were split into separate signals. Thus, it was impossible to elucidate the structure of $[2,6\text{-}B_{10}H_8O2CCH_3{*}H^{fac}]^0$ by only applying the 1D ^{11}B spectra. To decipher these spectra, the ^{11}B-^{11}B COSY spectra and theoretical calculations were used (Figure S5). This type of spectrum indicates that, contrary to the case of the $[B_{10}H_{11}]^-$ anion, the proton in this anion does not migrate along one of the equatorial belts but is localized on one of the edges. Based on the calculated data, the most stable isomer Iso1 was found. In this isomer, the proton was localized on the edge opposite the substituted position. The structure of this isomer is described in more detail below. Similar to the initial anion $[2,6\text{-}B_{10}H_8O_2CCH_3]^-$, the B97 functional worked well for the prediction of NMR spectra. The RMSD value for the B97/IGLO-III level of theory was equal to 3.5 (Table S5).

As in the case of $[B_{10}H_{11}]^-$, the anion $[2,6\text{-}B_{10}H_8O_2CCH_3{*}H^{fac}]^0$ has two non-equivalent apical positions. The first of them is bonded with additional proton atoms H^{fac} and the second one is not. Signal from the boron atom of the apical position B10, which was not bonded with H^{fac}, shifted into a low field and appeared at 19.4 ppm. Similar observations were made in the case of the apical position B10 of the $[B_{10}H_{11}]^-$ anion. In addition, the existence of this shift was proved by theoretical calculation. According to theoretical calculation, the signal of the apical position B10 appeared at 13.4 ppm. The chemical shift from the apical position B1, which bonded with H^{fac}, shifted in the high field and appeared at −26.9 ppm. The boron atoms in the substituted positions were also not equivalent in the case of a protonated cluster. The signal at 0.2 ppm correlated with the B10 position of the boron cluster in the ^{11}B-^{11}B COSY NMR spectra, thus it corresponded to the B6-substituted position, which is opposite to the apical position with H^{fac}. Another substituted position was found at −1.6 ppm. At −15.4 ppm, the signal from the B4,5 positions appeared. Signals from the B3 and B7 positions were relatively close, appearing at −19.0 and −20.1 ppm, respectively. Signals from the B9 and B8 positions had the greatest negative chemical shift value at −26.1 and −27.8 ppm, respectively.

In the ^{1}H NMR spectra of $[2,6\text{-}B_{10}H_8O_2CCH_3{*}H^{fac}]^0$, the signal of the methyl group was observed at 2.45 ppm (Figure S3). Signals from the exo-polyhedral cluster were in the range 5.16–0.33 ppm. It is noteworthy that the signal at 5.16 ppm corresponded to a hydrogen atom connected with a boron atom in the B10 position. This assumption is based on the result of ^{1}H NMR spectra modeling. According to the results of the calculations, the proton in the downfield region belonged to the given atom. The signal from H^{fac}, according to theoretical calculations, was in the downfield region at 0.00 ppm. In the experimental spectra, the most upfield signal was at 0.33 ppm. In the ^{13}C NMR spectra of $[2,6\text{-}B_{10}H_8O_2CCH_3{*}H^{fac}]^0$, the signal of the carbonyl group carbon atom was observed at 194.0 ppm and the signal of the methyl group carbon atom was observed at 20.10 ppm (Figure S4). In the IR-spectra, the B-H bands were shifted from 2490 cm^{-1} in the initial $[2,6\text{-}B_{10}H_8O_2CCH_3]^-$ to 2540 cm^{-1} in $[2,6\text{-}B_{10}H_8O_2CCH_3{*}H^{fac}]^0$ (Figure S6).

2.5. Structure Elucidation

The structure of $[2,6\text{-}B_{10}H_8O_2CCH_3{*}H^{fac}]^0$ was examined using DFT calculation. Several possible isomers were calculated (Figures 6 and S7, Table S4). Calculations were performed in the gas phase and considering the solvation effect. In both cases, the isomers differed significantly in terms of energy. The Iso1 and Iso2 isomers were the most thermodynamically stable (Figure 6). In these isomers, an extra proton, in addition to the equatorial boron atoms, was localized on the apical boron atom. The Iso1 in which the proton was localized on the edge opposite the substituted position had the highest negative Gibbs energy among all isomers. The energetic barrier for the migration of the proton atom from Iso1 to Iso2 was equal to 30 kJ·mol^{-1}in the gas phase and 25 kJ·mol^{-1} in dichloromethane solution. Isomers with the localized proton only on equatorial boron atoms were the least thermodynamically stable and had the smallest negative Gibbs energy values. The energy barrier between Iso1 and Iso3 was 50 kJ·mol^{-1} for both the gas and dichloromethane phase. This energy barrier was less than that of the anion $[B_{10}H_{11}]^-$. Thus, Iso1 was the most stable isomer among the other possible structures of $[2,6\text{-}B_{10}H_8O_2CCH_3{}^*H^{fac}]^0$.

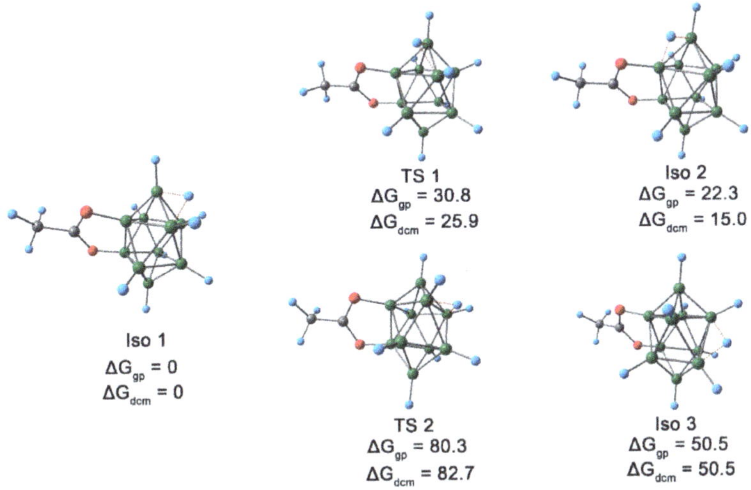

Figure 6. Optimized structures of the main isomers of $[2,6\text{-}B_{10}H_8O_2CCH_3{}^*H^{fac}]^0$. The atomic designations are given in Figure 1. ΔG_{gp}—relative Gibbs energy of isomer in gas phase. ΔG_{dcm}—relative Gibbs energy of isomer in dichlormethane solution.

The geometric parameters of this isomer were discussed in detail (Table S1). The bond length between H^{fac} and the apical boron atom was 1.28 Å, whereas the analogous parameter for the bond between H^{fac} and the equatorial boron atom was 1.42 Å. The bond length between equatorial boron atoms and the apical boron atom that bonded with the proton increased compared to the same parameter in the original cluster. This parameter increased from 1.72 to 1.84 Å. The phenomena of B-H^{fac} contacts were investigated using Wiberg bond order indices and QTAIM analysis (Figure 7). Applying the QTAIM method, a molecular graph of the electron density distribution in $[2,6\text{-}B_{10}H_8O_2CCH_3{}^*H^{fac}]^0$ was obtained. The molecular graph indicated that H^{fac} has interactions with apical and equatorial boron atoms. The B_{ap}–H^{fac} interaction was characterized by a greater value of the Wiberg bond order index compared to that of B_{eq}–H^{fac}. The B_{ap}–H^{fac} interaction was also characterized by greater values of $\rho(r)$, total energy at the bond critical point, and the delocalization index. These data indicate that H^{fac} is predominantly bonded with the boron atom in the apical position. An analogous investigation was carried out for the case of the $[B_{10}H_{11}]^-$ anion. For this anion, H^{fac} only bonded with the boron atom in the apical position. This contact was characterized by a longer B_{ap}-H^{fac} bond and smaller values of the main bond descriptors. Thus, in the case of $[2,6\text{-}B_{10}H_8O_2CCH_3{}^*H^{fac}]^0$ B_{ap}–H^{fac} has greater covalent interaction than the $[B_{10}H_{11}]^-$ anion.

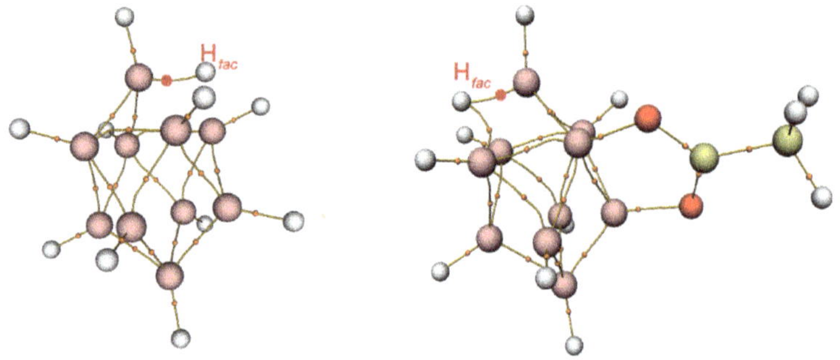

Figure 7. Molecular graph showing the results of the topological analysis of the electron density distribution in the model structures of $[B_{10}H_{11}]^-$ and $[2,6\text{-}B_{10}H_8O_2CCH_3{*}H^{fac}]^0$. The atomic designations are given in Figure 2.

The values of the main bond descriptors indicated that the B_{ap}–H^{fac} interactions were significantly weaker than those of the covalent interactions between boron atoms and exo-polyhedral hydrogen atoms. It is noteworthy that the values of the main bond descriptors of B-H and B-O bonds were greater in the case of $[2,6\text{-}B_{10}H_8O_2CCH_3{*}H^{fac}]^0$ than in $[2,6\text{-}B_{10}H_8O_2CCH_3]^-$. This indicates that the strength of these contacts increases. A possible reason for these phenomena may be the reduction in the total charge in the system compared to that of the initial anion $[2,6\text{-}B_{10}H_8O_2CCH_3]^-$. The atomic charges on the boron atoms are reduced and the electron repulsion in the exo-polyhedral bonds is also reduced.

3. Materials and Methods

3.1. Computational Details

The full geometry optimization of all model structures was carried out at the ωB97X-D3/6–31++G(d,p) level of theory [69,70] with the help of the ORCA 4.2.1 program package (Mülheim an der Ruhr, Germany) [71] (the atom-pairwise dispersion correction with the zero-damping scheme was utilized [70]). The convergence tolerances for the geometry optimization procedure were as follows: energy change = 5.0×10^{-6} Eh, maximal gradient = 3.0×10^{-4} Eh/Bohr, RMS gradient = 1.0×10^{-4} Eh/Bohr, maximal displacement = 4.0×10^{-3} Bohr, and RMS displacement = 2.0×10^{-3} Bohr. Spin-restricted approximation for the model structures with closed electron shells was used. Symmetry operations were not applied during the geometry optimization procedure for any of the model structures. The Hessian matrices were calculated numerically for all optimized model structures in order to prove the location of correct minima on the potential energy surfaces (no imaginary frequencies for all reactants, intermediates, and final products; only one imaginary frequency for transition states). The connectivity of each reaction step was also confirmed using the intrinsic reaction coordinate (IRC) calculation from the transition states [72–74]. Solvent effects were considered using the Solvation Model based on Density (SMD) [75]. The natural bond orbital (NBO) method was emplyed, using the NBO7 program package (Madison, WI, USA) [76]. Topological analysis of the electron density distribution, using the Quantum Theory of Atoms in Molecules (QTAIM) formalism developed by Bader [12], was employed with the Multiwfn program, version 3.7 (Beijing, China) [77]. The Cartesian atomic coordinates for all optimized equilibrium model structures are presented in the Supplementary Materials. Visualization of the optimized structures was carried out with the help of ChemCraft program version 1.7 (Ivanovo, Russia) [78]. In the case of the molecular graph showing the results of the topological analysis of the electron density distribution visualization, the Multiwfn program (version 3.7) was employed [77].

3.2. IR Spectra

The IR spectra of the prepared compounds were recorded on an Infralyum FT 02 Fourier transform spectrometer (Lumex Instruments Research and Production Company, Fraserview Place, Vancouver, BC, Canada) in the region of 300–4000 cm^{-1} and with a resolution of 1 cm^{-1}. Samples were prepared as dichloromethane CH_2Cl_2 solution.

3.3. NMR Spectra

The NMR (^1H, ^{11}B, ^{13}C) spectra of the solutions of the studied compounds in CD3CN were recorded on a Bruker Avance II 300 spectrometer (Ettlingen, Germany) operating at 300.3, 96.32, and 75.49 MHz, respectively, using an internal deuterium lock. Tetramethylsilane and boron trifluoride etherate were used as external references.

3.4. Protonation of $((C_6H_5)_4P)[2,6\text{-}B_{10}H_8O_2CCH_3]$

CF_3SO_3OH (0.018 mL, 1.5 mmol) was added to a solution of $((C_6H_5)_4P)[2,6\text{-}B_{10}H_8O_2CCH_3]$ (20 mg, 0.3 mmol) in deuterated dichloromethane CD_2Cl_2 (0.5 mL), in a dry argon atmosphere. The resulting mixture was kept at room temperature for 10 min. IR-spectra (dichloromethane solution): υ(BH) 2590 cm^{-1}, 2555 cm^{-1}. $^{11}B\text{-}\{^1H\}$ NMR (CD_2Cl_2, ppm): 19.2 (s, B10, I = 1), 1.4 (d, B6, I = 1), −0.2 (d, B2, I = 1), −15.2 (d, B4, I = 1), −15.7 (d, B3, I = 1), −19.1 (d, B5, I = 1), −20.1 (d, B9, I = 1), −24.4 (d, B7, I = 1), −26.1 (d, B1, I = 1), −27.8 (d, B8, I = 1). 1H NMR (CD_2Cl_2, ppm): 11.75 (CF_3SO_3H), 7.97 (m, Ph_4P^+, I = 4), 7.80 (m, Ph_4P^+, I = 8), 7.66 (m, Ph_4P^+, I = 8), 5.17 (s, B(10)H, I = 1), 2.68 (s, B(3)H, I = 1), 2.50 (s, CH_3, I = 3), 2.03 (s, B(9)H, I = 1), 1.60 (s, B(4)H, I = 1), 1.32 (s, B(5)H, I = 1), 0.93 (s, B(7)H, I = 1), 0.62 (s, B(8)H, I = 1), 0.36 (s, H^{fac}, I = 1). ^{13}C NMR (CD_2Cl_2, ppm): 194.7 (O2CCH3), 140.6 (Ph_4P+), 139.8 (Ph_4P_+), 135.8 (Ph_4P^+), 122.7 (Ph_4P^+), 118.1 (CF_3SO_3H), 20.1 (O_2CCH_3).

4. Conclusions

The process of protonation of the carboxonium derivative $[2,6\text{-}B_{10}H_8O_2CCH_3]^-$ was investigated. In the case of the carboxonium derivatives of *closo*-decaborate anions, trifluoromethanesulfonic acid CF_3SO_3H was used as a proton donor. By considering the reaction mechanisms, the reason for the excess of CF_3SO_3H was established. The excess acid was required to stabilize the anion $CF_3SO_3^-$, which in turn caused the total protonation process to become exergonic. In contrast to the anion $[B_{10}H_{11}]^-$, in the case of $[2,6\text{-}B_{10}H_8O_2CCH_3{*}H^{fac}]^0$, the proton did not migrate along the equatorial belt and was localized on a facet opposite the B atoms bonded with the exo-polyhedral substituent. Based on theoretical modeling data, it was shown that the proton was predominantly bound to the apical boron atom. In addition, the $^{11}B\text{-}^1H$ NMR spectra of the final compound were quite complicated and difficult to decipher. Using theoretical modeling data, all signals were correlated with the positions of the boron atoms in the cluster. Theoretical calculations at the B97/IGLO-III level of theory corresponded perfectly with the experimental $^{11}B\text{-}^1H$ NMR spectra data. $[2,6\text{-}B_{10}H_8O_2CCH_3{*}H^{fac}]^0$ can be considered to be a promising synthon for the preparation of trisubstituted derivatives of closo-borate anion.

Supplementary Materials: The following supporting information can be downloaded at: https://www.mdpi.com/article/10.3390/ijms23084190/s1.

Author Contributions: Manuscript conception, I.N.K. and A.S.N.; writing and original draft preparation, I.N.K., A.S.N. and A.P.Z.; experimental protonation of $[2,6\text{-}B_{10}H_8O_2CCH_3]^-$, A.V.K. and I.N.K.; NMR analysis, N.A.S. and A.Y.B.; IR analysis, A.V.N.; Fukui function calculation and structure elucidation, I.N.K., A.V.K. and A.S.N.; mechanism investigation, visualization, A.A.K. and I.N.K.; editing, data analysis, and interpretation, I.N.K., A.P.Z., A.S.N. and K.Y.Z.; project administration I.N.K., K.Y.Z. and N.T.K.; supervision N.T.K. All authors have read and agreed to the published version of the manuscript.

Funding: This work was supported by the Russian Science Foundation (project no. 20-73-00326).

Institutional Review Board Statement: Not applicable.

Informed Consent Statement: Not applicable.

Data Availability Statement: Not applicable.

Acknowledgments: NMR Studies were performed within the framework of the State Assignment of the Kurnakov Institute (IGIC RAS) in the field of fundamental scientific research.

Conflicts of Interest: The authors declare no conflict of interest.

References

1. Pino-Rios, R.; Inostroza, D.; Tiznado, W. Neither too Classic nor too Exotic: One-Electron Na·B Bond in NaBH $^{3-}$ Cluster. *Angew. Chem. Int. Ed.* **2021**, *60*, 12747–12753. [CrossRef] [PubMed]
2. Su, B.; Cao, Z.-C.; Shi, Z.-J. Exploration of Earth-Abundant Transition Metals (Fe, Co, and Ni) as Catalysts in Unreactive Chemical Bond Activations. *Acc. Chem. Res.* **2015**, *48*, 886–896. [CrossRef] [PubMed]
3. Woodward, R.B.; Hoffmann, R. The Conservation of Orbital Symmetry. *Angew. Chem. Int. Ed.* **1969**, *8*, 781–853. [CrossRef]
4. Abu-Youssef, M.A.M.; Langer, V.; Barakat, A.; Haukka, M.; Soliman, S.M. Molecular, Supramolecular Structures Combined with Hirshfeld and DFT Studies of Centrosymmetric M(II)-azido {M = Ni(II), Fe(II) or Zn(II)} Complexes of 4-Benzoylpyridine. *Symmetry* **2021**, *13*, 2026. [CrossRef]
5. Gouda, M.; Ferjani, H.; El-Lateef, H.M.A.; Khalaf, M.M.; Shaaban, S.; Yousef, T.A. A Competition between Hydrogen, Stacking, and Halogen Bonding in *N*-(4-((3-Methyl-1,4-dioxo-1,4-dihydronaphthalen-2-yl)selanyl)phenyl)acetamide: Structure, Hirshfeld Surface Analysis, 3D Energy Framework Approach, and DFT Calculation. *Int. J. Mol. Sci.* **2022**, *23*, 2716. [CrossRef]
6. Kolomeychuk, F.M.; Safonova, E.A.; Polovkova, M.A.; Sinelshchikova, A.A.; Martynov, A.G.; Shokurov, A.V.; Kirakosyan, G.A.; Efimov, N.N.; Tsivadze, A.Y.; Gorbunova, Y.G. Switchable Aromaticity of Phthalocyanine via Reversible Nucleophilic Aromatic Addition to an Electron-Deficient Phosphorus(V) Complex. *J. Am. Chem. Soc.* **2021**, *143*, 14053–14058. [CrossRef]
7. Weinhold, F.; Landis, C.R. *Discovering Chemistry with Natural Bond Orbitals*; John Wiley & Sons, Inc.: Hoboken, NJ, USA, 2012; ISBN 9781118229101.
8. Lepetit, C.; Fau, P.; Fajerwerg, K.; Kahn, M.L.; Silvi, B. Topological analysis of the metal-metal bond: A tutorial review. *Coord. Chem. Rev.* **2017**, *345*, 150–181. [CrossRef]
9. Cortes-Guzman, F.; Bader, R. Complementarity of QTAIM and MO theory in the study of bonding in donor-acceptor complexes. *Coord. Chem. Rev.* **2005**, *249*, 633–662. [CrossRef]
10. Kumar, P.S.V.; Raghavendra, V.; Subramanian, V. Bader's Theory of Atoms in Molecules (AIM) and its Applications to Chemical Bonding. *J. Chem. Sci.* **2016**, *128*, 1527–1536. [CrossRef]
11. Bader, R.F.W.; Legare, D.A. Properties of atoms in molecules: Structures and reactivities of boranes and carboranes. *Can. J. Chem.* **1992**, *70*, 657–676. [CrossRef]
12. Bader, R.F.W. *Atoms in Molecules: A Quantum Theory*; Oxford University Press: Oxford, UK, 1990.
13. Savin, P.-D.A.; Nesper, R.; Wengert, S.; Fässler, T.F. ELF: The Electron Localization Function. *Angew. Chem. Int. Ed.* **1997**, *36*, 1808–1832. [CrossRef]
14. Morgante, P.; Peverati, R. The devil in the details: A tutorial review on some undervalued aspects of density functional theory calculations. *Int. J. Quant. Chem.* **2020**, *120*, e26332. [CrossRef]
15. Shen, Y.-F.; Xu, C.; Cheng, L.-J. Deciphering chemical bonding in BnHn^{2-}(n = 2–17): Flexible multicenter bonding. *RSC Adv.* **2017**, *7*, 36755–36764. [CrossRef]
16. Teixidor, C.V. The uniqueness of boron as a novel challenging element for drugs in pharmacology, medicine and for smart biomaterials. *Future Med. Chem.* **2013**, *5*, 617–619. [CrossRef] [PubMed]
17. Evamarie Hey-Hawkins, C.V.T. *Boron-Based Compounds: Potential and Emerging Applications in Medicine*; John Wiley & Sons Ltd.: Hoboken, NJ, USA, 2018; ISBN 978-1-119-27555-8.
18. Poater, J.; Solà, M.; Viñas, C.; Teixidor, F. Hückel's Rule of Aromaticity Categorizes Aromatic closo Boron Hydride Clusters. *Chem. A Eur. J.* **2016**, *22*, 7437–7443. [CrossRef] [PubMed]
19. Duchêne, L.; Remhof, A.; Hagemann, H.; Battaglia, C. Status and prospects of hydroborate electrolytes for all-solid-state batteries. *Energy Stor. Mater.* **2020**, *25*, 782–794. [CrossRef]
20. Kochnev, V.K.; Avdeeva, V.V.; Goeva, L.V.; Malinina, E.A.; Kuznetsov, N.T. The undecahydrodecaborate anion $B_{10}H_{11}$—As the starting reagent in exopolyhedral substitution and complexation: Theoretical and experimental prerequisites. *Russ. J. Inorg. Chem.* **2012**, *57*, 331–336. [CrossRef]
21. Huang, Z.; Wang, S.; Dewhurst, R.D.; Ignat'Ev, N.V.; Finze, M.; Braunschweig, H. Boron: Its Role in Energy—Related Processes and Applications. *Angew. Chem. Int. Ed.* **2019**, *59*, 8800–8816. [CrossRef]
22. Ali, F.; Hosmane, N.S.; Zhu, Y. Boron Chemistry for Medical Applications. *Molecules* **2020**, *25*, 828. [CrossRef]
23. Hu, K.; Yang, Z.; Zhang, L.; Xie, L.; Wang, L.; Xu, H.; Josephson, L.; Liang, S.H.; Zhang, M.-R. Boron agents for neutron capture therapy. *Coord. Chem. Rev.* **2020**, *405*, 213139. [CrossRef]
24. Vorobyeva, M.; Dymova, M.; Novopashina, D.; Kuligina, E.; Timoshenko, V.; Kolesnikov, I.; Taskaev, S.; Richter, V.; Venyaminova, A. Tumor Cell-Specific 2′-Fluoro RNA Aptamer Conjugated with *Closo*-Dodecaborate as A Potential Agent for Boron Neutron Capture Therapy. *Int. J. Mol. Sci.* **2021**, *22*, 7326. [CrossRef] [PubMed]

25. Lipscomb, W.N. Framework Rearrangement in Boranes and Carboranes. *Science* **1966**, *153*, 373–378. [CrossRef] [PubMed]
26. Dobrott, R.D.; Lipscomb, W.N. Structure of $Cu_2B_{10}H_{10}$. *J. Chem. Phys.* **1962**, *37*, 1779–1784. [CrossRef]
27. Schwalbe, C.H.; Lipscomb, W.N. Structures of $B_{20}H_{182}$- and $B_{20}H_{18}NO_3$—And conformations of the triethylammonium ion. *J. Am. Chem. Soc.* **1969**, *91*, 194–196. [CrossRef]
28. Lipscomb, W.N. The Boranes and Their Relatives. *Science* **1977**, *196*, 1047–1055. [CrossRef]
29. Hoffmann, R.; Lipscomb, W.N. Theory of Polyhedral Molecules. I. Physical Factorizations of the Secular Equation. *J. Chem. Phys.* **1962**, *36*, 2179–2189. [CrossRef]
30. Hoffmann, R.; Lipscomb, W.N. Theory of Polyhedral Molecules. III. Population Analyses and Reactivities for the Carboranes. *J. Chem. Phys.* **1962**, *36*, 3489–3493. [CrossRef]
31. Hoffmann, R.; Lipscomb, W.N. Boron Hydrides: LCAO—MO and Resonance Studies. *J. Chem. Phys.* **1962**, *37*, 2872–2883. [CrossRef]
32. Wade, K. Structural and Bonding Patterns in Cluster Chemistry. *Adv. Inorg. Chem. Radiochem.* **1976**, *18*, 1–66. [CrossRef]
33. Fox, M.A.; Nervi, C.; Crivello, A.; Batsanov, A.S.; Howard, J.A.K.; Wade, K.; Low, P.J. Structural, spectroscopic, electrochemical and computational studies of C,C′-diaryl-ortho-carboranes, 1-(4-XC_6H_4)-2-Ph-1,2-$C_2B_{10}H_{10}$ (X = H, F, OMe, NMe_2, NH_2, OH and O−). *J. Solid State Electrochem.* **2009**, *13*, 1483–1495. [CrossRef]
34. Nelyubin, A.V.; Selivanov, N.A.; Bykov, A.Y.; Klyukin, I.N.; Novikov, A.S.; Zhdanov, A.P.; Karpechenko, N.Y.; Grigoriev, M.S.; Zhizhin, K.Y.; Kuznetsov, N.T. Primary Amine Nucleophilic Addition to Nitrilium Closo-Dodecaborate $[B_{12}H_{11}NCCH_3]^-$: A Simple and Effective Route to the New BNCT Drug Design. *Int. J. Mol. Sci.* **2021**, *22*, 13391. [CrossRef] [PubMed]
35. Mingos, D.M.P. Polyhedral skeletal electron pair approach. *Acc. Chem. Res.* **1984**, *17*, 311–319. [CrossRef]
36. Copley, R.C.B.; Mingos, D.M.P. The novel structure of the $[Au_{11}(PMePh_2)_{10}]^{3+}$ cation: Crystal structures of $[Au_{11}(PMePh_2)_{10}]$ $[C_2B_9H_{12}]_3$·4thf and $[Au_{11}(PMePh_2)_{10}][C_2B_9H_{12}]_3$ (thf = tetrahydrofuran). *J. Chem. Soc. Dalt. Trans.* **1996**, *17*, 479–489. [CrossRef]
37. Yan, Y.-K.; Mingos, D.M.P.; Kurmoo, M.; Li, W.-S.; Scowen, I.J.; McPartlin, M.; Coomber, A.T.; Friend, R.H. Synthesis, structure and physical properties of tetrathiafulvalenium salts of the ferracarborane complex commo-[3,3′-Fe{1-(C_4H_3S)-1,2-$C_2B_9H_{10}$}$_2$]$^-$. *J. Chem. Soc. Dalt. Trans.* **1995**, *17*, 2851–2860. [CrossRef]
38. Hertler, W.R.; Knoth, W.H.; Muetterties, E.L. Chemistry of Boranes. XXIV. Carbonylation of Derivatives of $B_{10}H_{10}^{2-}$ and $B_{12}H_{12}^{2-}$ with Oxalyl Chloride. *Inorg. Chem.* **1965**, *4*, 288–293. [CrossRef]
39. Miller, H.C.; Miller, N.E.; Muetterties, E.L. Chemistry of Boranes. XX. Syntheses of Polyhedral Boranes. *Inorg. Chem.* **1964**, *3*, 1456–1463. [CrossRef]
40. Chamberland, B.L.; Muetterties, E.L. Chemistry of Boranes. XVIII. Oxidation of $B_{10}H_{10}^{-2}$ and Its Derivatives. *Inorg. Chem.* **1964**, *231*, 1450–1456. [CrossRef]
41. Miller, H.C.; Hertler, W.R.; Muetterties, E.L.; Knoth, W.H.; Miller, N.E. Chemistry of Boranes. XXV. Synthesis andc Chemistry of Base Derivatives of $B_{10}H_{10}^{2-}$ and $B_{12}H_{12}^{2-}$. *Inorg. Chem.* **1965**, *4*, 1216–1221. [CrossRef]
42. Golub, I.; Filippov, O.; Belkova, N.; Epstein, L.; Shubina, E. The Reaction of Hydrogen Halides with Tetrahydroborate Anion and Hexahydro-*closo*-hexaborate Dianion. *Molecules* **2021**, *26*, 3754. [CrossRef]
43. Černý, R.; Brighi, M.; Murgia, F. The Crystal Chemistry of Inorganic Hydroborates. *Chemistry* **2020**, *2*, 805–826. [CrossRef]
44. Zhao, X.; Yang, Z.; Chen, H.; Wang, Z.; Zhou, X.; Zhang, H. Progress in three-dimensional aromatic-like closo-dodecaborate. *Coord. Chem. Rev.* **2021**, *444*, 214042. [CrossRef]
45. Kamin, A.A.; Juhasz, M.A. Exhaustive Cyanation of the Dodecaborate Dianion: Synthesis, Characterization, and X-ray Crystal Structure of $[B12(CN)12]^{2-}$. *Inorg. Chem.* **2019**, *59*, 189–192. [CrossRef]
46. Lacroix, M.R.; Liu, Y.; Strauss, S.H. Hydrated Metal Ion Salts of the Weakly Coordinating Fluoroanions PF_6^-, TiF_6^{2-}, $B_{12}F_{12}^{2-}$, $Ga(C_2F_5)_4^-$, $B(3,5\text{-}C_6H_3(CF_3)_2)_4^-$, and $Al(OC(CF_3)_3)_4^-$. In Search of the Weakest HOH···F Hydrogen Bonds. *Inorg. Chem.* **2019**, *58*, 14900–14911. [CrossRef] [PubMed]
47. Klyukin, I.N.; Novikov, A.S.; Zhdanov, A.P.; Zhizhin, K.Y.; Kuznetsov, N.T. QTAIM Analysis of Mono-Hydroxy Derivatives of closo-Borate Anions $[B_nH_{n-1}OH]^2$ (n = 6, 10, 12). *Russ. J. Inorg. Chem.* **2019**, *64*, 1825–1828. [CrossRef]
48. Kononova, E.; Klemenkova, Z. The electronic structure of nido-B10H14 and [6-Ph-nido-6-CB_9H_{11}] in terms of Bader's theory (AIM). *J. Mol. Struct.* **2013**, *1036*, 311–317. [CrossRef]
49. Mindich, A.L.; Bokach, N.A.; Dolgushin, F.M.; Haukka, M.; Lisitsyn, L.A.; Zhdanov, A.P.; Zhizhin, K.Y.; Miltsov, S.A.; Kuznetsov, N.T.; Kukushkin, V.Y. 1,3-Dipolar Cycloaddition of Nitrones to a Nitrile Functionality in closo-Decaborate Clusters: A Novel Reactivity Mode for the Borylated C≡N Group. *Organometallics* **2012**, *31*, 1716–1724. [CrossRef]
50. Burianova, V.K.; Mikherdov, A.S.; Bolotin, D.S.; Novikov, A.S.; Mokolokolo, P.P.; Roodt, A.; Boyarskiy, V.P.; Suslonov, V.V.; Zhdanov, A.P.; Zhizhin, K.Y.; et al. Electrophilicity of aliphatic nitrilium closo-decaborate clusters: Hyperconjugation provides an unexpected inverse reactivity order. *J. Organomet. Chem.* **2018**, *870*, 97–103. [CrossRef]
51. Bolotin, D.S.; Burianova, V.K.; Novikov, A.S.; Demakova, M.Y.; Pretorius, C.; Mokolokolo, P.P.; Roodt, A.; Bokach, N.A.; Suslonov, V.V.; Zhdanov, A.P.; et al. Nucleophilicity of Oximes Based upon Addition to a Nitrilium closo-Decaborate Cluster. *Organometallics* **2016**, *35*, 3612–3623. [CrossRef]
52. Stogniy, M.Y.; Erokhina, S.A.; Sivaev, I.B.; Bregadze, V.I. Nitrilium derivatives of polyhedral boron compounds (boranes, carboranes, metallocarboranes): Synthesis and reactivity. *Phosph. Sulf. Silicon Relat. Elem.* **2019**, *194*, 983–988. [CrossRef]
53. Golub, I.E.; Filippov, O.A.; Belkova, N.V.; Epstein, L.M.; Shubina, E.S. The Mechanism of Halogenation of Decahydro-closo-Decaborate Dianion by Hydrogen Chloride. *Russ. J. Inorg. Chem.* **2021**, *66*, 1639–1648. [CrossRef]

54. Shore, S.G.; Hamilton, E.J.M.; Bridges, A.N.; Bausch, J.; Krause-Bauer, J.A.; Dou, D.; Liu, J.; Liu, S.; Bin, D.; Hall, H.; et al. The Solid State Structure of $[B_{10}H_{11}]^-$ and Its Dynamic NMR Spectra in Solution. *Inorg. Chem.* **2003**, *42*, 1175–1186. [CrossRef] [PubMed]
55. Kochnev, V.K.; Avdeeva, V.V.; Malinina, E.A.; Kuznetsov, N.T. Theoretical study of H2 elimination from $[B_nH_{n+1}]$ monoanions (n = 6–9, 11). *Russ. J. Inorg. Chem.* **2014**, *59*, 1268–1275. [CrossRef]
56. Sun, Y.; Zhang, J.; Zhang, Y.; Liu, J.; Van Der Veen, S.; Duttwyler, S. The closo-Dodecaborate Dianion Fused with Oxazoles Provides 3D Diboraheterocycles with Selective Antimicrobial Activity. *Chem. A Eur. J.* **2018**, *24*, 10364–10371. [CrossRef] [PubMed]
57. Cao, K.; Zhang, C.-Y.; Xu, T.-T.; Wu, J.; Wen, X.-Y.; Jiang, W.-J.; Chen, M.; Yang, J. Synthesis of Polyhedral Borane Cluster Fused Heterocycles via Transition Metal Catalyzed B-H Activation. *Molecules* **2020**, *25*, 391. [CrossRef]
58. Zhu, T.-C.; Xing, Y.-Y.; Sun, Y.; Duttwyler, S.; Hong, X. Directed B–H functionalization of the closo-dodecaborate cluster via concerted iodination—deprotonation: Reaction mechanism and origins of regioselectivity. *Org. Chem. Front.* **2020**, *7*, 3648–3655. [CrossRef]
59. Voinova, V.V.; Selivanov, N.A.; Plyushchenko, I.V.; Vokuev, M.F.; Bykov, A.Y.; Klyukin, I.N.; Novikov, A.S.; Zhdanov, A.P.; Grigoriev, M.S.; Rodin, I.A.; et al. Anion—Novel Members of Diboroheterocycle Class. *Molecules* **2021**, *26*, 248. [CrossRef]
60. Klyukin, I.; Zhdanov, A.; Matveev, E.; Razgonyaeva, G.; Grigoriev, M.; Zhizhin, K.; Kuznetsov, N. Synthesis and reactivity of closo -decaborate anion derivatives with multiple carbon-oxygen bonds. *Inorg. Chem. Commun.* **2014**, *50*, 28–30. [CrossRef]
61. Laskova, J.; Ananiev, I.; Kosenko, I.; Serdyukov, A.; Stogniy, M.Y.; Sivaev, I.B.; Grin, M.A.; Semioshkin, A.; Bregadze, V.I. Nucleophilic addition reactions to nitrilium derivatives $[B_{12}H_{11}NCCH_3]^-$ and $[B_{12}H_{11}NCCH_2CH_3]$. Synthesis and structures of closo-dodecaborate-based iminols, amides and amidines. *Dalt. Trans.* **2022**, *51*, 3051–3059. [CrossRef]
62. Avdeeva, V.V.; Polyakova, I.N.; Goeva, L.V.; Malinina, E.A.; Kuznetsov, N.T. $[2,6(9)\text{-}B_{10}H_8(O)_2CCH_3]^-$ and [2,7(8)- $B_{10}H_8(OC(O)CH_3)_2]^{2-}$ derivatives in synthesis of position isomers of the $[B_{10}H_8(OC(O)CH_3)(OH)]^{2-}$ anion with the 2,6(9)- and 2,7(8)-arrangement of functional groups. *Russ. J. Inorg. Chem.* **2014**, *59*, 1247–1258. [CrossRef]
63. Klyukin, I.; Kubasov, A.; Limarev, I.; Zhdanov, A.; Matveev, E.; Polyakova, I.; Zhizhin, K.; Kuznetsov, N. The new approach to formation of exo boron-oxygen bonds from the decahydro-closo-decaborate(2-) anion. *Polyhedron* **2015**, *101*, 215–222. [CrossRef]
64. Oller, J.; Pérez, P.; Ayers, P.W.; Vöhringer-Martinez, E. Global and local reactivity descriptors based on quadratic and linear energy models forα,β-unsaturated organic compounds. *Int. J. Quant. Chem.* **2018**, *118*, 1–13. [CrossRef]
65. Gómez, T.; Fuentealba, P.; Robles-Navarro, A.; Cárdenas, C. Links among the Fukui potential, the alchemical hardness and the local hardness of an atom in a molecule. *J. Comput. Chem.* **2021**, *42*, 1681–1688. [CrossRef] [PubMed]
66. Khanum, G.; Ali, A.; Shabbir, S.; Fatima, A.; Alsaiari, N.; Fatima, Y.; Ahmad, M.; Siddiqui, N.; Javed, S.; Gupta, M. Vibrational Spectroscopy, Quantum Computational and Molecular Docking Studies on 2-[(1H-Benzimidazol-1-yl)-methyl]benzoic Acid. *Crystals* **2022**, *12*, 337. [CrossRef]
67. Klyukin, I.N.; Vlasova, Y.S.; Novikov, A.S.; Zhdanov, A.P.; Hagemann, H.R.; Zhizhin, K.Y.; Kuznetsov, N.T. B-F bonding and reactivity analysis of mono- and perfluoro-substituted derivatives of closo-borate anions (6, 10, 12): A computational study. *Polyhedron* **2022**, *211*, 115559. [CrossRef]
68. Klyukin, I.; Vlasova, Y.; Novikov, A.; Zhdanov, A.; Zhizhin, K.; Kuznetsov, N. Theoretical Study of *closo*-Borate Anions $[B_nH_n]^{2-}$ (n = 5–12): Bonding, Atomic Charges, and Reactivity Analysis. *Symmetry* **2021**, *13*, 464. [CrossRef]
69. Hehre, W.J.; Ditchfield, R.; Pople, J.A. Self—Consistent Molecular Orbital Methods. XII. Further Extensions of Gaussian—Type Basis Sets for Use in Molecular Orbital Studies of Organic Molecules. *J. Chem. Phys.* **1972**, *56*, 2257–2261. [CrossRef]
70. Grimme, S.; Antony, J.; Ehrlich, S.; Krieg, H. A consistent and accurate ab initio parametrization of density functional dispersion correction (DFT-D) for the 94 elements H-Pu. *J. Chem. Phys.* **2010**, *132*, 154104–154119. [CrossRef]
71. Neese, F. The ORCA program system. *Wiley Interdiscip. Rev. Comput. Mol. Sci.* **2012**, *2*, 73–78. [CrossRef]
72. Gonzalez, C.; Schlegel, H.B. Improved algorithms for reaction path following: Higher-order implicit algorithms. *J. Chem. Phys.* **1991**, *95*, 5853–5860. [CrossRef]
73. Gonzalez, C.; Schlegel, H.B. An improved algorithm for reaction path following. *J. Chem. Phys.* **1989**, *90*, 2154–2161. [CrossRef]
74. Gonzalez, C.; Schlegel, H.B. Reaction path following in mass-weighted internal coordinates. *J. Phys. Chem.* **1990**, *94*, 5523–5527. [CrossRef]
75. Marenich, A.V.; Cramer, C.J.; Truhlar, D.G. Universal Solvation Model Based on Solute Electron Density and on a Continuum Model of the Solvent Defined by the Bulk Dielectric Constant and Atomic Surface Tensions. *J. Phys. Chem. B* **2009**, *113*, 6378–6396. [CrossRef] [PubMed]
76. Glendening, E.D.; Landis, C.R.; Weinhold, F. NBO 7.0: New vistas in localized and delocalized chemical bonding theory. *J. Comput. Chem.* **2019**, *40*, 2234–2241. [CrossRef] [PubMed]
77. Lu, T.; Chen, F. Multiwfn: A multifunctional wavefunction analyzer. *J. Comput. Chem.* **2011**, *33*, 580–592. [CrossRef] [PubMed]
78. Adrienko, G.A. Chemcraft, Version 1.8 (Build 530). 2017. Available online: http://www.chemcraftprog.com (accessed on 9 April 2022).

International Journal of
Molecular Sciences

MDPI

Article

Vibrational Spectra and Molecular Vibrational Behaviors of Dibenzyl Disulfide, Dibenzyl Sulphide and Bibenzyl

Ziyi Wang, Ruimin Song, Weigen Chen *, Jianxin Wang, Pinyi Wang, Zhixian Zhang, Xinyuan Zhang and Fu Wan

State Key Laboratory of Power Transmission Equipment & System Security and New Technology, Chongqing University, Chongqing 400044, China; wzycsust@163.com (Z.W.); ruiminsong@cqu.edu.cn (R.S.); wang.jianxin@cqu.edu.cn (J.W.); wang.pinyi@outlook.com (P.W.); zhang.zhixian@cqu.edu.cn (Z.Z.); 18846116694@163.com (X.Z.); fuwan@cqu.edu.cn (F.W.)

* Correspondence: weigench@cqu.edu.cn

Abstract: The vibration spectroscopy (Raman and infrared) of widely concerned molecules in sulfur corrosion phenomenon (Dibenzyl Disulfide, Dibenzyl Sulphide, and Bibenzyl) is detailedly analyzed based on density functional theory and experimental measurement. The dominant conformations of these molecules are determined according to Boltzmann distribution in relative Gibbs free energy. Additionally, noncovalent interaction analysis is conducted to indicate intramolecular interaction. Vibration normal mode is assigned based on potential energy distribution, which comprehensively reveals the molecular vibrational behaviors. Conformations weighted spectra are obtained and compared with experimentally measured spectra. We found that experimental spectra are in good agreement with the theoretical spectra in B3LYP-D3(BJ)/6-311G** level with a frequency correction factor. Furthermore, the divergence among these molecules is discussed. The vibrational behavior of the methylene group in the molecule shows a trend with the presence of the sulfur atom.

Keywords: Raman spectroscopy; infrared spectroscopy; corrosive sulfur; density functional theory; vibration assignment

1. Introduction

Nowadays, the sulfur corrosion problem of power equipment has received widespread concern [1–5]. The failures caused by corrosive sulfur greatly affect the safety and stability of the power system and result in substantial economic losses. Dibenzyl Disulfide (DBDS) is regarded as the most important corrosive sulfur. The corrosion mechanism was proposed by Mitsubishi Electric Corporation as shown in Figure 1 [6]. The intermediate product Dibenzyl Sulphide (DBS) is also corrosive, and the final product of this reaction is Bibenzyl (BiBz) and cuprous sulfide. The DBDS, DBS, and BiBz are considered as indicators in power equipment sulfur corrosion phenomenon. The study on their physical and chemical properties will contribute to the development of advanced corrosive resistant materials and the realization of corrosive state monitoring.

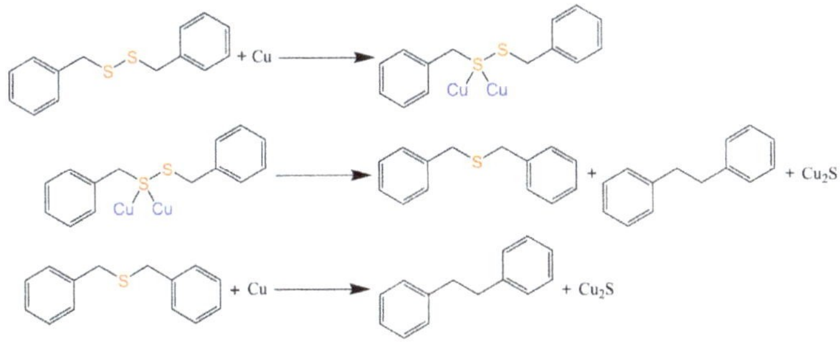

Figure 1. The corrosion mechanism of DBDS in power equipment.

Numerous efforts have been made to study properties of the DBDS, DBS, and BiBz, including chemical reactivity [7,8], synergistic effect [9,10], detection methods [6,11], etc. [12,13]. In this paper, the vibrational spectra and molecular vibrational behaviors are analyzed detailedly. The vibrational spectroscopy includes infrared and Raman spectroscopy, which is based on periodic changes of dipolmoments (IR) or polarizabilities (Raman) caused by molecular vibrations behaviors. IR and Raman spectra are both widely used methods for composition analysis and trace detection [14–16].

In addition to experimentally measured vibrational spectra, density functional theory (DFT) provides a solid foundation of theoretically calculated spectra. Considering the presence of a large number of fundamentals and overtones bands, it is difficult to interpret experimental spectra of large molecule [17]. However, the conformational distribution and theoretical spectrum of molecules can be obtained based on DFT. Then, a potential energy distribution (PED) based analysis can be performed to achieve vibration assignment, which gives a clear and crispy concept about the structural–vibrational properties of the molecule. A lot of studies have recently appeared in the literature concerning the calculation of spectroscopic properties by DFT methods [17–22].

2. Experimental Section

A commercially available Fourier transform infrared spectroscopy (FTIR, Nicolet iS50 FT-IR) was utilized for experimental test. The DBDS (CAS:150-60-7, 98%, Aladdin, China), DBS (CAS:538-74-9, 99%, Aladdin, China), and BiBz (CAS:103-29-7, 99%, Aladdin, China) samples were mixed with KBr (CAS:7758-02-3, 99.5%, Aladdin, China) and ground in a agate mortar for 1 min. Then, the mixed sample was pressurized by a tablet press at 20 MPa for 2 min for transmission testing. The IR spectra were measured between 400 cm^{-1} to 4000 cm^{-1} at a 4 cm^{-1} resolution. The spectra were an average of 32 scans.

A confocal Raman spectrometer mentioned in our previous paper [23] was used to detect experimental spectra of the DBDS, DBS, and BiBz. There are three kinds of gratings (600 lines/mm, 1200 lines/mm, and 1800 lines/mm) built in the spectrometer. The grating with minor grooves number is designed for wide spectral range detection, although the spectral resolution will decrease. In order to compare with theoretical spectra in full spectral range, 600 lines/mm grating was adopted. The other test parameters of the measurement are listed below: the slit width was 50 um, the integration time was 1 s, and the integration number was 20 times. The Raman platform was calibrated with monocrystalline silicon before each test. The Raman shift of the first-order silicon peak was adjusted to 520.7 cm^{-1}. The result of Raman spectrum comes from the average of three measurements.

The Raman/IR spectra of DBDS, DBS, and BiBz were plotted in Section 4 after normalization and de-baseline operations.

3. Molecular Structure and Intramolecular Interaction

3.1. Conformation Determination

Flexible molecules always have multiple conformations, such as DBDS, DBS, and BiBz studied in this paper. The proportion of different conformations was calculated according to Boltzmann distribution in relative Gibbs free energy. The molecular conformation search program Molclus [24] is used for conformation determination in this paper.

Firstly, a batch of initial conformations of DBDS, DBS, and BiBz were generated by rotating chemical bond via Molclus. The chemical bond within the benzene ring was ignored; only C-C bonds, C-S bonds, and S-S bonds between benzene rings are considered. All the considered bonds rotate every 120°. The DBDS with five rotation bonds generates a total of 3^5 initial conformations. Similarly, DBS generates 3^4 initial conformations. BiBz generates 3^3 initial conformations.

All initial conformations generated above were pre-optimized geometrically by Gaussian [25] in a semi-empirical method PM7 [26]. The optimized conformations with the energy difference within 0.25 kcal/mol and the geometric difference within 0.1 Å will be classified as the same conformation. The conformations with relatively lower energy were applied further geometric optimization and frequency analysis in B3LYP-D3(BJ)/6-311G** [27–31] level. For an authentic conformation distribution, a double hybrid functional revDSD-PBEP86-D3(BJ) [32] in conjunction with may-cc-PVTZ [33] was introduced to re-calculate the single point energy. The single point can be added with thermal correction to find the precise Gibbs free energy. The thermal correction is obtained in frequency analysis based on rigid-rotor harmonic oscillator (RRHO) model [34] under 298.15K and 1 atm. The results are shown in Table 1.

Table 1. Energy and conformation proportion of DBDS, DBS, and BiBz.

Molecule	Conformation	Single Point Energy (a.u.)	Thermal Correction (a.u.)	Gibbs Free Energy (a.u.)	Proportion
BiBz	1	−541.04114860	0.198497	−540.8426516	55.17%
	2	−541.03965980	0.197204	−540.8424558	44.83%
DBS	1	−938.9390301	0.196498	−938.7425321	91.45%
	2	−938.9372458	0.197657	−938.7395888	4.50%
	3	−938.9338298	0.194142	−938.7396878	4.05%
DBDS	1	−1336.847129	0.195781	−1336.651348	53.12%
	2	−1336.847175	0.195965	−1336.65121	45.89%
	3	−1336.844026	0.197072	−1336.646954	0.51%
	4	−1336.843633	0.196717	−1336.646916	0.49%

Conformations in Table 1 with a proportion less than 5% were ignored. The geometric structures and surface electrostatic potential of several dominant conformations were visualized via VMD [35], as shown in Figure 2. The electrostatic potential involved was evaluated by Multiwfn, based on the highly effective algorithm proposed in ref. [36].

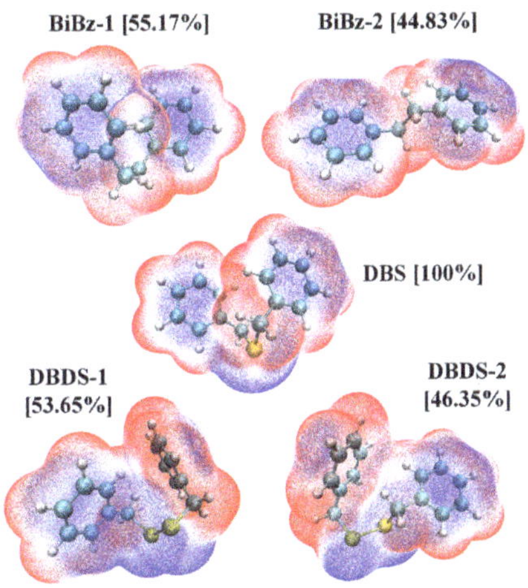

Figure 2. The geometric structure and surface electrostatic potential of the dominant conformation.

It can be found from the surface electrostatic potential distribution that the negative charge region mainly occurs around the sulfur atom and the benzene ring planar center. It is difficult to explain the geometric structure only by the surface electrostatic potential distribution due to the existence of various intramolecular interaction. Further NCI analysis is thus needed.

3.2. Noncovalent Interaction Analysis of the Dominant Conformation

To reveal the nature of these dominant conformations with low free energy, the noncovalent interaction (NCI) [37] method was applied for weak molecular interactions analysis.

The NCI method uses Reduced Density Gradient (RDG) to distinguish regions with noncovalent interactions in the system. The expression of RDG is as follows, where ρ is the electron density.

$$\mathrm{RDG} = \frac{1}{2 \cdot \sqrt[3]{3\pi^2}} \cdot \frac{|\nabla \rho|}{\rho^{4/3}} \tag{1}$$

The noncovalent interactions areas can be determined by RDG value. However, the strength and type of interaction are determined by ρ and $sign(\lambda_2)$, respectively. The $sign(\lambda_2)$ is the second largest eigenvalue of the electron density Hessian matrix. Overlapping the color map of $sign(\lambda_2) \cdot \rho$, as shown in Figure 3, on the RDG value isosurface, means the areas, intensity, and type of the noncovalent interactions can be visualized at the same time.

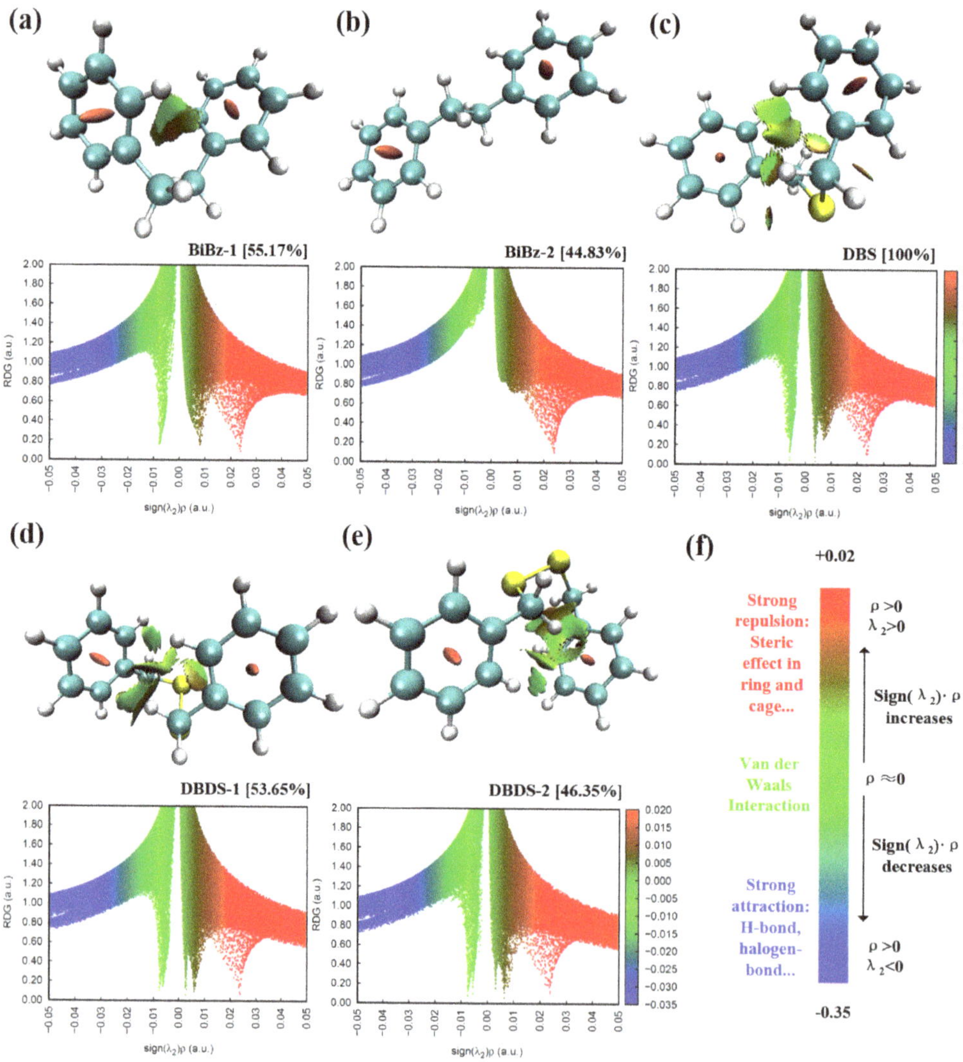

Figure 3. NCI analysis of dominant conformation. (**a**) RDG isosurface map and scatter plot of BiBz-1, (**b**) Bibz-2, (**c**) DBS, (**d**) DBDS-1, and (**e**) DBDS-2. (**f**) color map and chemical explanation.

The isosurface map and scatter plot of five dominant conformations were visualized via Multiwfn [38,39] and VMD [35]. As described in Figure 3f, the red area in the isosurface map and the scatter plot indicates the repulsion between molecules, such as steric effect. The green area indicates the van der Waals interaction, while the blue area indicates hydrogen bonds and halogen bonds that have a significant attraction.

There are two common features in the five dominant conformations. Firstly, significant steric effect is observed in the two benzene rings. This steric effect is shown as a red spindle-shaped area in the benzene ring on the RDG isosurface map, and as a red spike on the right side in the scatter plot. Secondly, there is no notable attraction in these conformations.

For BiBz-1, the van der Waals interaction is formed between the two benzene rings. The molecular structure of BiBz-2 is stretched, the two benzene rings are far apart, and no significant van der Waals effect is observed. This was shown as two spikes near the zero point on the scatter plot. The spikes can be positive or negative as the sign of $sign(\lambda_2) \cdot \rho$ is relatively unstable as the electron density in the van der Waals area is small. The BiBz-1 formed by two benzene rings attracting each other is more stable and has a lower single point energy due to intermolecular forces. However, as BiBz-2 has a lower thermal correction, the Gibbs free energy and proportion of these two conformations are not remarkably different.

For DBS, in addition to the zone between benzene rings, the van der Waals effect also occurs between the sulfur atom and the hydrogen atom on the benzene ring. However, this kind of H-S interaction is distinguished from hydrogen bonds (shown in blue). The RDG isosurface between the two benzene rings in DBDS is much more complicated, but in conclusion, it is still van der Waals interaction. These broken isosurfaces were shown as many fine spikes near the zero point on the scatter plot in Figure 3d,e.

4. Vibration Analysis and Vibrational Spectra

4.1. PED Based Vibrational Assignment

Each characteristic group in molecules has its own unique vibration behavior. However, these groups will be coupled with the molecular systems, which change frequency and intensity. Non-linear molecules have 3n−6 normal vibration modes, where n is the number of atoms in the molecule. The normal modes are the coupling of multiple groups vibration. It is difficult to directly identify which groups and vibration types contribute to the normal mode.

PED (potential energy distribution) analysis is an effective method that decomposes the normal vibration mode into group characteristic vibrations. The contribution proportion of the group vibration can be obtained, and the vibrational features of normal mode can be revealed. In this paper, PED based vibrational assignment of DBDS, DBS, and BiBz was performed via VEDA4 program [40].

The non-linear molecule DBDS consists of 30 atoms, it has a total of 84 normal modes. Therefore, 84 linearly independent internal coordinates are introduced to describe molecular vibration. The internal coordinates and related atoms of the DBDS-1 conformation are shown in Table 2. Atom numbers are listed in Figure 4. Additionally, the internal coordinates and related information of other four conformations can be found in Supplementary Material. The vibration type in VEDA4 program are divided into stretching (ν), bending (δ), and torsion (τ/γ), which correspond to specific internal coordinates. The stretching vibration involves two atoms. Bending vibration involves three atoms and torsion vibration involves four atoms. Among them, torsion can be further divided into two types: τ and γ. In many cases, the two definitions can describe the same atoms movement, but not always. τ ABCD means the dihedral angle between the ABC and BCD planes. γ ABCD means the angle between the AD vector and the BCD plane (reffed from VEDA4 user document).

Table 2. Internal coordinates and related atoms in DBDS-1.

Stretching	Related Atoms	Bending	Related Atoms	Torsion	Related Atoms
ν1	C(1)-H(7)	δ30	C(22)-C(25)-C(27)	τ58	H(7)-C(1)-C(2)-C(3)
ν2	C(2)-H(8)	δ31	C(3)-C(12)-S(15)	τ59	H(8)-C(2)-C(1)-C(6)
ν3	C(4)-H(9)	δ32	H(7)-C(1)-C(2)	τ60	H(9)-C(4)-C(5)-C(6)
ν4	C(5)-H(10)	δ33	H(8)-C(2)-C(1)	τ61	H(10)-C(5)-C(6)-C(1)
ν5	C(6)-H(11)	δ34	H(9)-C(4)-C(5)	τ62	H(11)-C(6)-C(1)-C(2)
ν6	C(12)-H(13)	δ35	H(10)-C(5)-C(6)	τ63	H(13)-C(12)-S(15)-S(16)
ν7	C(12)-H(14)	δ36	H(11)-C(6)-C(1)	τ64	H(14)-C(12)-S(15)-S(16)
ν8	C(17)-H(18)	δ37	H(13)-C(12)-S(15)	τ65	H(18)-C(17)-S(16)-S(15)
ν9	C(17)-H(19)	δ38	H(14)-C(12)-H(13)	τ66	H(19)-C(17)-S(16)-S(15)
ν10	C(21)-H(24)	δ39	H(18)-C(17)-S(16)	τ67	H(24)-C(21)-C(23)-C(27)
ν11	C(22)-H(26)	δ40	H(19)-C(17)-H(18)	τ68	H(26)-C(22)-C(25)-C(27)
ν12	C(23)-H(28)	δ41	H(24)-C(21)-C(23)	τ69	H(28)-C(23)-C(27)-C(25)
ν13	C(25)-H(29)	δ42	H(26)-C(22)-C(25)	τ70	H(29)-C(25)-C(27)-C(23)
ν14	C(27)-H(30)	δ43	H(28)-C(23)-C(27)	τ71	H(30)-C(27)-C(25)-C(22)
ν15	C(2)-C(1)	δ44	H(29)-C(25)-C(27)	τ72	C(2)-C(1)-C(6)-C(5)
ν16	C(22)-C(25)	δ45	H(30)-C(27)-C(25)	τ73	C(22)-C(25)-C(27)-C(23)
ν17	C(1)-C(6)	δ46	C(2)-C(1)-C(6)	τ74	C(1)-C(6)-C(5)-C(4)
ν18	C(5)-C(6)	δ47	C(1)-C(6)-C(5)	τ75	C(6)-C(5)-C(4)-C(3)
ν19	C(27)-C(23)	δ48	C(6)-C(5)-C(4)	τ76	C(4)-C(3)-C(12)-C(15)
ν20	C(4)-C(3)	δ49	C(4)-C(3)-C(12)	τ77	C(5)-C(4)-C3)-C(12)
ν21	C(5)-C(4)	δ50	C(5)-C(4)-C(3)	τ78	C(20)-C(21)-C(23)-C(27)
ν22	C(20)-C(21)	δ51	C(20)-C(21)-C(23)	τ79	C(25)-C(27)-C(23)-C(21)
ν23	C(21)-C(23)	δ52	C(21)-C(23)-C(27)	τ80	C(3)-C(12)-S(15)-S(16)
ν24	C(25)-C(27)	δ53	C(25)-C(27)-C(23)	τ81	C(12)-S(15)-S(16)-C(17)
ν25	C(3)-C(12)	δ54	C(17)-C(20)-C(22)	τ82	S(16)-C(17)-C(20)-C(21)
ν26	C(17)-C(20)	δ55	C(12)-S(15)-S(16)	τ83	S(15)-S(16)-C(17)-C(20)
ν27	S(15)-C(12)	δ56	S(16)-C(17)-C(20)	γ84	C(17)-C(21)-C(22)-C(20)
ν28	S(16)-C(17)	δ57	S(15)-S(16)-C(17)		
ν29	S(15)-S(16)				

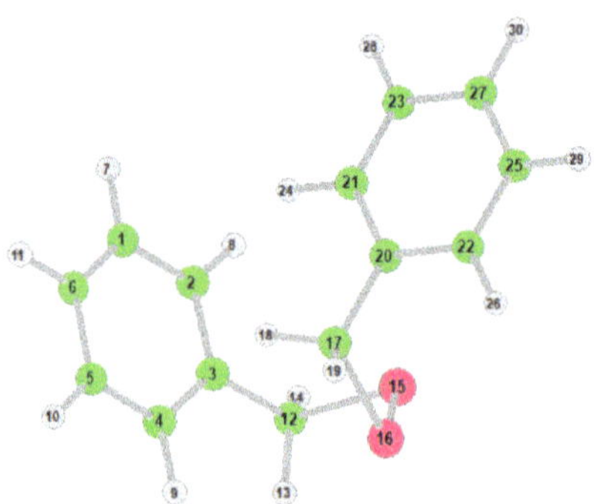

Figure 4. The atomic number of DBDS in VEDA4.

The internal coordinates of DBDS-1 conformation consist of 29 kinds of stretching, 28 kinds of bending, and 27 kinds of torsion. The contribution of different internal coordinates to the normal vibration mode is shown in Tables 3 and 4. Similarly, the detailed table of other conformations is provided in Supplementary Material. It should be noted that the internal coordinates contribution less than 10% are ignored. The normal modes of DBDS-1 are divided into low frequency (<400 cm^{-1}), middle frequency (400~2800 cm^{-1}), and high frequency (>2800 cm^{-1}). The scaled frequency was corrected by a factor of 0.9640 [41].

Table 3. Assignment of high frequency normal mode of DBDS-1.

Normal Mode	Scaled Frequency (cm^{-1})	Vibration Assignment (>10%)
#1	3077.87	%14ν(1) + %11ν(3) + %25ν(4) + %34ν(5)
#2	3077.32	%40ν(10) + %20ν(12) + %16ν(14)
#3	3071.02	%38ν(10) + %31ν(13) + %24ν(14)
#4	3068.73	%27ν(1) + %41ν(3) + %15ν(4)
#5	3061.73	%23ν(1) + %22ν(2) + %24ν(3) + %23ν(5)
#6	3061.38	%23ν(11) + %22ν(12) + %25ν(13) + %21ν(14)
#7	3053.17	%22ν(2) + %20ν(3) + %42ν(4) + %16ν(5)
#8	3051.51	%36ν(11) + %40ν(12) + %14ν(14)
#9	3046.70	%33ν(1) + %47ν(2) + %12ν(5)
#10	3045.33	%33ν(11) + %14ν(12) + %30ν(13) + %21ν(14)
#11	3024.71	%88ν(8) + %11ν(9)
#12	3007.35	%77ν(6) + %23ν(7)
#13	2946.40	%22ν(6) + %77ν(7)
#14	2942.76	%11ν(8) + %88ν(9)

Table 4. Assignment of middle frequency normal mode of DBDS-1.

Normal Mode	Scaled Frequency (cm^{-1})	Vibration Assignment (>10%)
#15	1590.90	%30ν(16) + %10ν(19)
#16	1588.63	%30ν(15) + %11ν(18)
#17	1571.78	%28ν(19) + %21ν(22)
#18	1569.58	%27ν(18) + %23ν(20)
#19	1476.26	%16δ(41) + %15δ(42) + %16δ(43) + %18δ(44)
#20	1474.92	%19δ(32) + %15δ(33) + %15δ(34) + %16δ(35)
#21	1437.59	%11δ(40) + %20δ(45)
#22	1434.55	%10ν(21) + %10δ(35) + %24δ(36)
#23	1420.17	%53δ(40) + %12τ(66)
#24	1415.30	%60δ(38) + %12τ(63) + %16τ(64)
#25	1313.79	%27δ(41) + %24δ(42) + %11δ(44) + %10δ(45)
#26	1311.88	%11δ(32) + %25δ(33) + %26δ(34) + %10δ(35) + %10δ(36)
#27	1296.69	%20ν(16) + %22ν(22) + %17ν(23) + %16ν(24)
#28	1295.12	%18ν(15) + %17ν(17) + %24ν(20) + %16ν(21)
#29	1225.70	%12δ(39) + %18δ(40) + %12τ(65) + %16τ(66)
#30	1220.57	%20δ(38) + %11τ(63) + %14τ(64)
#31	1190.12	%22ν(25) + %14ν(26)
#32	1188.81	%12ν(25) + %22ν(26)
#33	1162.78	%15δ(41) + %16δ(42) + %14δ(43) + %11δ(44)
#34	1160.99	%13δ(32) + %16δ(33) + %15δ(34) + %13δ(35)
#35	1141.18	%18δ(32) + %19δ(35) + %38δ(36)
#36	1140.39	%19δ(43) + %19δ(44) + %37δ(45)
#37	1126.57	%33δ(39) + %10τ(65) + %34τ(66)
#38	1123.89	%35δ(37) + %10τ(63) + %31τ(64)
#39	1062.59	%13ν(16) + %15ν(23) + %10δ(39) + %10δ(42) + %13δ(45)
#40	1060.74	%13ν(15) + %15ν(21) + %12δ(36) + %11δ(37)
#41	1015.83	%19ν(19) + %24ν(24) + %13δ(30) + %11δ(43) + %10δ(44)
#42	1014.97	%24ν(17) + %21ν(18) + %11δ(35) + %15δ(46)
#43	985.97	%10ν(19) + %11ν(24) + %26δ(30) + %13δ(52) + %18δ(53)
#44	985.52	%11ν(17) + %10ν(18) + %25δ(46) + %19δ(47) + %13δ(48)
#45	967.00	%22τ(58) + %15τ(61) + %26τ(62) + %16τ(72)
#46	964.61	%20τ(69) + %18τ(70) + %30τ(71) + %10τ(73)
#47	948.26	%17τ(58) + %16τ(59) + %15τ(60) + %26τ(61)
#48	943.31	%19τ(67) + %17τ(68) + %23τ(69) + %26τ(70)
#49	897.82	%21τ(67) + %19τ(68) + %24τ(71)
#50	894.49	%20τ(59) + %23τ(60) + %24τ(62)
#51	864.65	%14δ(39) + %11τ(63) + %20τ(65)
#52	862.53	%17δ(37) + %20τ(63) + %11τ(65)
#53	826.45	%22τ(58) + %22τ(59) + %23τ(60) + %20τ(61)

Table 4. Cont.

Normal Mode	Scaled Frequency (cm^{-1})	Vibration Assignment (>10%)
#54	823.51	$\%21\tau(67) + \%24\tau(68) + \%21\tau(69) + \%21\tau(70)$
#55	797.93	$\%19\nu(26) + \%10\delta(52) + \%30\delta(53)$
#56	794.54	$\%18\nu(25) + \%30\delta(47)$
#57	750.92	$\%17\tau(58) + \%15\tau(61) + \%14\tau(62) + \%21\tau(75)$
#58	745.18	$\%14\tau(69) + \%13\tau(70) + \%11\tau(71) + \%20\tau(78) + \%13\gamma(84)$
#59	687.46	$\%10\tau(59) + \%10\tau(60) + \%15\tau(62) + \%10\tau(72) + \%19\tau(75)$
#60	685.91	$\%10\tau(68) + \%14\tau(71) + \%25\tau(78)$
#61	634.47	$\%50\nu(27)$
#62	628.24	$\%41\nu(28)$
#63	614.53	$\%23\delta(46) + \%33\delta(48) + \%19\delta(50)$
#64	613.15	$\%13\nu(28) + \%20\delta(30) + \%27\delta(52)$
#65	558.03	$\%20\delta(51) + \%14\delta(56)$
#66	551.85	$\%16\delta(31) + \%19\delta(50)$
#67	468.56	$\%11\nu(27) + \%17\nu(29) + \%22\tau(74)$
#68	462.98	$\%30\nu(29) + \%14\gamma(84)$
#69	454.94	$\%51\nu(29)$

Modes #1~#14 are high frequency normal modes. All of them are contributed by stretching vibration, and there is a large frequency gap between #14 and #15. They have completely different characteristics in the frequency band. This is because the force constant of the stretching vibration is larger. The stretching vibration energy is greater than bending and torsion vibration energy. Therefore, the characteristic frequencies corresponding to the stretching vibration are all located in the high frequency region. Generally, almost all molecules have abundant vibration peaks corresponding to stretching vibrations in the high frequency region. The main feature of the low frequency zone is that no stretching contribution is assigned, while the middle frequency area is composed of various vibration contributions. The spectrum characteristics in the low frequency region is ignored in the following analysis. As the measurement range of FTIR only covers 400–4000 cm^{-1}, and the range of Raman spectroscopy in the low frequency zone is seriously affected by Rayleigh scattering.

The normal modes always appear in pairs. This is because DBDS is a quasi-symmetric molecule. Although the DBDS-1 conformation does not have molecular symmetry, its molecular composition is still symmetrical. For example, the vibration mode of one benzene ring must have the corresponding vibration mode of another benzene ring.

4.1.1. Methylene(-CH_2) Group

In the high frequency zone, the asymmetric stretching vibration of the two -CH2 groups corresponds to the #11 and #12 normal modes. The two mode peaks appear at 3007 cm^{-1} and 3025 cm^{-1} respectively, due to the vibration coupling (the calculated Raman and IR spectra are shown in Figures S4 and S5, respectively). The infrared activity of #12 mode is weak, almost invisible, in the IR spectrum. The symmetric stretching vibration of the two -CH2 groups corresponds to the #13 and #14 normal modes. These two modes appear as the same peak at 2946 cm^{-1}. The frequency of -CH2 symmetrical vibration is often smaller than that of asymmetrical vibration [21].

In the middle frequency zone, the scissoring vibration of -CH2 appears at 1416 cm^{-1}, corresponding to mode #23 and #24. When wagging vibration of -CH2 appears at 1223 cm^{-1}, corresponding to mode #29 and #30. High Raman activity and infrared activity are simultaneously exhibited in wagging vibration. The twisting vibration of -CH2 corresponds to #37 and #38 at 1126 cm^{-1}. The rocking vibration of -CH2 corresponds to #51 and #52 at 863 cm^{-1}. The activity of these four modes is relatively weak.

For the -CH2 group, the order of normal vibration frequency from large to small is: asymmetric stretch, symmetric stretch, scissoring, wagging, twisting, and rocking.

4.1.2. Phenyl(-C_5H_6) Group

In the high frequency zone, the #1 and #2 normal modes correspond to the in-phase stretching vibration of the C-H bond on the benzene ring. For both #1 and #2 modes, the five C-H bonds on the benzene ring will vibrate in the synchronous phase. Two modes appear as the same peak at 3077 cm^{-1}. This characteristic peak is the strongest peak in the Raman spectrum, but not in the infrared spectrum. The eight #3~#10 normal modes correspond to the asynchronous phase vibration of the C-H bonds on the benzene ring, and the Raman activity and infrared activity of these eight modes are quite different. Modes #3 and #4 appear at 3069 cm^{-1} in the IR spectrum. Modes #5 and #6 appear at 3062 cm^{-1} in both the IR and Raman spectrum. Modes #7 and #8 appear as a peak at 3051 cm^{-1} in the Raman spectrum.

In the middle frequency zone, the stretching vibration of the carbon skeleton of the benzene ring corresponds to modes #15~#18, #27, and #28. The frequency of these stretching vibration modes is much smaller than the C-H bond stretching vibration. This shows that the vibration force constant of the carbon skeleton is much smaller than the force constant of the C-H bond. The bending vibration of the hydrogen atoms on the benzene ring corresponds to #19~#22, #25, #26, #33~#36, #39, and #40. Modes #31 and #32 have very strong coupling characteristics. All atoms except disulfide bonds are involved, and both apparent carbon skeleton stretching and hydrogen atoms bending occurred.

The #41 mode (frequency < 1016 cm^{-1}) began to exhibit bending of the benzene ring carbon skeleton The bending and stretching vibrations of the carbon skeleton of the benzene ring simultaneously appear in the #41 and #42 modes. The #43, #44, #55, #56, and #63~#66 modes correspond to bending vibration of benzene ring carbon skeleton. High Raman activity is exhibited in these two modes, appearing as peak at 986 cm^{-1}, which is also the characteristic peak used for Raman spectroscopy based DBDS detection in our study in progress.

From the #45 mode (frequency < 967 cm^{-1}), the out-of-plane bending of hydrogen atoms on benzene ring occurs. Modes #45~#50, #53 and #54, and #57~#60 correspond to this type of vibration. Modes #67~#69 correspond to the deformation vibration of the benzene ring couple with the stretching vibration of disulfide bond. The other out-of-plane vibration of the benzene ring appeared in the low frequency zone.

The #61 and #62 modes are composed of multiple vibrations, mainly including the in-plane bending, out-of-plane bending of the benzene ring, and the stretching vibration of C-S bond.

4.2. Weighted Vibrational Spectra

Different conformations have their own vibrational spectra. The weighed spectrum was calculated by different weighted conformations spectra according to their proportion. Frequency calculation is carried out in the optimal conformations; there is no imaginary frequency observed in the result. A frequency correction factor (0.9640) for B3LYP/6-311G** [41] was applied to the vibration spectra of DBDS, DBS, and BiBz due to the overestimated frequency of anharmonic effect.

Take the Raman spectra of DBDS as an example to illustrate the role of weighted spectrum. Figure 5 shows the Raman spectra of two DBDS conformations in the high frequency zone. The most significant difference between the two conformations is that DBDS-1 has a characteristic peak at 3025 cm^{-1}, but DBDS-2 has a characteristic peak at 2970 cm^{-1}.

Based on the vibration assignment in Section 4.1, the peak at 3025 cm^{-1} is assigned to asymmetrical stretching vibration of two -CH_2 groups in DBDS-1. The peak at 2970 cm^{-1} is assigned to symmetric stretching of -CH_2 group in DBDS-2. The Raman spectra of different conformations are different, but they are all superimposed in the weighted curve, as shown in Figure 5's blue line. The vibration spectroscopy is revealed much better when based on conformation weighted curve.

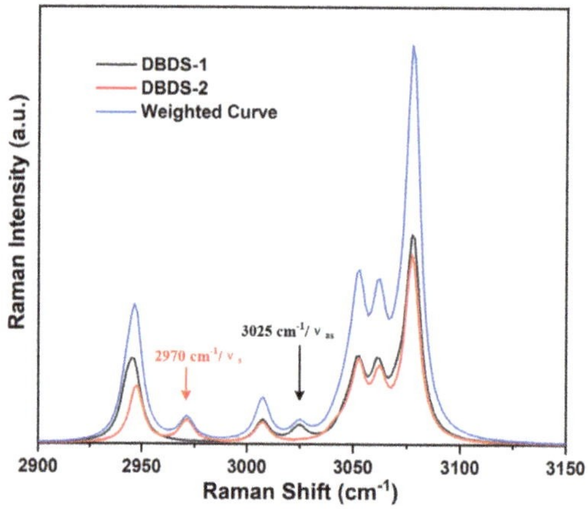

Figure 5. Theoretically calculated conformations Raman spectra and weighted curve of DBDS.

It should be noted that the Raman spectrum obtained by theoretical calculation is a spectrum of Raman activity (unit in $Å^4$/amu). In order to compare with the experimental spectrum, the Raman activity is converted into the Raman intensity, depending on the frequency of incident light and ambient temperature [42]:

$$I_i = \frac{C(v_o - v_i)^4 S_i}{v_i B_i} \quad (2)$$

$$B_i = 1 - exp(-\frac{hcv_i}{kT}) \quad (3)$$

where S_i, I_i, and v_i are the Raman activity, Raman intensity, and vibration frequency of ith vibration mode respectively. v_0 is the frequency of incident light. T is the ambient temperature, h is Planck's constant, c is the speed of light, k is Boltzmann's constant, and C is an arbitrary constant. The conversion process above was conducted via Multiwfn code [38]. The laser source wavelength used in our experimental platform is 532 nm, and room temperature was set as 298.15 K.

4.3. Comparison of Theoretical and Experimental Spectra

The experimentally measured and theoretically calculated spectra of DBDS, DBS, and BiBz are provided in Supplementary Material. It can be found from the Figures S4–S9 that the experimental Raman spectrum is in good agreement with the theoretical spectrum. Generally, the B3LYP hybrid functional combining with 6-311G** basis set overestimates the vibrational frequency in the high vibrational frequency zone but underestimates the frequency in the low frequency zone. This phenomenon is consistent with previous studies [43]. Dual scaling factor in the high frequency zone and low frequency zone is optional for agreement improvement.

In order to quantitatively describe the consistency between the theoretical spectrum and the experimental spectrum, a linear fitting operation is performed to each characteristic peak of DBDS. The abscissa of peak data is the experimental frequency, and the ordinate is the theoretical frequency. The peak data fitting line close to $y = x$ indicates a better agreement between the experimental value and the theoretical value. As shown in Figure 6, the fitting lines of Raman spectrum and infrared spectrum both show good linearity. The goodness of fit reaches 0.99979 and 0.99981, respectively. A good agreement is revealed at B3LYP-D3(BJ)/6-311G** level as the slope of the fitted line is close to 1.

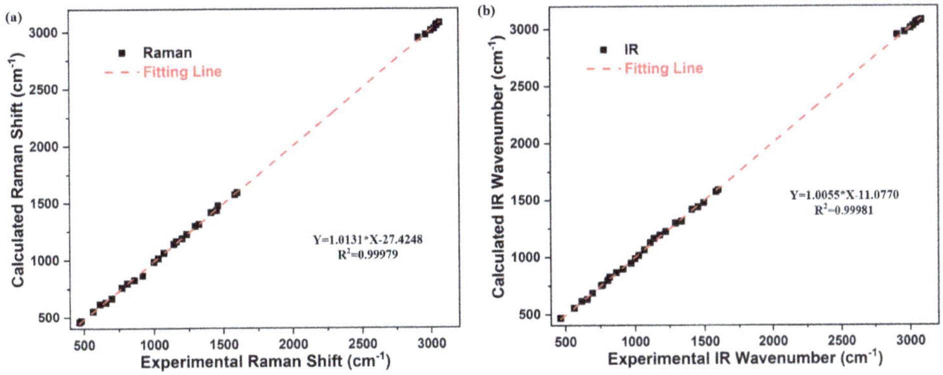

Figure 6. Linear fitting between theoretical and experimental Raman shift.

The mode identification is shown in Tables 5 and 6. Most experimental Raman peaks can be identified to corresponding normal modes within a reasonable deviation. There are also some normal vibration modes which are not measured in the experimental spectrum due to the weak Raman activity or infrared activity.

Table 5. Comparison of experimental and theoretical Raman spectrum.

Experimental Raman Peak (cm^{-1})	Corresponding Calculated Peak (cm^{-1})	Mode Identification	Deviation (cm^{-1})
467	456	#69	11
481	467	#67 + #68	14
504	-	Overtone of low frequency vibration	-
567	552	#65 + #66	15
616	615	#63 + #64	1
659	630	#61 + #62 (DBDS-1)	29
701	666	#61 + #62 (DBDS-2)	35
772	760	#57 (DBDS-2)	12
811	796	#55 + #56	15
860	824	#53 + #54	36
920	864	#51 + #52	56
1002	985	#43 + #44	17
1033	1015	#41 + #42	18
1074	1062	#39 + #40	12
-	1125	#37 + #38	-
1143	1140	#35 + #36	3
1162	1162	#33 + #34	0
1202	1189	#31 + #32	13
1234	1225	#29 + #30	9
1295	1296	#27 + #28	−1
1324	1312	#25 + #26	12
1338	-	Double frequency band of 659	-
1411	1416	#23 + #24	−5
1452	1432	#21 + #22	20
1463	1475	#19 + #20	−12
1495	-	Combination of 1002 and 481 under Fermi resonance	-
1526	-	Combination of 1002 and 481 under Fermi resonance	-
1584	1571	#17 + #18	13
1600	1590	#15 + #16	10
2908	2946	#13 + #14(DBDS-1)	−38
2964	2972	#13 (DBDS-2)	−8
3002	3008	#12	−6
3028	3025	#11 (DBDS-1)	3
3042	3053	#7 + #8	−11
3049	3062	#5 + #6	−13
3068	3077	#1 + #2	−9

Table 6. Comparison of experimental and theoretical IR spectrum.

Experimental IR Peak (cm^{-1})	Corresponding Calculated Peak (cm^{-1})	Mode Identification	Deviation (cm^{-1})
467	467	#67 + #68	0
505	-	Overtone of low frequency vibration	-
563	555	#65 + #66	8
569	-	Overtone of low frequency vibration	-
617	615	#63 + #64	2
658	633	#61 + #62	25
695	687	#59 + #60	8
757	751	#57 + #58 (DBDS-1)	6
766	758	#57 (DBDS-2)	8
803	796	#55 + #56	7
816	826	#53 + #54	−10
864	864	#51 + #52	0
912	895	#49 + #50	17
971	947	#47 + #48	24
1001	986	#43 + #44	15
1028	1015	#41 + #42	13
1068	1062	#39 + #40	6
1112	1126	#37 + #38	−14
1139	1162	#33 + #34	−23
1181	1190	#31 + #32	−9
1198	-	Combination of 695 and 563	-
1223	1223	#29 + #30	0
1291	1296	#27 + #28	−5
1336	1313	#25 + #26	23
1409	1416	#23 + #24	−7
1452	1436	#21 + #22	16
1494	1475	#19 + #20	19
1582	1570	#17 + #18	12
1599	1588	#15 + #16	11
2909	2946	#13 + #14 (DBDS-1)	−37
2964	2972	#13 (DBDS-2)	−8
3006	3007	#12	−1
3026	3025	#11 (DBDS-1)	1
3043	3047	#9 + #10	−4
3052	3062	#5 + #6	−10
3068	3070	#3 + #4	−2
3083	3077	#1 + #2	6

Different conformations contribute to different experimental peaks. For example, in the both Raman and infrared experimental spectra, the peak at 2964 cm^{-1} is only generated by the #13 normal mode of DBDS-2, while the #13 normal mode of DBDS-1 contribute to the peak at 2908 cm^{-1} together with the #14 mode. The characteristic peaks generated by a single conformation are marked with parentheses in Tables 5 and 6.

In addition to the fundamental frequency band in the vibrational spectroscopy, there are many anharmonic phenomenon such as overtone, combination band, Fermi resonance, etc. Which contribute to several characteristic peaks in experiment spectrum that are not observed in the theoretical spectrum. For example, two peaks are observed in experimental Raman spectrum at 1495 cm^{-1} and 1526 cm^{-1}. While the adjacent peaks at 1463 cm^{-1} and 1584 cm^{-1} correspond to the #19, #20, #17, and #18 mode, respectively. There is no corresponding theoretical normal mode for these two peaks (1495 and 1526) assignment. They are actually a combination band of peaks at 1002 cm^{-1} and 481 cm^{-1}. The peak at 1002 cm^{-1} (#43 + #44) corresponds to in-plane bending vibration of benzene ring carbon skeleton, while the peak at 481 cm^{-1} (#67 + #68) is corresponding to the out-of-plane deformation vibration of the benzene ring. This kind of highly correlated vibration is prone to coupling and generating combined frequency band. Fermi resonance most often occurs between fundamental and overtone (combination) excitations, if they are nearly coincident in energy. The two peaks (1495 and 1526) are thus split under Fermi resonance with the action of strong fundamental frequency band (1600 cm^{-1}).

4.4. Comparison of DBDS, DBS, and BiBz

DBDS, DBS, and Bibz have remarkable similarities in molecular structure. The only difference is the number of sulfur atoms between the two benzyl groups. The electronegativity of sulfur atoms is stronger than that of carbon atoms and hydrogen atoms. The sulfur atom will cause the redistribution of electron as discussed in Section 3.1, which changes the molecular vibration frequency. The spectra comparison of DBDS, DBS, and BiBz is shown in Figure 7. In order to reveal the similarity and difference of the vibrational behaviors of the three molecules, the six vibrational behaviors of the methylene groups is analyzed.

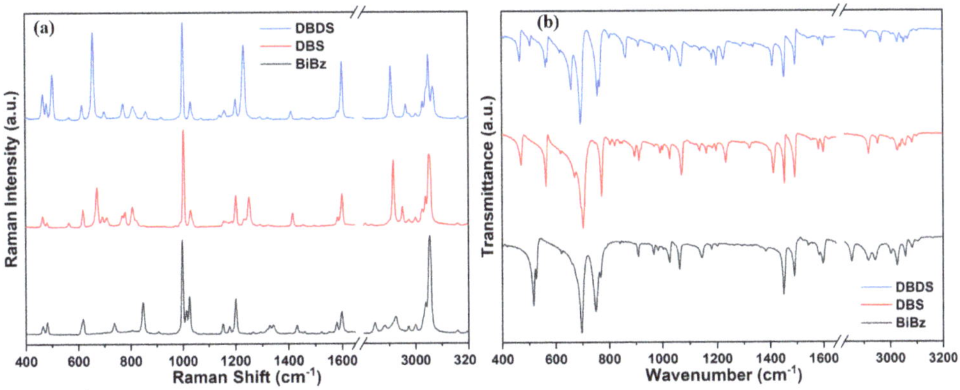

Figure 7. Experimental measured Raman (**a**) and IR (**b**) spectra of DBDS, DBS, and BiBz.

The frequency distributions of these six vibrations are generally consistent. The order from largest to smallest is asymmetric stretching, symmetric stretching, scissoring, wagging, twisting, and rocking, according to the frequency magnitude.

The characteristic vibration frequency of six different vibration behaviors in the DBDS, DBS, and BiBz is shown in Table 7. The trend is plotted in Figure 8 and the frequency is averaged by two dual vibrations. It can be found that, for the stretching vibration located in the high frequency region, when the sulfur atom in the middle of the benzyl group is lost, the stretching vibration frequency tends to decrease, while the four bending vibration frequencies in the middle frequency region tend to increase.

The most interesting aspect of these graphs is the low rocking vibration value of BiBz-2, which shows a different trend from several other conformations. This is because the frequency difference between the two dual vibration is very large. BiBz-2 is the only molecule with C_{2h} symmetry among the five dominant conformations. The rocking vibration at 748 cm^{-1} and 964 cm^{-1} are synchronous rocking and asynchronous rocking, respectively. The synchronous rocking at 748 cm^{-1} keeps symmetry during the vibration based on the center plane of two benzene rings. The polariseability change rate of this vibration is zero, thus the Raman activity is zero. Exhibiting extremely low frequency.

Table 7. The characteristic vibration frequency of methylene groups in different molecular.

Conformation	Stretching Vibration (cm^{-1})		Bending Vibration (cm^{-1})			
	Asymmetric (ν_s)	Symmetric (ν_{as})	Scissoring (δ)	Wagging (ω)	Twisting (τ)	Rocking (ρ)
DBDS-1	3007/3025	2943/2946	1415/1420	1221/1226	1124/1127	863/865
DBDS-2	3007/3041	2947/2971	1417/1431	1220/1227	1112/1122	878/894
DBS	2990/2991	2944	1412/1416	1223/1237	1130/1155	878/918
BiBz-1	2956/2968	2919/2924	1438/1443	1279/1324	1126/1170	921/977
BiBz-2	2950/2972	2920/2929	1436/1452	1251/1316	1122/1248	748/964

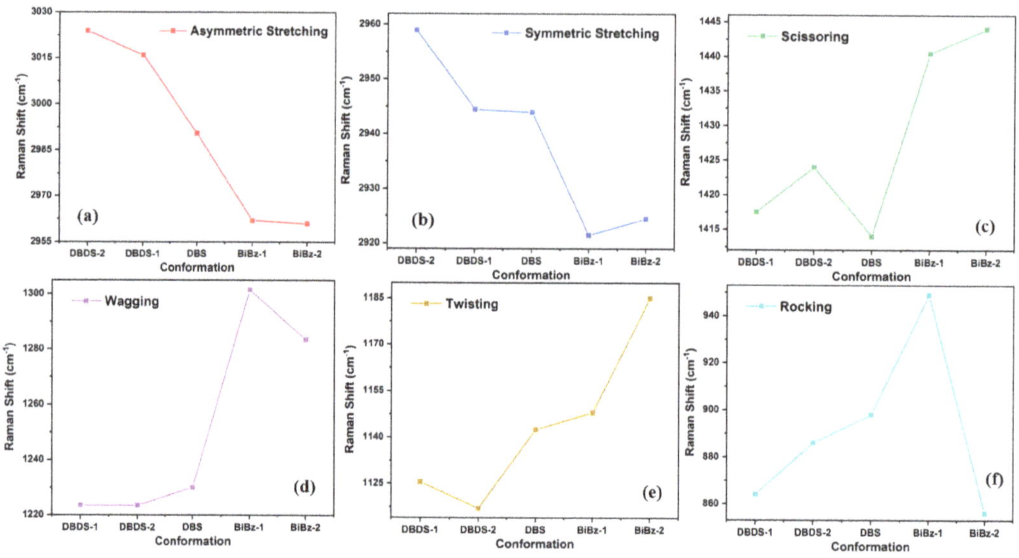

Figure 8. Tendency of different vibration behavior frequencies in the dominant conformations of DBDS, DBS, and BiBz. (**a**) Asymmetric Stretching; (**b**) Symmetric Stretching; (**c**) Scissoring; (**d**) Wagging; (**e**) Twisting; (**f**) Rocking.

5. Conclusions

In this paper, the vibration spectroscopy (Raman and infrared) and molecular vibrational behaviors of widely concerned molecules in sulfur corrosion phenomenon (Dibenzyl Disulfide, Dibenzyl Sulphide, and Bibenzyl) is detailedly analyzed. First, the proportion of different conformations are calculated according to Boltzmann distribution in relative Gibbs free energy via Molclus program. In order to obtain an accurate Gibbs free energy, the single-point energy is calculated in revDSD-PBEP86-D3(BJ)/may-cc-PVTZ and thermodynamic correction is calculated in the B3LYP-D3(BJ)/6-311G** level, respectively. Five dominant conformations of these three molecules were obtained, and their intramolecular interactions were detailedly discussed by NCI analysis.

PED based vibrational assignment of DBDS, DBS, and BiBz was performed via VEDA4 program, the normal vibration modes are assigned. Conformations weighted spectra is compared with experimentally measured spectra. It can be found that experimental spectra are in good agreement with the theoretical spectra in the B3LYP-D3(BJ)/6-311G** level with a frequency correction factor (0.9640). The characteristic peak in experimental spectra is also identified to corresponding modes within a reasonable deviation. The overtone due to the anharmonic effect is also pointed out. The divergence among these molecules is discussed. It can be found that, for the stretching vibration of methylene group located in the high frequency region, when the sulfur atom in the middle of the benzyl group lost, the stretching vibration frequency tends to decrease, while the four bending vibration frequencies in the middle frequency region tend to increase.

In summary, the results presented in this paper provide a theoretical and experimental guidance for understanding the vibrational spectra and molecular vibrational behaviors of these molecules. We also believe that this article will lay a solid foundation for the development of advanced corrosive resistant materials and realization of the corrosive state monitoring.

Supplementary Materials: The following supporting information can be downloaded at: https://www.mdpi.com/article/10.3390/ijms23041958/s1.

Author Contributions: Z.W.: conceptualization, methodology, software, validation, visualization, investigation, and writing—original draft preparation; R.S.: conceptualization, methodology, and writing—review and editing; J.W. and P.W.: conceptualization; Z.Z. and X.Z.: writing—review and editing; W.C.: supervision and project administration; and F.W.: funding acquisition. All authors have read and agreed to the published version of the manuscript.

Funding: This research was funded by the National Natural Science Foundation of China under grant number 51977017.

Data Availability Statement: The data presented in this study are available on request from the corresponding author.

Acknowledgments: We Would like to thank Gong at Analytical and Testing Center of Chongqing University for their assistance with FTIR test.

Conflicts of Interest: The authors declare no conflicts of interest.

References

1. Scatiggio, F.; Tumiatti, V.; Maina, R.; Tumiatti, M.; Pompili, M.; Bartnikas, R. Corrosive sulfur induced failures in oil-filled electrical power transformers and shunt reactors. *IEEE Trans. Power Deliv.* **2009**, *24*, 1240–1248. [CrossRef]
2. Scatiggio, F.; Tumiatti, V.; Maina, R.; Tumiatti, M.; Pompili, M.; Bartnkas, R. Corrosive sulfur in insulating oils: Its detection and correlated power apparatus failures. *IEEE Trans. Power Deliv.* **2007**, *23*, 508–509. [CrossRef]
3. Dahlund, M. *Copper Sulphide in Transformer Insulation*; CIGRE WG A2-32, Final Report, No. 378; CIGRE: Paris, France, 2009.
4. Yuan, Y.; Gao, X.; Zhou, J.; Liu, G.; Kuang, X.; Yang, L.; Liao, R. A review: Research on corrosive sulphur in electrical power equipment. *High Volt.* **2021**, doi:10.1049/hve2.12155. [CrossRef]
5. Cong, H.; Pan, H.; Qian, D.; Zhao, H.; Li, Q. Reviews on sulphur corrosion phenomenon of the oil–paper insulating system in mineral oil transformer. *High Volt.* **2020**, *6*, 193–209. [CrossRef]
6. Toyama, S.; Tanimura, J.; Yamada, N.; Nagao, E.; Amimoto, T. Highly sensitive detection method of dibenzyl disulfide and the elucidation of the mechanism. *IEEE Trans. Dielectr. Electr. Insul.* **2009**, *16*, 509–515. [CrossRef]
7. Kamishima, S.; Ito, T.; Morishima, Y. Change in corrosivity of insulating oil caused by oxidative deterioration of the oil. *IEEE Trans. Dielectr. Electr. Insul.* **2012**, *19*, 505–509. [CrossRef]
8. Lukic, J.M.; Milosavljevic, S.B.; Orlovic, A.M. Degradation of the insulating system of power transformers by copper sulfide deposition: Influence of oil oxidation and presence of metal passivator. *Ind. Eng. Chem. Res.* **2010**, *49*, 9600–9608. [CrossRef]
9. Cong, H.; Pan, H.; Hu, X.; Zhang, M.; Li, Q. Micro-mechanism study on synergistic degradation of the oil-paper insulation with dibenzyl disulfide, hexadecyl mercaptan and benzothiophene. *High Volt.* **2021**, *6*, 448–459. [CrossRef]
10. Jaber, A.; Mehanna, N.; Oweimreen, G.; Abulkibash, A. The effect of DBDS, DBPC, BTA and DBP combinations on the corrosion of copper immersed in mineral transformer oil. *IEEE Trans. Dielectr. Electr. Insul.* **2016**, *23*, 1–7. [CrossRef]
11. Tumiatti, V.; Roggero, C.; Tumiatti, M.; Di Carlo, S.; Maina, R.; Kapila, S. IEC 62697-2012: State of the art methods for quantification of DBDS and other corrosive sulfur compounds in unused and used insulating liquids. *IEEE Trans. Dielectr. Electr. Insul.* **2012**, *19*, 1633–1641. [CrossRef]
12. Du, D.; Tang, C.; Zhang, J.; Hu, D. Effects of hydrogen sulfide on the mechanical and thermal properties of cellulose insulation paper: A molecular dynamics simulation. *Mater. Chem. Phys.* **2020**, *240*, 122153. [CrossRef]
13. Naicker, S.A.; Moodley, M. A computational study of the adsorption of corrosive sulphur on Ag surfaces. *J. Mater. Sci.* **2021**, *56*, 13386–13399. [CrossRef]
14. Zhang, Y.; Katayama, Y.; Tatara, R.; Giordano, L.; Yu, Y.; Fraggedakis, D.; Sun, J.G.; Maglia, F.; Jung, R.; Bazant, M.Z.; et al. Revealing electrolyte oxidation via carbonate dehydrogenation on Ni-based oxides in Li-ion batteries by in situ Fourier transform infrared spectroscopy. *Energy Environ. Sci.* **2020**, *13*, 183–199. [CrossRef]
15. Jianxin, W.; Weigen, C.; Zhixian, Z.; Fu, W.; Feng, Z.; Ruimin, S.; Yingying, W.; Shoufei, G. Fiber-enhanced Raman spectroscopy for highly sensitive H_2 and SO_2 sensing with a hollow-core anti-resonant fiber. *Opt. Express* **2021**, *29*, 32296–32311.
16. Song, R.; Chen, W.; Yang, D.; Shi, H.; Zhang, R.; Wang, Z. Aging Assessment of Oil–Paper Insulation Based on Visional Recognition of the Dimensional Expanded Raman Spectra. *IEEE Trans. Instrum. Meas.* **2021**, *70*, 1–10. [CrossRef]
17. Inaoka, S.; Iwata, K.; Saha, S. Towards the critical understanding of selected vibrational features in biologically important dicyano aromatic conjugated molecules: Importance of electron donating/withdrawal groups and geometry associated with dicyano group. *Spectrochim. Acta Part A Mol. Biomol. Spectrosc.* **2020**, *224*, 117419.
18. Wu, Y.; Lan, H.; Ning, W.; Chen, F.; Yan, G.; Cai, K. Structural dynamics and vibrational feature of N-Acetyl-D-glucosamine in aqueous solution. *Spectrochim. Acta Part A Mol. Biomol. Spectrosc.* **2021**, *259*, 119918.

19. Muthunatesan, S.; Ragavendran, V. A study of vibrational spectra and investigations of charge transfer and chemical bonding features of 2-chloro benzimidazole based on DFT computations. *Spectrochim. Acta Part A Mol. Biomol. Spectrosc.* **2015**, *134*, 148–154. [CrossRef] [PubMed]
20. Balachandran, V.; Santhi, G.; Karpagam, V.; Rastogi, V. Structural features of the 2-amino-5-nitrobenzophenone by means of vibrational spectroscopy HF and DFT, first order hyperpolarizability, NBO, HOMO–LUMO and theromodynamic properties. *Spectrochim. Acta Part A Mol. Biomol. Spectrosc.* **2014**, *118*, 835–846. [CrossRef]
21. Sert, Y.; El-Emam, A.A.; Al-Abdullah, E.S.; Al-Tamimi, A.M.S.; Çırak, Ç.; Ucun, F. Use of vibrational spectroscopy to study 4-benzyl-3-(thiophen-2-yl)-4, 5-dihydro-1H-1, 2, 4-triazole-5-thione: A combined theoretical and experimental approach. *Spectrochim. Acta Part A Mol. Biomol. Spectrosc.* **2014**, *126*, 280–290. [CrossRef]
22. Kose, E.; Bardak, F.; Atac, A.; Karabacak, M.; Cipiloglu, M. Determination of structural and vibrational spectroscopic features of neutral and anion forms of dinicotinic acid by using NMR, infrared and Raman experimental methods combined with DFT and HF. *Spectrochim. Acta Part A Mol. Biomol. Spectrosc.* **2013**, *114*, 38–45. [CrossRef]
23. Chen, W.; Gu, Z.; Zou, J.; Wan, F.; Xiang, Y. Analysis of furfural dissolved in transformer oil based on confocal laser Raman spectroscopy. *IEEE Trans. Dielectr. Electr. Insul.* **2016**, *23*, 915–921. [CrossRef]
24. Lu, T. Molclus Program. Version 1.9.9.5. Available online: http://www.keinsci.com/research/molclus.html (accessed on 25 July 2021).
25. Frisch, M.J.; Trucks, G.W.; Schlegel, H.B.; Scuseria, G.E.; Robb, M.A.; Cheeseman, J.R.; Scalmani, G.; Barone, V.; Petersson, G.A.; Nakatsuji, H.; et al. *Gaussian 16, Revision A.03*; Gaussian, Inc.: Wallingford, CT, USA, 2016.
26. Stewart, J.J. Optimization of parameters for semiempirical methods VI: More modifications to the NDDO approximations and re-optimization of parameters. *J. Mol. Model.* **2013**, *19*, 1–32. [CrossRef]
27. Stephens, P.J.; Devlin, F.J.; Chabalowski, C.F.; Frisch, M.J. Ab initio calculation of vibrational absorption and circular dichroism spectra using density functional force fields. *J. Phys. Chem.* **1994**, *98*, 11623–11627. [CrossRef]
28. Grimme, S.; Antony, J.; Ehrlich, S.; Krieg, H. A consistent and accurate ab initio parametrization of density functional dispersion correction (DFT-D) for the 94 elements H-Pu. *J. Chem. Phys.* **2010**, *132*, 154104. [CrossRef]
29. Grimme, S.; Ehrlich, S.; Goerigk, L. Effect of the damping function in dispersion corrected density functional theory. *J. Comput. Chem.* **2011**, *32*, 1456–1465. [CrossRef]
30. Hariharan, P.; Pople, J. Accuracy of AH n equilibrium geometries by single determinant molecular orbital theory. *Mol. Phys.* **1974**, *27*, 209–214. [CrossRef]
31. Petersson, A.; Bennett, A.; Tensfeldt, T.G.; Al-Laham, M.A.; Shirley, W.A.; Mantzaris, J. A complete basis set model chemistry. I. The total energies of closed-shell atoms and hydrides of the first-row elements. *J. Chem. Phys.* **1988**, *89*, 2193–2218. [CrossRef]
32. Santra, G.; Sylvetsky, N.; Martin, J.M. Minimally empirical double-hybrid functionals trained against the GMTKN55 database: revDSD-PBEP86-D4, revDOD-PBE-D4, and DOD-SCAN-D4. *J. Phys. Chem. A* **2019**, *123*, 5129–5143. [CrossRef]
33. Papajak, E.; Zheng, J.; Xu, X.; Leverentz, H.R.; Truhlar, D.G. Perspectives on basis sets beautiful: Seasonal plantings of diffuse basis functions. *J. Chem. Theory Comput.* **2011**, *7*, 3027–3034. [CrossRef]
34. Lu, T.; Chen, Q. Shermo: A general code for calculating molecular thermochemistry properties. *Comput. Theor. Chem.* **2021**,*1200*, 113249. [CrossRef]
35. Humphrey, W.; Dalke, A.; Schulten, K. VMD—Visual Molecular Dynamics. *J. Mol. Graph.* **1996**, *14*, 33–38. [CrossRef]
36. Zhang, J.; Lu, T. Efficient evaluation of electrostatic potential with computerized optimized code. *Phys. Chem. Chem. Phys.* **2021**, *23*, 20323–20328. [CrossRef]
37. Johnson, E.R.; Keinan, S.; Mori-Sánchez, P.; Contreras-García, J.; Cohen, A.J.; Yang, W. Revealing noncovalent interactions. *J. Am. Chem. Soc.* **2010**, *132*, 6498–6506. [CrossRef]
38. Lu, T.; Chen, F. Multiwfn: A multifunctional wavefunction analyzer. *J. Comput. Chem.* **2012**, *33*, 580–592. [CrossRef] [PubMed]
39. Lu, T.; Chen, Q. Interaction Region Indicator: A Simple Real Space Function Clearly Revealing Both Chemical Bonds and Weak Interactions. *Chem.-Methods* **2021**, *1*, 231–239. [CrossRef]
40. Jamróz, M.H. Vibrational energy distribution analysis (VEDA): Scopes and limitations. *Spectrochim. Acta Part A Mol. Biomol. Spectrosc.* **2013**, *114*, 220–230. [CrossRef] [PubMed]
41. Kashinski, D.; Chase, G.; Nelson, R.; Di Nallo, O.; Scales, A.; VanderLey, D.; Byrd, E. Harmonic vibrational frequencies: Approximate global scaling factors for TPSS, M06, and M11 functional families using several common basis sets. *J. Phys. Chem. A* **2017**, *121*, 2265–2273. [CrossRef] [PubMed]
42. Liu, Z.; Lu, T.; Chen, Q. Vibrational Spectra and Molecular Vibrational Behaviors of All-Carboatomic Rings, cyclo [18] carbon and Its Analogues. *Chem.—Asian J.* **2021**, *16*, 56–63. [CrossRef] [PubMed]
43. Halls, M.D.; Velkovski, J.; Schlegel, H.B. Harmonic frequency scaling factors for Hartree-Fock, S-VWN, B-LYP, B3-LYP, B3-PW91 and MP2 with the Sadlej pVTZ electric property basis set. *Theor. Chem. Accounts* **2001**, *105*, 413–421. [CrossRef]

International Journal of
Molecular Sciences

MDPI

Article

New Light on an Old Story: Breaking Kasha's Rule in Phosphorescence Mechanism of Organic Boron Compounds and Molecule Design

Dan Deng [1,2], Bingbing Suo [1,2,*] and Wenli Zou [1,2,*]

1 Institute of Modern Physics, Northwest University, Xi'an 710127, China; dengdan@stumail.nwu.edu.cn
2 Shaanxi Key Laboratory for Theoretical Physics Frontiers, Xi'an 710127, China
* Correspondence: bsuo@nwu.edu.cn (B.S.); zouwl@nwu.edu.cn (W.Z.)

Abstract: In this work, the phosphorescence mechanism of (E)-3-(((4-nitrophenyl)imino)methyl)-2H-thiochroman-4-olate-BF2 compound (S-BF2) is investigated theoretically. The phosphorescence of S-BF2 has been reassigned to the second triplet state (T_2) by the density matrix renormalization group (DMRG) method combined with the multi-configurational pair density functional theory (MCPDFT) to approach the limit of theoretical accuracy. The calculated radiative and non-radiative rate constants support the breakdown of Kasha's rule further. Our conclusion contradicts previous reports that phosphorescence comes from the first triplet state (T_1). Based on the revised phosphorescence mechanism, we have purposefully designed some novel compounds in theory to enhance the phosphorescence efficiency from T_2 by replacing substitute groups in S-BF2. Overall, both S-BF2 and newly designed high-efficiency molecules exhibit anti-Kasha T_2 phosphorescence instead of the conventional T_1 emission. This work provides a useful guidance for future design of high-efficiency green-emitting phosphors.

Keywords: radiative rate constants; non-radiative rate constants; DMRG; MCPDFT; anti-Kasha's rule

1. Introduction

Phosphorescence emitters are widely used in organic light-emitting diodes (OLEDs) as photoelectric devices in visualizing, imaging, sensing, and so on due to their 100% internal quantum efficiency (IQE) in theory by fully utilizing triplet excitons of 75% [1–4]. Traditionally, the most efficient phosphorescence materials are inorganic or organometallic compounds containing heavy atoms like iridium (Ir), platinum (Pt), and gold (Au) [5–8] which may facilitate intersystem crossing (ISC) between the higher-lying singlet and triplet excited states (S_m and T_n; m, n = 1, 2, ...) due to strong spin-orbit coupling (SOC) effects induced by heavy atoms [9]. Unfortunately, these kinds of materials are usually expensive, poisonous, or environmentally unfriendly, and therefore pure organic phosphors may be potential alternatives to substitute organometallic luminophores. Since organic phosphorescence materials are relatively cheap, biocompatible, easy processing, and versatile, they have received extensive attention [10,11]. However, the development of efficient and metal-free room-temperature phosphorescent (RTP) materials is rather challenging [10,12–17] because phosphorescent molecules are highly sensitive to temperature and oxygen molecules in the atmosphere. For example, their triplet excited states are extremely unstable at ambient environment and the non-radiative decay and T-T quench may easily take place through thermal collision process and triplet-triplet energy transfer, respectively. Moreover, without the help of heavy atoms, their tiny SOC constants lead to extremely weak ISC from S_m to T_n. Consequently, metal-free RTP materials are rarely applicable.

In recent years, luminescent organoboron compounds have become the focus of investigations because of their promising luminescent properties [2,18–22]. In 2017, Yu et al. [23]

synthesized the (E)-3-(((4-nitrophenyl)imino)methyl)-2H-thiochroman-4-olate-BF2 compound (S-BF2; see Figure 1b) based on the parent molecule (E)-2-(((4-nitrophenyl)imino)-methyl)-naphthalen-1-olate-BF2 (C-BF2; Figure 1a), and explored the photoluminescence (PL) mechanism experimentally and theoretically. Although their theoretical proof by time-dependent density functional theory (TDDFT) seems rough, their study opened the door for a new class of laser materials to utilize previously untapped triplet phosphors. In 2018, Paul et al. [24] attributed the luminescence of S-BF2 to the T_1 state by means of CASSCF(4,4)/NEVPT2 calculations because the widely used TDDFT method, according to them, "cannot"describe multi-reference character in the S_1 and T_1 states. Soon afterwards, Lv et al. [25] also revealed the phosphorescence mechanism of S-BF2 from the T_1 state by TDDFT.

All of the above authors assigned the phosphorescence of S-BF2 to the T_1 state, but unfortunately, there were some insufficiencies in their studies. In the two TDDFT studies [23,25], for example,

- The vertical emission energy of T_1 was in poor agreement with the experimental one.
- To improve the agreement of T_1, Lv et al. tried to use the adiabatic energy rather than the vertical emission one, but only an unusual exchange-correlation (XC) functional could obtain a reasonable result [25].
- Due to Kasha's rule [26], the probability of T_2 (as well as higher triplet states) was not considered at all.

Paul et al. contradicted the applicability of TDDFT in S-BF2 [24], but some of their views violates the standard model of quantum chemistry. Before thoroughly exploring the phosphorescence mechanism in S-BF2, here we comment on the work due to Paul et al. [24] and explain the contradiction between CASSCF(4,4)/NEVPT2 and TDDFT.

- Paul et al. claimed that TDDFT fails to get multi-reference S_1 and T_1 [24], which is conditional. In principle, a successful TDDFT calculation requires that (1) the reference state (usually S_0) by density functional theory (DFT) is single-reference characterized, and (2) the interested excited states may be accessed by one or more single-excitations from S_0. Therefore, an excited state with some multi-reference character may be well calculated by TDDFT only if it is single-excitation dominated from a single-reference S_0 (for example, see the third reason on page 4518 of Reference [27]), and of course it is also important to choose suitable XC functionals for some systems like transition metal compounds and charge-transfer (CT) excitations.
- CASSCF and NEVPT2 have a different theoretical basis from TDDFT, and both single-reference and multi-reference states may be described in a unified framework. A crucial question in nearly all the multi-configurational methods is whether adequate static correlations can be captured by the active space, and otherwise multi-configurational calculations merely reproduce single-configurational results or even worse. The NEVPT2 emission wavelength of $T_1 \rightarrow S_0$ by Paul et al. is 564 nm [24], at first glance being in good agreement with the experimental phosphorescence peak at 575 nm [23], but it is not clear whether all the important π and π^* orbitals have been included in their quite small active space of (4e,4o).

Due to the above doubts, the phosphorescence mechanism of S-BF2 remains unclear, and it is necessary to explore the emissive mechanism by more accurate approaches.

This paper is structured as follows. In Section 2, we introduce the theoretical methods performed in this study, i.e., TDDFT and high-precision DMRG-SCF/MCPDFT, and provide the fundamental definitions of radiative and non-radiative rate constants. In Section 3, the theoretical results and discussions are grouped into three aspects: (1) the reliability of the cheap TDDFT method is verified with the help of theoretical limit emission energies of S-BF2 by DMRG-SCF/MCPDFT, (2) the PL pathway of S-BF2 is analyzed based on the radiative and non-radiative rates by TDDFT and a revised phosphorescent mechanism is suggested, and (3) a series of new molecules with higher-performance phosphorescence efficiency have been designed theoretically according to the new mechanism. Some conclusions are drawn in Section 4.

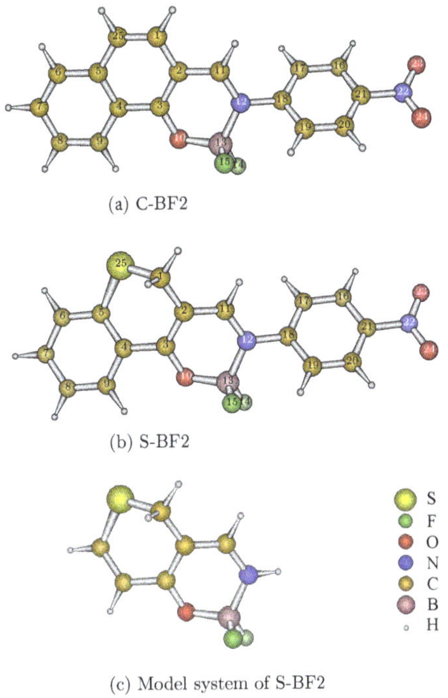

Figure 1. Structures of (**a**) C-BF2, (**b**) S-BF2, and (**c**) the model system.

2. Computational Methods

2.1. TDDFT

Geometry optimizations of all the systems (C-BF2, S-BF2 and designed molecules) at the ground and excited states were performed by means of DFT and TDDFT [28], respectively. Vibrational frequencies were also calculated after optimization to ensure that these structures are stable points. To find a suitable XC functional, a series of XC functionals have been examined combined with the basis set 6-311G(d,p) [29,30], including BLYP [31, 32], B3LYP [33,34], PBE0 [35,36], CAM-B3LYP [37], ωB97XD [38], recently developed SCAN0 [39–41], and so on (see the Supplementary Materials). The experimental spectra of S-BF2 were measured in dichloromethane (CH_2Cl_2), so the solvent environment was implicitly simulated by the polarizable continuum model (PCM) [42]. The above DFT and TDDFT calculations were performed using the GAUSSIAN 16 [43] program package, whereas the calculations involving the SCAN0 functional were performed with an in-house BDF program package [44–46] (see the Supplementary Materials for the details of SCAN0). The TDDFT/6-311G(d,p) results by different XC functionals may be found in Table S1, and the simulated absorption spectra of S-BF2 have been plotted in Figure S1.

2.2. ONIOM(DMRG-SCF/MCPDFT:TDDFT)

To verify the CASSCF(4,4)/NEVPT2 results of Paul et al. [24], we performed the two-layer ONIOM (QM_{high}:QM_{low}) combination approach of Morokuma [47,48] as demonstrated in Figure 1b,c, where the geometry of S-BF2 is optimized at the TDDFT(PBE0)/6-311G(d,p) level in CH_2Cl_2 solvent for T_n (n = 1 and 2; see Section S6 in the Supplementary

Materials). The ONIOM total energy or vertical emission energy of the real system (i.e., S-BF2) in CH_2Cl_2 solvent is calculated by

$$E_{real}^{CH_2Cl_2}(QM_{High} : QM_{Low}) \approx E_{model}^{gas}(QM_{High}) - E_{model}^{gas}(QM_{Low}) + E_{real}^{CH_2Cl_2}(QM_{Low}) \quad (1)$$

where the low level quantum mechanical method (QM_{Low}) is TDDFT/6-311G(d,p); two common XC functionals B3LYP and PBE0 are used in this work. In the model system part of calculation by a high level quantum mechanical method (QM_{High}), the minimal full valence active space consists of 54 electrons in 43 orbitals by neglecting all the core orbitals, deep-lying 2*s* or 3*s* valence orbitals of N, O, F, and S, and high-lying H 1*s* orbitals, i.e., (54e,43o), which is far beyond the capability of CASSCF, and therefore the density matrix renormalization group (DMRG) method is performed instead through self-consistent field orbital optimization iteratively (DMRG-SCF; also loosely called DMRG-CASSCF) with up to 1000 renormalized states (M = 1000). The basis set is def2-TZVPP(-f) [49] by eliminating all the f-functions from the original one, and the Cholesky decomposition (CD) of the two-electron integrals [50] is used to speed up the calculations with controlled accuracy. Since neither DMRG-NEVPT2 nor DMRG-CASPT2 is feasible for this large active space to capture dynamic correlations, the DMRG based multi-configurational pair-density functional theory (MCPDFT) [51] is carried out subsequently with three different on-top density functionals ftBLYP, ftPBE, and ftrevPBE [52]. It has been found in the literatures that MCPDFT with a suitable active space could perfectly reproduce the CASPT2 results of organic compounds [52,53], d-block transition metal molecules [54,55], and even actinide-containing systems [56]. The OPENMOLCAS [57,58] program package is used for DMRG-SCF(54e,43o)/MCPDFT calculations by driving the BLOCK program [59,60] as the DMRG solver.

2.3. Radiative and Non-Radiative Rates

In order to figure out the radiative and non-radiative photochemical decay processes, the minimal energy crossing points (MECPs) [61] were optimized using the sobMECP procedure [61,62], and the radiative, internal conversion (IC), and ISC rate constants were calculated by the MOMAP program package [63]. A molecule in excited electronic state loses its energy through different relaxation processes, which may be grouped into radiative and non-radiative ones. The radiative process is defined as the emission of photons in a transition between different electronic states, including (spin-conserving) fluorescence and (spin-nonconserving) phosphorescence emissions. On the contrary, there are no emitted photons in the non-radiation process, including (spin-conserving) IC and (spin-nonconserving) ISC [10,64].

The radiative rate (in s^{-1}) is estimated via [10]

$$K_r = \frac{3}{2} f (\Delta E)^2 \quad (2)$$

where the oscillator strength f (in atomic unit; *a.u.*) is defined by

$$f = \frac{2}{3} \mu_t^2 \Delta E \quad (3)$$

In Equations (2) and (3), μ_t is the transition dipole moment in *a.u.*, ΔE is the emission energy in cm^{-1} (Equation (2)) or *a.u.* (Equation (3)).

The non-radiative rate is given by the fermi Golden Rule.

$$K_{nr} = 2\pi \frac{1}{\hbar} \sum_f |\mathbf{H}_{if}^1|^2 \delta(E_i - E_f) \quad (4)$$

where $\mathbf{H}_{if}^1$ is the matrix element of the first-order Hamiltonian, being spin-orbit coupling (SOC) matrix element for the ISC rate and non-adiabatic coupling matrix element for the

IC rate, the δ function provides the energy conservation for the non-radiative transition, and i, f represents the initial and final states, respectively.

The SOC constant is an important parameter to calculate K_{isc} and K_r in phosphorescent transitions. In this study, the SOC effects were calculated perturbatively by the TDDFT(PBE0) + SOC method implemented in the BDF program package, where the one-center molecular mean-field approximation was applied to the two-electron SO integrals.

3. Results and Discussion

3.1. Absorption and Emission Energies of S-BF2

In order to verify the previous viewpoint that the phosphorescence of S-BF2 at 575 nm comes from T_1 [23–25], the emission energies and wavelengths of T_1 and T_2 are calculated by the DMRG-SCF/MCPDFT based ONIOM combination approach, and the results are summarized in Table 1. It can be seen that the experimental phosphorescence wavelength of S-BF2 at 575 nm [23] may be assigned to the T_2 state instead of T_1, which is contrary to all the early results [23–25]. This opposing assignment may be attribute to some charge-transfer character [24] and near-degeneracy correlation [65] in the excited states, which may be well described by DMRG/MCPDFT with an adequately large active space [53,66,67].

Table 1. Emission energies (in eV) and emission wavelengths (in parentheses; nm) of T_1 and T_2 of S-BF2 by ONIOM (QM_{High}:QM_{Low}).

System	QM_{High} $^{(a)}$	QM_{Low} $^{(b)}$	T_1	T_2
Model	A		1.75 (708)	2.40 (516)
	B		1.74 (714)	2.38 (520)
	C		1.77 (702)	2.42 (512)
Real (S-BF2)	A	a	1.87 (664)	2.07 (598)
	A	b	1.93 (643)	2.16 (574)
	B	a	1.85 (669)	2.06 (603)
	B	b	1.92 (647)	2.14 (579)
	C	a	1.88 (659)	2.10 (592)
	C	b	1.94 (638)	2.18 (569)
	Expt. $^{(c)}$		2.16 (575)	

$^{(a)}$ DMRG-SCF(54e,43o)/MCPDFT with ftBLYP (A), ftPBE (B), or ftrevPBE (C). $^{(b)}$ TDDFT with B3LYP (a) or PBE0 (b). $^{(c)}$ Reference [23].

As for the applicability of DFT/TDDFT to S-BF2, two issues need to be paid attention. It is well known that the accuracy of DFT/TDDFT calculations is generally affected by the amount of multi-reference character in S_0 and CT character in excited states [68]. The S_0 state of S-BF2 has single-reference character at its equilibrium geometry [24]. At the T_n geometries, however, DMRG-SCF(54e,43o) predicts some multi-reference character in S_0 of the model system. In addition, the analysis of spatial extent in CT excitations [69] implemented in the MULTIWFN program [70] shows that T_1 and T_2 are basically local excitations (see Section S3 in the Supplementary Materials). Consequently, DFT/TDDFT may be still applicable for S-BF2 if double-excitations are negligible, but some hybrid XC functionals are preferred to describe both the slight multi-reference character in S_0 and the modest CT character in the excited states of S-BF2.

Table S1 in the Supplementary Materials collects the emission energies and the corresponding phosphorescence wavelengths of T_1 and T_2 at their respective geometries, either in gas phase or in CH_2Cl_2 solution. On the whole, the energy of T_1 state by most of the functionals is too low whereas some common functionals support phosphorescence from the T_2 state instead of T_1. Among these functionals, PBE0 exhibits better performance than the others in emission spectra by comparing the wavelengths with the DMRG-SCF(54e,43o)/MCPDFT ones. Therefore, the PBE0 functional will be used in the following study.

At the TDDFT(PBE0) level of theory, the absorption and phosphorescence spectra of S-BF2 are calculated at the S_0 and T_n (n = 1 and 2) geometries, respectively, as collected in Table 2. In the absorption spectrum, there are two theoretical peaks at 435 ($S_0 \rightarrow S_1$) and

375 nm ($S_0 \rightarrow S_2$), which are merely 5 nm blue-shifted compared with the experimental ones at 440 and 380 nm [23] and are better than the CASSCF(4,4)/NEVPT2 results of 410 and 340 nm [24]. Since the oscillator strength (f) of 0.632 in the transition $S_0 \rightarrow S_2$ is much larger than that of 0.269 in $S_0 \rightarrow S_1$, S-BF2 is much easier to be excited to S_2 than to S_1. In the phosphorescence spectrum, the experimental peak at 575 nm [23] has been reassigned to the theoretical $T_2 \rightarrow S_0$ transition at 559 nm in this study, which has much stronger transition dipole moment (and f as well) than that of $T_1 \rightarrow S_0$. To confirm the phosphorescence coming from the higher-lying T_2 state instead of T_1, radiative and non-radiative rates need to be investigated further in the next subsection.

Table 2. Excitation energy (in eV), wavelength (in nm), and oscillator strength of S-BF2 in CH_2Cl_2 solution.

State	E	λ	f	Configuration [(a)] (%)
Absorption Spectrum				
S_1	2.85	435	0.269	H→L (93)
Expt. [(b)]	2.82	440		
S_2	3.30	375	0.632	H-1→L (97)
Expt. [(b)]	3.26	380		
Phosphorescence Spectrum				
T_1	1.64	758	1.31×10^{-7}	H→L (84), H-1→L (11)
T_2	2.22	559	1.63×10^{-5}	H-1→L (53), H→L (38)
Expt. [(b)]	2.16	575		

[(a)] H for HOMO and L for LUMO. [(b)] Reference [23].

3.2. Radiative and Non-radiative Processes of S-BF2

To deepen the understanding of the PL mechanism in S-BF2, some radiative and non-radiative rate constants are calculated as given in Table 3, including the ISC rate (K_{isc}), the IC rate (K_{ic}), and the radiative rate (K_r). In the transition $S_1 \rightarrow S_0$, K_{ic} of 5.05×10^9 s^{-1} is much faster than K_r of 7.37×10^7 s^{-1}, and therefore the fluorescence probability is relatively weak. In addition, an instantaneous IC process from S_2 to S_1 may increase the population on S_1 further as found in the previous studies [23,24]. On the other hand, K_{isc} of 2.24×10^9 s^{-1} in $S_1 \rightarrow T_2$ is comparable to K_{ic} in $S_1 \rightarrow S_0$ and is far beyond K_{isc} of 1.10×10^7 s^{-1} in $S_1 \rightarrow T_1$. The K_{isc} rate of $S_1 \rightarrow T_1$ is 2.26×10^9 s^{-1} in Reference [24], being about 200 times larger than our 1.10×10^7 s^{-1} because of the underestimated energy gap between S_1 and T_1 by CASSCF/NEVPT2. In our study, the corresponding ISC efficiency (Φ_{isc}) is 47.0% in $S_1 \rightarrow T_2$, being about 120 times larger than that of 0.4% in $S_1 \rightarrow T_1$ (see Table S12 in the Supplementary Materials), which explains the considerable population on the T_2 state. Obviously $K_r = 3.25 \times 10^3$ s^{-1} surpasses $K_{isc} = 1.33 \times 10^1$ s^{-1} in $T_2 \rightarrow S_0$. For comparison, $K_r = 1.20 \times 10^1$ s^{-1} in $T_1 \rightarrow S_0$ is two orders of magnitude smaller than that in $T_2 \rightarrow S_0$, not to mention the agreement of emission wavelength with the experimental one.

Table 3. Radiative and non-radiative rate constants (in s^{-1}) of S-BF2.

Rate	$S_2 \rightarrow S_1$	$S_1 \rightarrow S_0$	$S_1 \rightarrow T_1$	$T_1 \rightarrow S_0$	$S_1 \rightarrow T_2$	$T_2 \rightarrow S_0$
K_{ic}	2.80×10^{13}	5.05×10^9				
K_{isc}			1.10×10^7	5.12×10^4	2.24×10^9	1.33×10^1
K_r		7.37×10^7		1.20×10^1		3.25×10^3

According to the Marcus theory, the rate of the internal conversion process $T_2 \rightarrow T_1$ is estimated to be about 10^{10} s^{-1}, seeming to hinder the $T_2 \rightarrow S_0$ radiation processes with the rate of 3.25×10^3 s^{-1}. However, it is worth noting that the $K_{isc}(S_1 \rightarrow T_2)$ rate of 2.24×10^9 s^{-1} is so large that the population loss on T_2 due to $K_{ic}(T_2 \rightarrow T_1)$ can be compensated instantaneously. As a consequence, the phosphorescence of the transition $T_2 \rightarrow S_0$ may be observed although its radiative rate is much smaller than $K_{ic}(T_2 \rightarrow T_1)$. On the other side, the non-radiative rate in T_1 is much faster than the paltry radiative one with a seriously underestimated emission energy. Compared with $T_1 \rightarrow S_0$, $T_2 \rightarrow S_0$ has a much larger phosphorescence radiative rate with the better agreement of emission wavelength, and is more suitable for the assignment of the experimental phosphorescence [23]. Consequently, it can be inferred that the phosphorescence process is more likely to come from the T_2 state according to the above theoretical results, distinctly breaking Kasha's rule (i.e., fluorescence/phosphorescence usually comes from the lowest excited state [26]).

Although the ISC process is often negligible in the non-radiative transition between the states with different spin multiplicities, it frequently occurs in phosphorescence and may be heavily promoted by remarkable SOC and narrow energy gap [9]. In order to quantitatively unravel the strong non-radiative decay process in $S_1 \rightarrow T_2$, the minimum energy crossing points (MECPs) [61] are optimized and the SOC constants at MECPs (ξ_{MECP}) are calculated. The energy profile of non-radiative decay pathway has been plotted in Figure 2. After S-BF2 being excited to the S_2 state in the Franck-Condon region of S_0, it relaxes rapidly to the minimum of S_2 at 2.94 eV, vertically lying only 0.09 eV above S_1, and then converts to S_1 by conquering a small barrier at MECP1. From the S_1 state at 2.51 eV, there are two non-radiative channels to triplet states (T_1 via MECP2 with a larger barrier of 0.58 eV and T_2 via MECP3 with a smaller barrier of 0.02 eV), and the corresponding SOC constants at MECPs between S_1 and T_n are 1.09 and 9.18 cm^{-1}, respectively. So the quite large K_{isc} rate in $S_1 \rightarrow T_2$ may be attributed to both the low energy barrier and the significant SOC constant at MECP3. For comparison, we have noticed that in C-BF2 the SOC constants between S_1 and T_n are tiny (see Table S8 in the Supplementary Materials), so the ISC process and the subsequent phosphorescence emission are hard to occur. Along with the emission of $T_2 \rightarrow S_0$, there is also a fierce competitive processes, i.e., the internal conversion from T_2 to T_1 through MECP4.

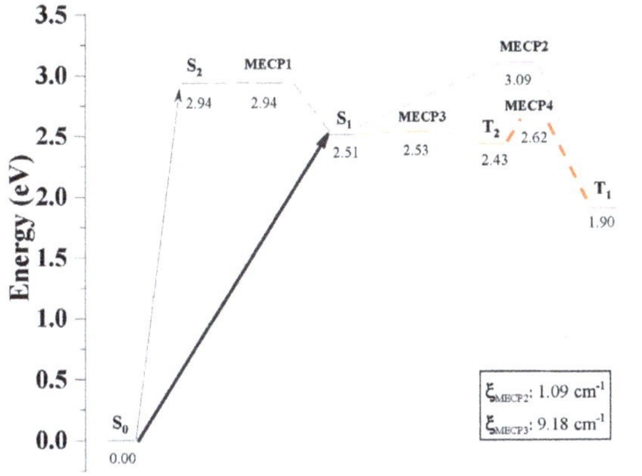

Figure 2. Energies of S-BF2 at the ground and exited states as well as the minimum energy crossing points (MECPs). ξ_{MECP} represents the SOC constant at MECP.

In summary, our quantitative results of S-BF2 suggest a revised phosphorescence mechanism $S_0 \rightarrow S_2 \rightarrow S_1 \rightarrow T_2 \rightarrow S_0$ as demonstrated in Figure 3. At first, S-BF2 in S_0 is excited to S_1 and S_2 but in the latter state there is an IC process to S_1. Then an efficient ISC process takes place through the channel $S_1 \rightarrow T_2$, and finally anti-Kasha phosphorescence occurs directly in the emission $T_2 \rightarrow S_0$. The anti-Kasha phenomenon was found experimentally only in very scarce organic compounds [64,71], e.g., the fluorescence of azulene derivatives [72] and the phosphorescence of N966 [73], ClBDBT [74], and CzCbDBT [75], whereas S-BF2 is a new case according to the above analysis.

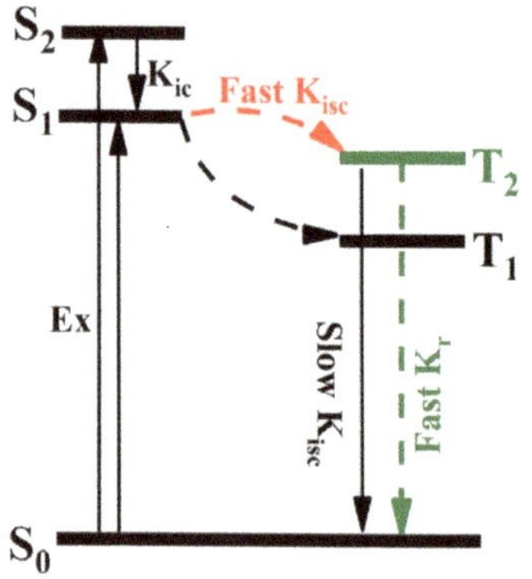

Figure 3. Revised phosphorescence mechanism of S-BF2. K_{ic}, K_{isc} and K_r refer to internal conversion rate, intersystem crossing rate, and radiative rate, respectively.

3.3. *Newly Designed S-BF2 Derivatives and Their Photoluminescence Properties*

The second purpose in this study is to improve S-BF2 theoretically for better phosphorescent performance. The nitrophenyl group in S-BF2 has little effect on the electronic configurations and excitation energies of the S-BF2 frame, and therefore we try to replace the nitrophenyl group with other substituent groups to increase (decrease) the luminous efficiency from T_2 (T_1). Since the phenyl, fluorene, carbazole, dibenzofuran, and dibenzothiophene groups are often used as phosphor in experimental synthesis [12,14], they are adopted in this study, leading to S-BF2*, S-BF2_C, S-BF2_N, S-BF2_O, and S-BF2_S, respectively (*cf.* Figure 4). Again, all these molecules are calculated at the TDDFT(PBE0)/6-311G(d,p) level in CH_2Cl_2 solution.

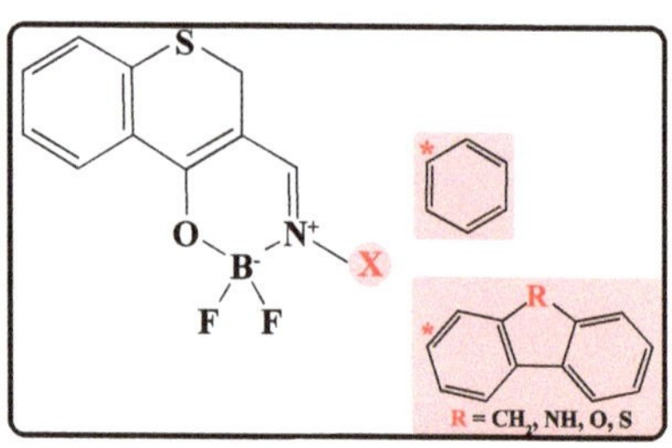

Figure 4. Designed molecules based on S-BF2. The pink part shows the substituents.

The properties of absorption spectra of the designed molecules are listed in Table S11 in the Supplementary Materials. On the whole, these molecules are directly excited to S_1 because of its larger oscillator strengths, so the IC process from from S_2 to S_1 may be ignored. Compared with S-BF2, the $K_{isc}(S_1 \rightarrow T_2)$ rates of the new compounds increase dramatically but except S-BF2_S due to larger SOC constants and smaller energy differences, whereas the $K_{isc}(S_1 \rightarrow T_1)$ ones decrease by one to three orders of magnitude (see Table S10 in the Supplementary Materials), so the population on T_1 become irrelevant. At the T_2 states, the emission energies of the new compounds are about 0.1 eV larger than that of S-BF2, resulting in one to three orders of magnitude reductions in $K_{isc}(T_2 \rightarrow S_0)$; the only exception is S-BF2* where $K_{isc}(T_2 \rightarrow S_0)$ is five times the S-BF2 one because of larger reorganization energy and stronger SOC constant (see Table S10). At the T_1 states, in contrast, the $K_{isc}(T_1 \rightarrow S_0)$ rates rise considerably with the increases of reorganization energy and reductions of emission energy. On the whole, the K_{isc} rates in $T_2 \rightarrow S_0$ are one to seven orders of magnitude weaker than those in $T_1 \rightarrow S_0$ so the phosphorescence emission from the latter transition may be difficult to detect. This result may be confirmed further by the radiative rates given in Table 4: the $K_r(T_1 \rightarrow S_0)$ rates of the new molecules are reduced by several times compared with the S-BF2 one due to the reductions of oscillator strengths, so the parasitic light from $T_1 \rightarrow S_0$ can be inhibited effectively. On the contrary, the K_r rates in $T_2 \rightarrow S_0$ become three to twenty-seven times the S-BF2 one; the only exception is S-BF2* since its oscillator strength of T_2 is nearly unchanged relative to the S-BF2 one. According to the energy gap law [76], the non-radiative rate of $T_2 \rightarrow T_1$ may be significantly reduced as the energy gap increases. Compared with the $K_{isc}(S_1 \rightarrow T_2)$ rates with the order of magnitude between nine and ten (see Table S10), the decay of population on T_2 caused by the much smaller $K_{ic}(T_2 \rightarrow T_1)$ rates can be compensated faster (*cf.* Table S9), and therefore the $T_2 \rightarrow S_0$ emissions in the new molecules are expected to be detected experimentally. In the phosphorescence spectra summarized in Table 4, the phosphorescence wavelengths from T_2 of the new molecules are only 1–7 nm red-shifted relative to the one of S-BF2, being still in the range of green light wave band as in S-BF2.

Table 4. Emission energies, wavelengths, oscillator strengths, radiative rates, and electron configurations of the new molecules in CH_2Cl_2 solution.

Molecule	State	E (eV)	λ (nm)	f	K_r (s^{-1})	Configuration [a] (%)
S-BF2*	T_1	1.44	864	1.65×10^{-7}	1.20×10^{1}	H→L (95)
	T_2	2.36	526	2.00×10^{-5}	4.40×10^{3}	H-1→L (78), H→L(15)
S-BF2_C	T_1	1.18	1050	6.74×10^{-8}	3.42×10^{0}	H→L (94)
	T_2	2.34	530	4.05×10^{-4}	8.76×10^{4}	H-1→L (89)
S-BF2_N	T_1	1.21	1025	6.82×10^{-8}	3.66×10^{0}	H→L (92)
	T_2	2.35	527	5.47×10^{-5}	1.19×10^{4}	H-2→L (89)
S-BF2_O	T_1	1.33	931	7.10×10^{-8}	4.57×10^{0}	H→L (92)
	T_2	2.33	533	1.25×10^{-4}	2.68×10^{4}	H-1→L (90)
S-BF2_S	T_1	1.35	917	7.44×10^{-8}	4.71×10^{0}	H→L (92)
	T_2	2.33	533	8.31×10^{-5}	1.77×10^{4}	H-2→L (89)

[a] H for HOMO and L for LUMO.

As expected, the PL efficiency of S-BF2 may be significantly improved by replacing the nitrophenyl group with fluorene, carbazole, dibenzofuran, or dibenzothiophene but with a simpler PL process $S_0 \rightarrow S_1 \rightarrow T_2 \rightarrow S_0$, where the energy dissipation between S_2 and S_1 has been avoided. Experimental syntheses of these new molecules (especially S-BF2_C and S-BF2_N) are highly desirable to verify the phosphorescence mechanism with enhanced PL efficiency.

4. Conclusions

In the present paper the PL mechanism of S-BF2 have been systematically and thoroughly investigated. By fully considering the static correlations through DMRG-SCF, dynamic correlations through MCPDFT, and substituents and environments through ONIOM, the previous phosphorescent transition $T_1 \rightarrow S_0$ has been reassigned to $T_2 \rightarrow S_0$, and the cheaper TDDFT calculations with suitable functionals may also be confirmed. The computed radiative and non-radiative rate constants as well as MECPs suggest a different PL pathway, namely $S_0 \rightarrow S_2 \rightarrow S_1 \rightarrow T_2 \rightarrow S_0$, and support the breakdown of Kasha's rule. Based on the new phosphorescence mechanism, we have purposefully designed some novel compounds to enhance the phosphorescence efficiency from the T_2 state and reduce the stray light from the T_1 state by changing the substitutes, which provide a valuable guidance for the design of high-efficiency green-emitting phosphors.

Supplementary Materials: The following are available at https://www.mdpi.com/article/10.3390/ijms23020876/s1.

Author Contributions: Conceptualization, D.D.; methodology, D.D. and W.Z.; validation, D.D., B.S. and W.Z.; formal analysis, D.D.; investigation, D.D.; writing—original draft preparation, D.D.; writing—review and editing, D.D., B.S. and W.Z.; visualization, D.D.; supervision, B.S and W.Z.; funding acquisition, B.S. and W.Z. All authors have read and agreed to the published version of the manuscript.

Funding: This research was financially supported by National Natural Science Foundation of China (NSFC, Grant Nos. 21873077 and 22073072) and the Double First-class University Construction Project of Northwest University.

Institutional Review Board Statement: Not applicable.

Informed Consent Statement: Not applicable.

Data Availability Statement: The data presented in this study and not reported in the Supplementary Materials are available on request from the corresponding author.

Conflicts of Interest: The authors declare no conflict of interest.

Abbreviations

The following abbreviations are used in this manuscript:

CT	Charge-Transfer
DFT	Density Functional Theory
DMRG	Density Matrix Renormalization Group
IC	Internal Conversion
ISC	Intersystem Crossing
MCPDFT	Multi-Configurational Pair-Density Functional Theory
MECP	Minimum Energy Crossing Point
PL	Photoluminescence
SOC	Spin-Orbit Coupling
TDDFT	Time-Dependent Density Functional Theory

References

1. Chen, H.W.; Lee, J.H.; Lin, B.Y.; Chen, S.; Wu, S.T. Liquid crystal display and organic light-emitting diode display: Present status and future perspectives. *Light-Sci. Appl.* **2018**, *7*, 17168. [CrossRef]
2. Li, D.; Zhang, H.; Wang, Y. Four-coordinate organoboron compounds for organic light-emitting diodes (OLEDs). *Chem. Soc. Rev.* **2013**, *42*, 8416–8433. [CrossRef]
3. Fateminia, S.A.; Mao, Z.; Xu, S.; Yang, Z.; Chi, Z.; Liu, B. Organic nanocrystals with bright red persistent room-temperature phosphorescence for biological applications. *Angew. Chem. Int. Ed.* **2017**, *56*, 12160–12164. [CrossRef]
4. Reineke, S.; Baldo, M.A. Room temperature triplet state spectroscopy of organic semiconductors. *Sci. Rep.* **2014**, *4*, 3797. [CrossRef]
5. Massuyeau, F.; Faulques, E.; Latouche, C. New insights to simulate the luminescence properties of Pt (II) complexes using quantum calculations. *J. Chem. Theory Comput.* **2017**, *13*, 1748–1755. [CrossRef] [PubMed]
6. Younker, J.M.; Dobbs, K.D. Correlating experimental photophysical properties of iridium (III) complexes to spin–orbit coupled TDDFT predictions. *J. Phys. Chem. C* **2013**, *117*, 25714–25723. [CrossRef]
7. Tang, M.C.; Leung, M.Y.; Lai, S.L.; Ng, M.; Chan, M.Y.; Wing-Wah Yam, V. Realization of thermally stimulated delayed phosphorescence in arylgold (III) complexes and efficient gold (III) based blue-emitting organic light-emitting devices. *J. Am. Chem. Soc.* **2018**, *140*, 13115–13124. [CrossRef] [PubMed]
8. Li, E.Y.T.; Jiang, T.Y.; Chi, Y.; Chou, P.T. Semi-quantitative assessment of the intersystem crossing rate: An extension of the El-Sayed rule to the emissive transition metal complexes. *Phys. Chem. Chem. Phys.* **2014**, *16*, 26184–26192.
9. Marian, C.M. Spin–orbit coupling and intersystem crossing in molecules. *Wiley Interdiscip. Rev. Comput. Mol. Sci.* **2012**, *2*, 187–203. [CrossRef]
10. Baryshnikov, G.; Minaev, B.; Ågren, H. Theory and calculation of the phosphorescence phenomenon. *Chem. Rev.* **2017**, *117*, 6500–6537. [CrossRef] [PubMed]
11. Mukherjee, S.; Thilagar, P. Recent advances in purely organic phosphorescent materials. *Chem. Commun.* **2015**, *51*, 10988–11003. [CrossRef]
12. Zhao, W.; He, Z.; Lam, J.W.; Peng, Q.; Ma, H.; Shuai, Z.; Bai, G.; Hao, J.; Tang, B.Z. Rational molecular design for achieving persistent and efficient pure organic room-temperature phosphorescence. *Chem* **2016**, *1*, 592–602. [CrossRef]
13. Xiao, L.; Fu, H. Enhanced Room-Temperature Phosphorescence through Intermolecular Halogen/Hydrogen Bonding. *Chem–Eur. J.* **2019**, *25*, 714–723. [CrossRef] [PubMed]
14. Yuasa, H.; Kuno, S. Intersystem crossing mechanisms in the room temperature phosphorescence of crystalline organic compounds. *Bull. Chem. Soc. Jpn.* **2018**, *91*, 223–229. [CrossRef]
15. Hirata, S. Intrinsic analysis of radiative and room-temperature nonradiative processes based on triplet state intramolecular vibrations of heavy atom-free conjugated molecules toward efficient persistent room-temperature phosphorescence. *J. Phys. Chem. Lett.* **2018**, *9*, 4251–4259. [CrossRef] [PubMed]

16. He, Z.; Zhao, W.; Lam, J.W.; Peng, Q.; Ma, H.; Liang, G.; Shuai, Z.; Tang, B.Z. White light emission from a single organic molecule with dual phosphorescence at room temperature. *Nat. Commun.* **2017**, *8*, 416. [CrossRef] [PubMed]
17. Ma, H.; Peng, Q.; An, Z.; Huang, W.; Shuai, Z. Efficient and long-lived room-temperature organic phosphorescence: Theoretical descriptors for molecular designs. *J. Am. Chem. Soc.* **2018**, *141*, 1010–1015. [CrossRef]
18. Pfister, A.; Zhang, G.; Zareno, J.; Horwitz, A.F.; Fraser, C.L. Boron polylactide nanoparticles exhibiting fluorescence and phosphorescence in aqueous medium. *ACS Nano* **2008**, *2*, 1252–1258. [CrossRef]
19. Zhang, G.; Evans, R.E.; Campbell, K.A.; Fraser, C.L. Role of boron in the polymer chemistry and photophysical properties of difluoroboron- dibenzoylmethane polylactide. *Macromolecules* **2009**, *42*, 8627–8633. [CrossRef]
20. Frath, D.; Azizi, S.; Ulrich, G.; Ziessel, R. Chemistry on Boranils: an entry to functionalized fluorescent dyes. *Org. Lett.* **2012**, *14*, 4774–4777. [CrossRef] [PubMed]
21. He, G.; Wiltshire, B.D.; Choi, P.; Savin, A.; Sun, S.; Mohammadpour, A.; Ferguson, M.J.; McDonald, R.; Farsinezhad, S.; Brown, A.; et al. Phosphorescence within benzotellurophenes and color tunable tellurophenes under ambient conditions. *Chem. Commun.* **2015**, *51*, 5444–5447. [CrossRef]
22. Torres Delgado, W.; Braun, C.A.; Boone, M.P.; Shynkaruk, O.; Qi, Y.; McDonald, R.; Ferguson, M.J.; Data, P.; Almeida, S.K.; de Aguiar, I.; et al. Moving Beyond Boron-Based Substituents To Achieve Phosphorescence in Tellurophenes. *ACS. Appl. Mater. Inter.* **2017**, *10*, 12124–12134. [CrossRef]
23. Yu, Z.; Wu, Y.; Xiao, L.; Chen, J.; Liao, Q.; Yao, J.; Fu, H. Organic phosphorescence nanowire lasers. *J. Am. Chem. Soc.* **2017**, *139*, 6376–6381. [CrossRef] [PubMed]
24. Paul, L.; Banerjee, A.; Paul, A.; Ruud, K.; Chakrabarti, S. Unraveling the Microscopic Origin of Triplet Lasing from Organic Solids. *J. Phys. Chem. Lett.* **2018**, *9*, 4314–4318. [CrossRef] [PubMed]
25. Lv, A.; Ye, W.; Jiang, X.; Gan, N.; Shi, H.; Yao, W.; Ma, H.; An, Z.; Huang, W. Room-Temperature Phosphorescence from Metal-Free Organic Materials in Solution: Origin and Molecular Design. *J. Phys. Chem. Lett.* **2019**, *10*, 1037–1042. [CrossRef]
26. Kasha, M. Characterization of electronic transitions in complex molecules. *Discuss. Faraday Soc.* **1950**, *9*, 14–19. [CrossRef]
27. Hait, D.; Head-Gordon, M. Orbital Optimized Density Functional Theory for Electronic Excited States. *J. Phys. Chem. Lett.* **2021**, *12*, 4517–4529. [CrossRef]
28. van Gisbergen, S.J.A.; Snijders, J.G.; Baerends, E.J. Implementation of time-dependent density functional response equations. *Comput. Phys. Commun.* **1999**, *118*, 119–138. [CrossRef]
29. Wong, M.W.; Gill, P.M.; Nobes, R.H.; Radom, L. 6-311G (MC)(d, p): A second-row analogue of the 6-311G (d, p) basis set: calculated heats of formation for second-row hydrides. *J. Phys. Chem.* **1988**, *92*, 4875–4880. [CrossRef]
30. Frisch, M.J.; Pople, J.A.; Binkley, J.S. Self-consistent molecular orbital methods 25. Supplementary functions for Gaussian basis sets. *J. Chem. Phys.* **1984**, *80*, 3265–3269. [CrossRef]
31. Becke, A.D. Density-functional exchange-energy approximation with correct asymptotic behavior. *Phy. Rev. A* **1988**, *38*, 3098. [CrossRef]
32. Lee, C.; Yang, W.; Parr, R.G. Development of the Colle-Salvetti correlation-energy formula into a functional of the electron density. *Phy. Rev. B* **1988**, *37*, 785. [CrossRef] [PubMed]
33. Raghavachari, K. Perspective on "Density functional thermochemistry. III. The role of exact exchange". *Theor. Chem. Acc.* **2000**, *103*, 361–363. [CrossRef]
34. Stephens, P.J.; Devlin, F.J.; Chabalowski, C.F.; Frisch, M.J. Ab initio calculation of vibrational absorption and circular dichroism spectra using density functional force fields. *J. Phys. Chem.* **1994**, *98*, 11623–11627. [CrossRef]
35. Adamo, C.; Barone, V. Toward reliable density functional methods without adjustable parameters: The PBE0 model. *J. Chem. Phys.* **1999**, *110*, 6158–6170. [CrossRef]
36. Ernzerhof, M.; Scuseria, G.E. Assessment of the Perdew–Burke–Ernzerhof exchange-correlation functional. *J. Chem. Phys.* **1999**, *110*, 5029–5036. [CrossRef]
37. Yanai, T.; Tew, D.P.; Handy, N.C. A new hybrid exchange–correlation functional using the Coulomb-attenuating method (CAM-B3LYP). *Chem. Phys. Lett.* **2004**, *393*, 51–57. [CrossRef]
38. Chai, J.D.; Head-Gordon, M. Long-range corrected hybrid density functionals with damped atom–atom dispersion corrections. *Phys. Chem. Chem. Phys.* **2008**, *10*, 6615–6620. [CrossRef]
39. Sun, J.; Ruzsinszky, A.; Perdew, J. Strongly Constrained and Appropriately Normed Semilocal Density Functional. *Phys. Rev. Lett.* **2015**, *115*, 036402. [CrossRef]
40. Hui, K.; Chai, J.D. SCAN-based hybrid and double-hybrid density functionals from models without fitted parameters. *J. Chem. Phys.* **2016**, *144*, 044114. [CrossRef]
41. Mezei, P.D.; Csonka, G.I.; Kállay, M. Simple modifications of the SCAN meta-generalized gradient approximation functional. *J. Chem. Theory Comput.* **2018**, *14*, 2469–2479. [CrossRef]
42. Tomasi, J.; Mennucci, B.; Cammi, R. Quantum mechanical continuum solvation models. *Chem. Rev.* **2005**, *105*, 2999–3094. [CrossRef]
43. Frisch, M.J.; Trucks, G.W.; Schlegel, H.B.; Scuseria, G.E.; Robb, M.A.; Cheeseman, J.R.; Scalmani, G.; Barone, V.; Petersson, G.A.; Nakatsuji, H.; et al. *Gaussian 16 Revision C.01*; Gaussian Inc.: Wallingford, CT, USA, 2016.
44. Liu, W.; Hong, G.; Dai, D.; Li, L.; Dolg, M. The Beijing four-component density functional program package (BDF) and its application to EuO, EuS, YbO and YbS. *Theor. Chem. Acc.* **1997**, *96*, 75–83. [CrossRef]

45. Li, Z.; Suo, B.; Zhang, Y.; Xiao, Y.; Liu, W. Combining spin-adapted open-shell TD-DFT with spin–orbit coupling. *Mol. Phys.* **2013**, *111*, 3741–3755. [CrossRef]
46. Zhang, Y.; Suo, B.; Wang, Z.; Zhang, N.; Li, Z.; Lei, Y.; Zou, W.; Gao, J.; Peng, D.; Pu, Z.; et al. BDF: A relativistic electronic structure program package. *J. Chem. Phys.* **2020**, *152*, 064113. [CrossRef] [PubMed]
47. Lundberg, M.; Sasakura, Y.; Zheng, G.; Morokuma, K. Case Studies of ONIOM(DFT:DFTB) and ONIOM(DFT:DFTB:MM) for Enzymes and Enzyme Mimics. *J. Chem. Theory Comput.* **2010**, *6*, 1413–1427. [CrossRef]
48. Chung, L.W.; Sameera, W.M.C.; Ramozzi, R.; Page, A.J.; Hatanaka, M.; Petrova, G.P.; Harris, T.V.; Li, X.; Ke, Z.; Liu, F.; et al. The ONIOM Method and Its Applications. *Chem. Rev.* **2015**, *115*, 5678–5796. [CrossRef] [PubMed]
49. Weigend, F.; Ahlrichs, R. Balanced basis sets of split valence, triple zeta valence and quadruple zeta valence quality for H to Rn: Design and assessment of accuracy. *Phys. Chem. Chem. Phys.* **2005**, *7*, 3297–3305. [CrossRef]
50. Aquilante, F.; Pedersen, T.B.; Lindh, R. Low-cost evaluation of the exchange Fock matrix from Cholesky and density fitting representations of the electron repulsion integrals. *J. Chem. Phys.* **2007**, *126*, 194106. [CrossRef]
51. Sharma, P.; Bernales, V.; Knecht, S.; Truhlar, D.G.; Gagliardi, L. Density matrix renormalization group pair-density functional theory (DMRG-PDFT): Singlet-triplet gaps in polyacenes and polyacetylenes. *Chem. Sci.* **2019**, *10*, 1716–1723. [CrossRef] [PubMed]
52. Hoyer, C.E.; Ghosh, S.; Truhlar, D.G.; Gagliardi, L. Multiconfiguration pair-density functional theory is as accurate as CASPT2 for electronic excitation. *J. Phys. Chem. Lett.* **2016**, *7*, 586–591. [CrossRef]
53. Ghosh, S.; Cramer, C.J.; Truhlar, D.G.; Gagliardi, L. Generalized-active-space pair-density functional theory: An efficient method to study large, strongly correlated, conjugated systems. *Chem. Sci.* **2017**, *8*, 2741–2750. [CrossRef]
54. Carlson, R.K.; Truhlar, D.G.; Gagliardi, L. Multiconfiguration pair-density functional theory: A fully translated gradient approximation and its performance for transition metal dimers and the spectroscopy of $Re_2Cl_8^{2-}$. *J. Chem. Theory Comput.* **2015**, *11*, 4077–4085. [CrossRef] [PubMed]
55. Wilbraham, L.; Verma, P.; Truhlar, D.G.; Gagliardi, L.; Ciofini, I. Multiconfiguration pair-density functional theory predicts spin-state ordering in iron complexes with the same accuracy as complete active space second-order perturbation theory at a significantly reduced computational cost. *J. Phys. Chem. Lett.* **2017**, *8*, 2026–2030. [CrossRef] [PubMed]
56. Adeyiga, O.; Suleiman, O.; Dandu, N.K.; Odoh, S.O. Ground-state actinide chemistry with scalar-relativistic multiconfiguration pair-density functional theory. *J. Chem. Phys.* **2019**, *151*, 134102. [CrossRef] [PubMed]
57. Galván, I.F.; Vacher, M.; Alavi, A.; Angeli, C.; Aquilante, F.; Autschbach, J.; Bao, J.J.; Bokarev, S.I.; Bogdanov, N.A.; Carlson, R.K.; et al. OpenMolcas: from source code to insight. *J. Chem. Theory Comput.* **2019**, *15*, 5925–5964. [CrossRef]
58. Aquilante, F.; Autschbach, J.; Baiardi, A.; Battaglia, S.; Borin, V.A.; Chibotaru, L.F.; Conti, I.; De Vico, L.; Delcey, M.; Fdez Galván, I.; et al. Modern quantum chemistry with [Open] Molcas. *J. Chem. Phys.* **2020**, *152*, 214117. [CrossRef]
59. Chan, G.K.L.; Sharma, S. The density matrix renormalization group in quantum chemistry. *Annu. Rev. Phys. Chem.* **2011**, *62*, 465–481. [CrossRef]
60. Sharma, S.; Chan, G.K.L. Spin-adapted density matrix renormalization group algorithms for quantum chemistry. *J. Chem. Phys.* **2012**, *136*, 124121. [CrossRef]
61. Harvey, J.N.; Aschi, M. Spin-forbidden dehydrogenation of methoxy cation: a statistical view. *Phys. Chem. Chem. Phys.* **1999**, *1*, 5555–5563. [CrossRef]
62. Lu, T. The sobMECP Program. Available online: http://sobereva.com/286 (accessed on 7 January 2022).
63. Niu, Y.; Li, W.; Peng, Q.; Geng, H.; Yi, Y.; Wang, L.; Nan, G.; Wang, D.; Shuai, Z. MOlecular MAterials Property Prediction Package (MOMAP) 1.0: A software package for predicting the luminescent properties and mobility of organic functional materials. *Mol. Phys.* **2018**, *116*, 1078–1090. [CrossRef]
64. Itoh, T. Fluorescence and phosphorescence from higher excited states of organic molecules. *Chem. Rev.* **2012**, *112*, 4541–4568. [CrossRef] [PubMed]
65. Mok, D.K.W.; Neumann, R.; Handy, N.C. Dynamical and nondynamical correlation. *J. Phys. Chem.* **1996**, *100*, 6225–6230. [CrossRef]
66. Ghosh, S.; Sonnenberger, A.L.; Hoyer, C.E.; Truhlar, D.G.; Laura, G. Multiconfiguration pair-density functional theory outperforms Kohn-Sham density functional theory and multireference perturbation theory for ground-state and excited-state charge transfer. *J. Chem. Theory Comput.* **2015**, *11*, 3643–3649. [CrossRef] [PubMed]
67. Presti, D.; Truhlar, D.G.; Gagliardi, L. Intramolecular charge transfer and local excitation in organic fluorescent photoredox catalysts explained by RASCI-PDFT. *J. Phys. Chem. C* **2018**, *122*, 12061–12070. [CrossRef]
68. Zhao, Y.; Truhlar, D.G. Density functional for spectroscopy: No long-range self-interaction error, good performance for Rydberg and charge-transfer states, and better performance on average than B3LYP for ground states. *J. Phys. Chem. A* **2006**, *110*, 13126–13130. [CrossRef] [PubMed]
69. Le Bahers, T.; Adamo, C.; Ciofini, I. A qualitative index of spatial extent in charge-transfer excitations. *J. Chem. Theory Comput.* **2011**, *7*, 2498–2506. [CrossRef]
70. Lu, T.; Chen, F. Multiwfn: A multifunctional wavefunction analyzer. *J. Comput. Chem.* **2012**, *33*, 580–592. [CrossRef] [PubMed]
71. Demchenko, A.P.; Tomin, V.I.; Chou, P.T. Breaking the Kasha rule for more efficient photochemistry. *Chem. Rev.* **2017**, *117*, 13353–13381. [CrossRef]

72. Eber, G.; Grüneis, F.; Schneider, S.; Dörr, F. Dual Fluorescence Emission of Azulene Derivatives in Solution. *Chem. Phys. Lett.* **1974**, *29*, 397–404. [CrossRef]
73. Paul, L.; Chakrabarti, S.; Ruud, K. Anomalous phosphorescence from an organometallic white-light phosphor. *J. Phys. Chem. Lett.* **2017**, *8*, 4893–4897. [CrossRef] [PubMed]
74. Paul, L.; Moitra, T.; Ruud, K.; Chakrabarti, S. Strong Duschinsky mixing induced breakdown of Kasha's rule in an organic phosphor. *J. Phys. Chem. Lett.* **2019**, *10*, 369–374. [CrossRef] [PubMed]
75. Feng, C.; Li, S.; Fu, L.; Xiao, X.; Xu, Z.; Liao, Q.; Wu, Y.; Yao, J.; Fu, H. Breaking Kasha's rule as a mechanism for solution-phase room-temperature phosphorescence from high-lying triplet excited state. *J. Phys. Chem. Lett.* **2020**, *11*, 8246–8251. [CrossRef] [PubMed]
76. Englman, R.; Jortner, J. The energy gap law for radiationless transitions in large molecules. *Mol. Phys.* **1970**, *18*, 145–164. [CrossRef]

MDPI AG
Grosspeteranlage 5
4052 Basel
Switzerland
Tel.: +41 61 683 77 34

International Journal of Molecular Sciences Editorial Office
E-mail: ijms@mdpi.com
www.mdpi.com/journal/ijms

www.ingramcontent.com/pod-product-compliance
Lightning Source LLC
Chambersburg PA
CBHW041018270926
51850CB00024B/1652
* 9 7 8 3 7 2 5 8 5 4 7 9 0 *